Mechanisms and Manifestations of Obesity in Lung Disease

Richard A. Johnston is serving in his personal capacity. The views expressed are his own and do not necessarily represent the views of the Centers for Disease Control and Prevention or the United States Government.

Mechanisms and Manifestations of Obesity in Lung Disease

Edited by

Richard A. Johnston

Benjamin T. Suratt

ACADEMIC PRESS
An imprint of Elsevier

Academic Press is an imprint of Elsevier
125 London Wall, London EC2Y 5AS, United Kingdom
525 B Street, Suite 1650, San Diego, CA 92101, United States
50 Hampshire Street, 5th Floor, Cambridge, MA 02139, United States
The Boulevard, Langford Lane, Kidlington, Oxford OX5 1GB, United Kingdom

Notices
Knowledge and best practice in this field are constantly changing. As new research and experience broaden our understanding, changes in research methods, professional practices, or medical treatment may become necessary.

Practitioners and researchers must always rely on their own experience and knowledge in evaluating and using any information, methods, compounds, or experiments described herein. In using such information or methods they should be mindful of their own safety and the safety of others, including parties for whom they have a professional responsibility.

To the fullest extent of the law, neither the Publisher nor the authors, contributors, or editors, assume any liability for any injury and/or damage to persons or property as a matter of products liability, negligence or otherwise, or from any use or operation of any methods, products, instructions, or ideas contained in the material herein.

Library of Congress Cataloging-in-Publication Data
A catalog record for this book is available from the Library of Congress

British Library Cataloguing-in-Publication Data
A catalogue record for this book is available from the British Library

ISBN: 978-0-12-813553-2

For information on all Academic Press publications
visit our website at https://www.elsevier.com/books-and-journals

www.elsevier.com • www.bookaid.org

Working together
to grow libraries in
developing countries

Publisher: Stacy Masucci
Acquisition Editor: Katie Chan
Editorial Project Manager: Barbara Makinster
Production Project Manager: Punithavathy Govindaradjane
Cover Designer: Matthew Limbert

Typeset by SPi Global, India

Contents

Contributors

Anurag Agrawal
CSIR Institute of Genomics & Integrative Biology, Delhi University, New Delhi, India

Maryellen C. Antkowiak
Department of Medicine, University of Vermont Larner College of Medicine, Burlington, VT, United States

Melinda A. Beck
Department of Nutrition, Gilling's School of Global Public Health, University of North Carolina, Chapel Hill, NC, United States

Sapna Bhatia
Department of Internal Medicine, University of New Mexico School of Medicine, Albuquerque, NM, United States

Ellen E. Blaak
Department of Human Biology, NUTRIM School of Nutrition and Translational Research in Metabolism, Maastricht University Medical Centre, Maastricht, The Netherlands

Richard N. Channick
Department of Medicine, Massachusetts General Hospital, Harvard Medical School, Boston, MA, United States

David Chapman
Translational Airways Group, School of Life Sciences, The University of Technology Sydney, Sydney, NSW, Australia

Rituparna Chaudhuri
CSIR Institute of Genomics & Integrative Biology, Delhi University, New Delhi, India

Veronica De Rosa
Istituto per l'Endocrinologia e l'Oncologia Sperimentale, Consiglio Nazionale delle Ricerche (IEOS-CNR), Napoli, Italy

Anne E. Dixon
Department of Medicine, Division of Pulmonary and Critical Care Medicine, University of Vermont, Burlington, VT, United States

Debbie M. Figueroa
Laboratory of Asthma and Lung Inflammation, Pulmonary Branch, Division of Intramural Research, NHLBI, NIH, Bethesda, MD, United States

Erick Forno
Department of Pediatrics, University of Pittsburgh School of Medicine; Division of Pulmonary Medicine, UPMC Children's Hospital of Pittsburgh, Pittsburgh, PA, United States

Gijs H. Goossens
Department of Human Biology, NUTRIM School of Nutrition and Translational Research in Metabolism, Maastricht University Medical Centre, Maastricht, The Netherlands

Elizabeth M. Gordon
Laboratory of Asthma and Lung Inflammation, Pulmonary Branch, Division of Intramural Research, NHLBI, NIH, Bethesda, MD, United States

William D. Green
Department of Nutrition, Gilling's School of Global Public Health, University of North Carolina, Chapel Hill, NC, United States

Chenjuan Gu
Department of Pulmonary and Critical Care Medicine, Ruijin Hospital, Shanghai Jiao Tong University School of Medicine, Shanghai, China

Richard A. Johnston
Division of Critical Care Medicine, Department of Pediatrics, McGovern Medical School at The University of Texas Health Science Center at Houston, Houston, TX, United States

Jonathan C. Jun
Division of Pulmonary and Critical Care, Department of Medicine, Johns Hopkins University, Baltimore, MD, United States

Erik A. Karlsson
Virology Unit, Institut Pasteur du Cambodge, Phnom Penh, Cambodia

Gregory King
The Woolcock Institute of Medical Research, The University of Sydney; Department of Respiratory Medicine, Royal North Shore Hospital, Sydney, NSW, Australia

Stewart J. Levine
Laboratory of Asthma and Lung Inflammation, Pulmonary Branch, Division of Intramural Research, NHLBI, NIH, Bethesda, MD, United States

Benjamin J. Marsland
Faculty of Biology and Medicine, University of Lausanne, Epalinges, Switzerland; Department of Immunology and Pathology, Monash University, Melbourne, VIC, Australia

Giuseppe Matarese
Istituto per l'Endocrinologia e l'Oncologia Sperimentale, Consiglio Nazionale delle Ricerche (IEOS-CNR); Treg Cell Lab, Dipartimento di Medicina Molecolare e Biotecnologie Mediche, Università degli Studi di Napoli "Federico II", Napoli, Italy

J. Justin Milner
Division of Biological Sciences, University of California, San Diego, CA, United States

Christina Pabelick
Department of Anesthesiology and Perioperative Medicine, Mayo Clinic, Rochester, MN, United States

Francesco Perna
Dipartimento di Medicina Clinica e Chirurgia, Università degli Studi di Napoli "Federico II", Napoli, Italy

Vsevolod Y. Polotsky
Division of Pulmonary and Critical Care, Department of Medicine, Johns Hopkins University, Baltimore, MD, United States

Y.S. Prakash
Department of Anesthesiology and Perioperative Medicine, Mayo Clinic, Rochester, MN, United States

Claudio Procaccini
Istituto per l'Endocrinologia e l'Oncologia Sperimentale, Consiglio Nazionale delle Ricerche (IEOS-CNR), Napoli, Italy

Deepa Rastogi
Albert Einstein College of Medicine, Bronx, NY, United States

Aman Rathore
Division of Pulmonary and Critical Care, Department of Medicine, Johns Hopkins University, Baltimore, MD, United States

Jennifer Rebeles
National Research Council Postdoctoral Research Associate, US Army Institute of Surgical Research, San Antonio, TX, United States

Stacey Schultz-Cherry
Department of Infectious Diseases, St. Jude Children's Research Hospital, Memphis, TN, United States

Stephanie A. Shore
Molecular and Integrative Physiological Sciences Program, Department of Environmental Health, Harvard T.H. Chan School of Public Health, Boston, MA, United States

Akshay Sood
Department of Internal Medicine, University of New Mexico School of Medicine, Albuquerque, NM, United States

Renee D. Stapleton
Department of Medicine, Division of Pulmonary and Critical Care Medicine, University of Vermont, Burlington, VT, United States

Benjamin T. Suratt
Department of Medicine, Division of Pulmonary and Critical Care Medicine, University of Vermont, Burlington, VT, United States

Michael A. Thompson
Department of Anesthesiology and Perioperative Medicine, Mayo Clinic, Rochester, MN, United States

Niki D.J. Ubags
Faculty of Biology and Medicine, University of Lausanne, Epalinges, Switzerland

Rens L.J. van Meijel
Department of Human Biology, NUTRIM School of Nutrition and Translational Research in Metabolism, Maastricht University Medical Centre, Maastricht, The Netherlands

Xianglan Yao
Laboratory of Asthma and Lung Inflammation, Pulmonary Branch, Division of Intramural Research, NHLBI, NIH, Bethesda, MD, United States

Haris Younas
Division of Pulmonary and Critical Care, Department of Medicine, Johns Hopkins University, Baltimore, MD, United States

Preface

During the last four decades, the global prevalence of obesity has increased dramatically among children, adolescents, and adults, and given current trends, 51% of adults in the United States are expected to be obese by the year 2030. The negative health and socioeconomic consequences of obesity are well-known. For example, a significant association exists between obesity and several chronic diseases, including cardiovascular and gastrointestinal diseases, certain forms of cancer, osteoarthritis, and type 2 diabetes. Furthermore, obesity is responsible for 20% of all deaths in the United States, and as of 2016, the annual direct costs for the medical treatment of obese adults are almost 210 billion dollars. Obesity also accounts for indirect costs such as absenteeism and presenteeism that total 66 billion dollars per year. For the foreseeable future, direct and indirect costs secondary to obesity are expected to surge and eventually be as much as one trillion dollars per year. In addition, as the prevalence of obesity continues to rise at an alarming rate, obesity-related morbidity and mortality will do the same. Thus, it is important to understand the relationship between obesity and a number of different diseases.

Globally, cardiovascular disease and type 2 diabetes are the number one and number seven causes of death, respectively, and the mechanistic bases for the relationship between obesity and these diseases have been well characterized. However, four of the top ten global causes of death in 2016 were lung diseases, including chronic obstructive pulmonary disease, lower respiratory tract infections, lung cancer, and tuberculosis. Until very recently, the impact of obesity on the development and/or course of lung diseases has been ignored. Even in the absence of any lung disease, obesity has negative effects on lung function in both children and adults. For example, obesity increases breathing frequency and decreases FEV_1, $FEF_{25\%-75\%}$, and lung compliance, which are functional changes that are frequently associated with lung disease in nonobese individuals. Taken together, these data suggest that the interactions between obesity and lung disease may have profound consequences for human health now and increasingly in the decades to come.

Obesity-hypoventilation syndrome was the first lung disease to be causally associated with obesity. Subsequently, investigators demonstrated that obesity was also a risk factor for and/or modified the course of a number of lung diseases, including obstructive sleep apnea, asthma, acute lung injury, chronic obstructive pulmonary disease (emphysema and chronic bronchitis), pulmonary hypertension, and lung infections (bacterial and viral). From the results of both human and animal studies, several hypotheses have emerged to explain the relationship between obesity and lung disease. Such mechanisms include adipose tissue dysfunction, chronic systemic inflammation, mechanical factors, elements of the metabolic syndrome, hypoxia, and physical inactivity. As obesity is typically associated with a confluence of these factors, it is improbable that any of these would affect lung disease in isolation. However, to date, chronic systemic inflammation has garnered the most attention. Obesity-induced chronic systemic inflammation arises when adipose tissue becomes infiltrated with macrophages, which in turn, along with adipocytes, release adipokines (e.g., acute-phase reactants, cytokines, hormones, and soluble cytokine receptors) into the systemic circulation. Because adipokines, which are largely proinflammatory in nature, have been previously shown to correlate directly with the severity of other obesity-related diseases, including cardiovascular disease and type 2 diabetes, investigators hypothesized that the same may also occur with lung disease. To that end, a number of investigators have demonstrated that specific adipokines (adiponectin, IL-13, IL-17A, IL-33, osteopontin, and TNF-α) may contribute to obesity-induced lung dysfunction in animal models. Besides adipokines, there is emerging evidence that mitochondria, lipoproteins, the metabolome, and the microbiome can influence the development and/or course of lung disease in both the presence and absence of obesity. Although these studies are currently in their infancy, a better understanding of the contribution of each of these factors to obesity-induced lung dysfunction is needed.

The impact of obesity on the development and/or course of lung disease has recently garnered much interest among physicians and scientists because of the obesity epidemic. Nevertheless, we continue to have a poor understanding of the biological mechanisms coupling obesity and lung disease. Given that (1) the prevalence of obesity continues to escalate at an alarming rate, (2) four of the top ten global causes of death are the result of lung disease, and (3) obese individuals with lung disease are often less or unresponsive to standard therapies, preclinical animal studies and clinical trials in this area are desperately needed in order to either prevent or mitigate the negative health and socioeconomic effects of obesity-associated lung disease. *Mechanisms and Manifestations of Obesity in Lung Disease* is the first textbook that provides an all-in-one comprehensive summary of the biological mechanisms by which obesity causes lung dysfunction in health and in disease. This textbook is an excellent resource for basic and clinical scientists interested in the emerging area of obesity and lung disease where very little progress has been made to either prevent or alleviate disease symptoms.

Adipose tissue metabolism and inflammation in obesity

Rens L.J. van Meijel, Ellen E. Blaak, Gijs H. Goossens

Department of Human Biology, NUTRIM School of Nutrition and Translational Research in Metabolism, Maastricht University Medical Centre, Maastricht, The Netherlands

ABBREVIATIONS

ATGL	adipose triglyceride lipase
ATM	adipose tissue macrophage
ATP	adenosine triphosphate
bFGF	basic fibroblast growth factor
CD	cluster of differentiation
COPD	chronic obstructive pulmonary disease
CRP	C-reactive protein
CXCL	C-X C motif ligand
DAMP	damage-associated molecular pattern
ECM	extracellular matrix
ERK1/2	extracellular signal-regulated kinase 1 and 2
FFA	free fatty acids
F_IO_2	inspired air fraction
HSL	hormone-sensitive lipase
IFN-γ	interferon-gamma
IKKI	κB kinase
IL	interleukin
IRS-1	insulin receptor substrate 1
JNK1/2	C-Jun N-terminal kinases 1/2
LPL	lipoprotein lipase
LPS	lipopolysaccharides
MCP-1	monocyte chemoattractant protein-1
MGL	monoacylglycerol lipase
MMP	matrix metalloproteinase
Myf5	myogenic factor 5
NLR	nucleotide-binding oligomerization domain-like receptor
NLRP3	NLR-leucine-rich family, pyrine domain-containing 3
p38-MAPK	p38 mitogen-activated protein kinase
PI-3K	phosphoinositide 3-kinase
pO_2	oxygen tension

Mechanisms and Manifestations of Obesity in Lung Disease. https://doi.org/10.1016/B978-0-12-813553-2.00001-4

RBC	red blood cell
T2DM	type 2 diabetes mellitus
TAG	triacylglycerides
TGF-β	transforming growth factor-β
Th1	type 1 T-helper
TIMP	tissue inhibitors of metalloproteinases
TLR4	Toll-like-receptor-4
TNF-α	tumor necrosis factor-α
Tregs	regulatory T-lymphocytes
UCP-1	uncoupling protein-1
VEGF	vascular endothelial growth factor
VEGFR2	vascular endothelial growth factor receptor 2

INTRODUCTION

Obesity is a chronic metabolic disorder characterized by a complex and multifactorial etiology, involving both external and internal factors such as lifestyle, the environment, stress, sleep, and (*epi*)genetics. Globally, obesity is a major cause of morbidity and mortality [1]. Worldwide, >1.9 billion adults are overweight, of whom at least 650 million are obese, defined as a body mass index (BMI) $\geq 30\,kg/m^2$ [2]. According to current estimates, the global prevalence of obesity will increase drastically in the near future [3]. This poses a major public health issue because obesity is associated with a variety of comorbidities and chronic diseases, including hypertension, dyslipidemia, type 2 diabetes, cardiovascular disease, cognitive decline, depression, obstructive sleep apnea syndrome (OSAS), skin problems, asthma, and several types of cancer [4–9].

Initiated by an energy imbalance in which energy intake exceeds expenditure for a prolonged period of time, adipose tissue mass may drastically increase. The progressive increase in adipose tissue mass induces adipose tissue dysfunction, which is closely related to the development of insulin resistance [10]. In fact, adipose tissue dysfunction is a key contributor to metabolic and endocrine derangements, characterized by dyslipidemia, hyperglycemia, hypertension, and low-grade systemic inflammation [11, 12]. In this chapter, we will first discuss the normal metabolic and immune functions of adipose tissue. Next, different types of adipose tissue as well as distinct adipose tissue depots where lipids can be stored will be discussed. Thereafter, an overview of the metabolic and immunological consequences of adipose tissue dysfunction in obesity will be provided. Finally, we shall briefly elaborate on the possible link between adipose tissue dysfunction and select lung diseases.

THE IMPORTANCE OF HEALTHY ADIPOSE TISSUE

Because obesity is associated with several comorbidities, the many benefits provided by healthy adipose tissue are often easily ignored. For example, adipose tissue protects delicate organs and certain regions of the body exposed to high levels of mechanical stress [13]. Furthermore, healthy adipose tissue has thermal insulation properties and is associated with normal reproductive function [14, 15]. However, the classical function of adipose tissue is to provide a long-term energy reserve, which can be used during food deprivation by mobilizing fatty acids to be utilized by other organs [16]. Therefore, the amount of adipose tissue can fluctuate, increasing following a prolonged period of positive energy balance and declining when energy expenditure exceeds intake for a substantial amount of time.

In the past, adipose tissue was considered an inert organ, which mainly served as a storage depot for excess energy in the form of triacylglycerides (TAG). Over the past several decades, however, a variety of functions have been attributed to adipose tissue, which is now recognized as a highly dynamic and metabolically active organ [17].

Adipose tissue contains different cell types roughly categorized as adipocytes and nonadipocyte cells. The latter includes fibroblasts, endothelial cells, and a diverse number of immune cells, which together comprise the stromal vascular fraction of adipose tissue [18]. A large proportion of cells present within adipose tissue are adipocytes, which are differentiated fibroblasts able to store large quantities of TAG. These adipocytes originate from mesenchymal stem cells and differentiate into preadipocytes. Subsequently, preadipocytes differentiate into mature adipocytes and acquire properties such as insulin sensitivity, the ability to store lipids, and the capacity to secrete proteins and other moieties, which are known as adipokines [19].

Adipose tissue is capable of storing excess energy in the form of TAG in the postprandial phase (i.e., after the completion of a meal). On the other hand, adipose tissue releases free fatty acids (FFA) when other tissues are in need of energy such as during states of fasting and exercise [20]. Meal intake results in increased circulating TAG concentrations, which are present in chylomicrons and very-low-density-lipoproteins and their remnants [21]. A large proportion of circulating TAG will be taken up from the circulation by adipose tissue as they pass through adipose tissue capillaries. To process TAG, adipocytes secrete an inactive form of the enzyme lipoprotein lipase (LPL), which is subsequently transported to the endothelial surface of capillaries present in the adipose tissue [22]. Once active, LPL hydrolyses circulating TAG into FFA and glycerol. The regulation of LPL activity is complex, involving both nutrients and hormones [23]. Within adipose tissue, insulin levels primarily determine LPL activity. As such, LPL activity is increased postprandially and decreased during fasting [24]. After hydrolysis, FFA can be taken up by adipocytes and converted into TAG. However, a substantial proportion of FFA may also enter the circulation, thereby increasing plasma FFA concentration [25]. Excess storage of TAG in adipocytes may result in enlarged adipocytes, which in turn results in expansion of adipose tissue mass, which will be discussed later. During energy deprivation (e.g., fasting), the major physiological role of adipose tissue is to provide FFA for cellular functions requiring energy. To release FFA, a catabolic process occurs within adipocytes. As such, TAG contained within intracellular lipid droplets are hydrolyzed into diacylglycerol, followed by monoacylglycerol, and ultimately glycerol by adipose triglyceride lipase (ATGL), hormone-sensitive lipase (HSL), and monoacylglycerol lipase (MGL), respectively [26]. During each step, a FFA is liberated from the glyceride residue and results in the formation of glycerol. Lipolysis is under both nutritional and hormonal control, and several bioactive molecules regulate the lipolytic rate of adipocytes, including catecholamines (e.g., norepinephrine) and natriuretic peptides (e.g., atrial natriuretic peptide) [23, 27]. Because insulin is a potent antilipolytic agent [28], postprandial insulin peaks inhibit lipolysis.

In addition to its role in lipid metabolism, adipose tissue is a metabolically active endocrine organ that regulates a multitude of physiological processes, including energy homeostasis, blood pressure, inflammation, angiogenesis, and food intake [29]. A key discovery that established adipose tissue as an endocrine organ was the identification of the hormone leptin, which is principally produced in and secreted by adipocytes and acts both centrally (i.e., hypothalamus) and in peripheral organs to regulate food intake and energy homeostasis [30]. To date, an enormous number of adipokines have been described, which can exert effect through paracrine-, autocrine-, and endocrine-mediated pathways [31]. Thus adipose tissue exerts many functions and is not merely a passive energy depot as originally thought. As such, adipose tissue exerts a key role in maintaining cardiometabolic health.

ADIPOSE TISSUE: AN ORGAN WITH DIFFERENT SHADES

Our knowledge about adipose tissue has profoundly increased over the last several decades. In most species, unilocular white adipocytes are by far the most abundant fat cells, and they dynamically regulate lipid turnover and storage. Brown adipocytes, on the other hand, are highly specialized cells that dissipate stored energy in the form of heat due to the presence of uncoupling protein (UCP)-1 within the many mitochondria in these cells. More specifically, UCP-1 catalyzes the leak of protons across the inner mitochondrial membrane, thereby "uncoupling" substrate oxidation from adenosine triphosphate (ATP) synthesis [32].

Infants have significant depots of brown fat, presumably to provide heat immediately after birth [33]. For many years, however, it was primarily thought that only white adipose tissue was present in adults, unless individuals were exposed to cold for a long period of time (e.g., outside workers in Scandinavia) or had certain pathologies (e.g., pheochromocytomas), which are rare adrenal tumors that produce excessive amounts of norepinephrine and epinephrine [34, 35]. However, it is now known that brown adipose tissue is present in normal weight adults, and becomes more active when catecholamine levels increase due to stimulation of lipolysis and subsequent UCP-1 activation in brown fat [36].

Recently, evidence has emerged that another type of adipocyte exists that resembles the classical brown adipocyte but has a different pattern of gene expression. These so-called "beige"/"brite" adipocytes originate from the lateral mesoderm, similar to white adipocytes, whereas brown adipocytes originate from the paraxial mesoderm and are characterized by myogenic marker 5 (Myf5) expression [37]. Many studies are currently ongoing to elucidate the physiological factors that activate brown and beige cells. In addition to the sympathetic nervous system, exposure to cold may provide an independent stimulus for brown adipose tissue activation [38]. Moreover, several circulating hormones (e.g., natriuretic peptides, thyroid hormones, and catecholamines) and bile acids, among other factors, have been suggested to stimulate the activity of brown adipose tissue in rodents and/or humans [39–41]. Interestingly, there seems to be remarkable plasticity among adipocytes because white, beige/brite, and brown adipocytes can transform into other adipocytes during chronic overfeeding or exposure to cold [42]. In this chapter, we shall focus on white adipose tissue, because progressive expansion of this fat depot in obesity is accompanied by the development of associated comorbidities.

LOCATION OF LIPID STORAGE DOES MATTER

The majority of lipids is stored in adipocytes. When large amounts of adipocytes cluster together, an adipose tissue depot is formed. In humans, we can distinguish subcutaneous and visceral white adipose tissue, with each of these depots subdivided into specific adipose tissue depots. Subcutaneous adipose tissue is mainly stored in the abdominal (upper-body) and gluteofemoral (lower-body) region. Visceral adipose tissue refers to the adipose tissue dispersed between the abdominal organs, including the omental, mesenteric, and retroperitoneal fat depots. Distribution of body fat is considered to be one of the strongest risk factors for the development of both metabolic as well as cardiovascular disorders [10]. The reason why some individuals store more fat within certain adipose tissue depots than others is a topic of current investigation. Body fat distribution seems to be influenced by numerous intrinsic as well as environmental factors, such as alcohol intake, smoking, and time of the onset of obesity. Besides, (*epi*)genetic factors appear to play a role in both adipose tissue mass gain and loss in specific fat depots [43].

Interestingly, it seems that cell-autonomous mechanisms underlie depot-specific differences in adipocyte physiology. This is further exemplified by the findings that preadipocytes express gene signatures specific for their depot of origin, and these cells continue to function distinctly after isolation and prolonged passage under identical conditions [44, 45]. Clearly, important intrinsic differences between adipose tissue depots exist. Upper-body fat mass, characterized by increased abdominal and visceral fat mass, is associated with an increased risk for cardiometabolic complications [46]. However, the cause is still under debate, yet different hypotheses may explain this association. In the presence of insulin resistance, the storage capacity of fat, particularly in subcutaneous abdominal adipose tissue, may be reduced and result in systemic lipid overflow and the deposition of droplets containing TAG within nonadipocyte cells in various tissues (e.g., ectopic fat accumulation). The accumulation of TAG can occur within (cardio)myocytes, hepatocytes, pancreatic cells, and other cell types [23]. Potentially, visceral fat can be considered as another ectopic fat depot. Ectopic fat storage and, more specifically, the accumulation, composition, and localization of bioactive lipid metabolites is an established risk factor for the development of insulin resistance and related complications [23, 47].

ADIPOSE TISSUE DYSFUNCTION IN OBESITY: A KEY DETERMINANT OF CARDIOMETABOLIC RISK

During a prolonged period of excess caloric intake that leads to weight gain, the structure and composition of adipose tissue changes dramatically. Obese adipose tissue is usually characterized by adipocyte enlargement (i.e., hypertrophy) (Figure 1.1), infiltration of different types of immune cells, and alterations in the extracellular

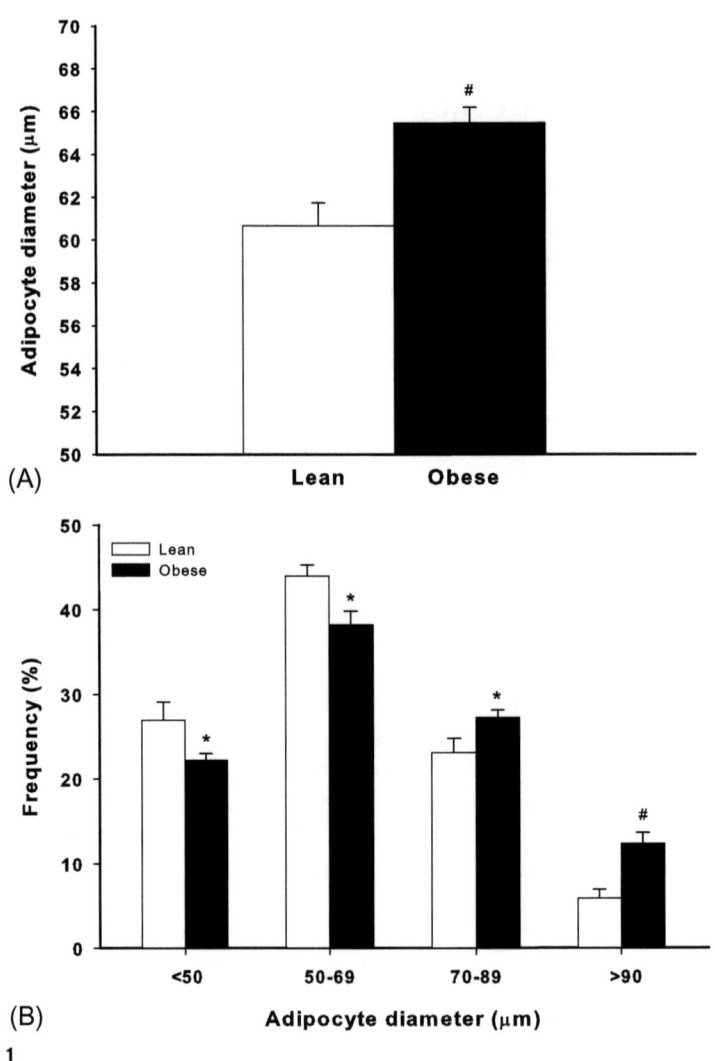

FIGURE 1.1

Mean adipocyte size (A) is usually higher in obese as compared with individuals (adipocyte hypertrophy) with (B) a shift toward a higher proportion of large/very large adipocytes. Values are mean ± SEM. *P < 0.05, #P < 0.01 versus lean [48].

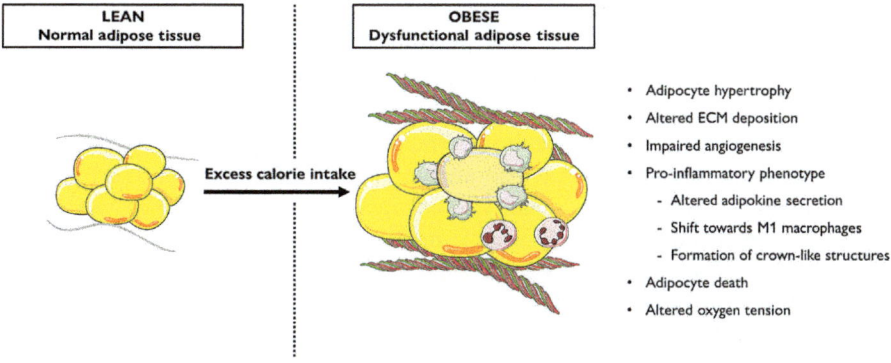

FIGURE 1.2

Obesity, which is the result of a prolonged positive energy balance, contributes to the development of dysfunctional adipose tissue characterized by adipocyte hypertrophy, a proinflammatory phenotype (i.e., a shift toward more M1 macrophages), adipocyte cell death, and an altered adipokine secretion pattern. Surrounding necrotic adipocytes, crown-like structures can be observed, which consist of phagocytic cells that remove cell debris. Furthermore, dysfunctional adipose tissue shows disproportionate deposition of ECM components. Together with impaired angiogenesis, this may limit further expansion of adipose tissue in obese individuals.

matrix (ECM) (Figure 1.2) [49]. An increase in adipose tissue mass can be achieved by hypertrophy or by recruiting new adipocytes from the resident pool of progenitors (i.e., hyperplasia). Although it is generally accepted that weight gain is the result of hypertrophy rather than hyperplasia of adipocytes, there is evidence that this may depend on the respective adipose tissue depots [50]. Nevertheless, stable isotope labeling from midcentury nuclear weapon testing has revealed that the number of adipocytes becomes fixed during childhood or early adulthood, yet obese individuals seem to have a greater number of adipocytes [51]. Furthermore, adipocyte number will not decrease once the number of adipocytes has been maximally achieved during childhood or early adulthood. Thus marked weight loss will reduce adipocyte size but not the overall number [51]. The changes that occur during substantial expansion of adipose tissue mass and that predispose to cardiometabolic diseases in obese individuals will be discussed later in more detail.

ADIPOSE TISSUE LIPID METABOLISM

To accommodate the need for additional fat storage under conditions of energy excess, adipose tissue enlarges. Adipose tissue has been hypothesized to have a limit to which it can expand [52]. During the progression of obesity, the increase in dietary lipids may exceed the storage capacity of adipose tissue and result in the ectopic deposition of lipids [53]. The deposition of abnormal quantities of lipids and lipid metabolites in a variety of cell types such as hepatocytes, myocytes, and beta cells interferes with cellular homeostasis. For example, ectopic fat storage, and more specifically, the accumulation, composition, and localization of bioactive lipid metabolites may interfere with insulin signaling and provoke apoptosis and inflammation, thereby compromising tissue function [54]. Lipodystrophic animals exemplify this phenomenon, often referred to as lipotoxicity [55]. In these particular animals, fat accumulates ectopically within insulin-sensitive tissues such as the liver and results in a plethora of symptoms such as insulin resistance, hepatic steatosis, and dyslipidemia [56]. Lipodystrophic humans are also unable to generate and maintain adipose tissue, which is characterized by scarcity of fat storage in subcutaneous tissue. Similar to lipodystrophic animals, lipodystrophic humans are severely insulin resistant, and thus most will develop overt type 2 diabetes [57].

Insulin resistance in obesity is often characterized by systemic lipid overflow, and more specifically, by increased circulating concentrations of TAG and FFA. As adipocytes become hypertrophic, a variety of both molecular and cellular alterations occur within adipocytes, which ultimately affects their capacity to store fat. An impaired postprandial lipid-buffering capacity of adipose tissue has been recognized as an important contributor to lipid overflow. This has been primarily characterized by impaired extraction of TAG in postprandial states and impaired regulation of lipolysis [58, 59]. In obesity, extraction of TAG appears to be impaired due to diminished insulin-mediated stimulation of LPL activity, which thereby increases spillover of FFA [23]. In addition, impaired insulin-mediated inhibition of lipolysis, together with the increased fat mass in obesity, often results in an increase in systemic concentrations of FFA, which has been thought to contribute to dyslipidemia in obesity [60]. The rise in the concentration of FFA is predominantly influenced by the activity of enzymes involved in lipolysis and the conformation of lipid droplet integrity proteins [59]. These proteins, called perilipins, are expressed within adipocytes on the surface of droplets containing TAG and prevent lipolytic enzymes from hydrolyzing the pool of TAG [61]. Nevertheless, obese individuals are characterized by a deficiency in perilipin expression, which leads to an increase in the basal rate of lipolysis and consequently circulating concentrations of FFA [62]. Furthermore, FFA have the capacity to inhibit the antilipolytic effects of insulin, thereby attenuating this negative feedback mechanism, further increasing systemic levels of FFA [63]. Several studies have demonstrated that an increase in FFA can mediate many adverse effects, in particular insulin resistance, impaired glucose tolerance, hepatic gluconeogenesis, and reduced insulin clearance by the liver [59]. Taken together, impaired regulation of the storage and release of lipids is associated with a variety of metabolic dysregulations, highlighting the importance of metabolic homeostasis in adipocytes.

ADIPOSE TISSUE INFLAMMATION

Adipose tissue expansion, which is characterized by adipocyte hypertrophy, leads to adipocyte necrosis and apoptosis, thereby contributing to the infiltration of macrophages into adipose tissue [64]. In obesity, adipocyte cell death may contribute to leukocyte recruitment as well as release of damage-associated molecular patterns (DAMP), which in turn lead to activation of inflammasomes within macrophages [65, 66]. Inflammasomes are multimeric complexes consisting of sensor and adaptor molecules and pro-caspase-1, which upon stimulation lead to the activation of inflammatory precursors, eventually causing inflammatory-mediated cell death [67]. In particular, the leucine-rich-containing family, pyrine domain-containing 3 inflammasome (NLRP3), is highly expressed in adipose tissue of obese mice, yet its role remains to be elucidated in obese humans [65]. Within macrophages, activated inflammasomes release caspase-1 that subsequently cleaves cytokine precursors, prointerleukin (IL)-1β and pro-IL-18 to biologically active IL-1β and IL-18, which in turn are secreted and amplify the inflammatory response in adipose tissue [68]. Indeed, a hallmark of adipose tissue expansion in obesity is the formation of so-called crown-like structures surrounding necrotic or dying adipocytes [64]. These crown-like structures are phagocytic cells, which remove adipocyte debris and incorporate lipids as well.

In normal weight individuals, adipose tissue macrophages (ATMs) account for approximately 10% of the cells found in adipose tissue [69]. Thus these cells are the most abundant immune cells in adipose tissue. The number of ATMs dramatically increases in obesity and accounts for 30%–40% of the cells present in adipose tissue, thereby illustrating the role of inflammatory processes in adipose tissue dysfunction [69]. Increased levels of FFA and lipopolysaccharide (LPS), which bind to Toll-like-receptor-4 (TLR4) and activate ATMs, may explain the increased abundance of ATMs in obese humans. Indeed, TLRs play an important role in immune responses. However, the importance of TLRs within the context of obesity and insulin resistance remains elusive. Several animal studies have demonstrated that both TLR2 and 4 contribute to the regulation of adipose tissue inflammation, and hence, insulin sensitivity [70, 71]. Initially, TLR2 and 4 were found to be receptors for lipoproteins derived from gram-negative and gram-positive bacteria, respectively [65]. Indeed, LPS is a well-known exogenous ligand for TLR4, which has been reported to be partly responsible for the low-grade inflammatory state in obesity, a phenomenon often referred to as metabolic endotoxemia [72]. However, mechanistic insights suggest that FFA can also potentially act as endogenous substrates for both receptors [73]. Subsequently, activation of TLRs might result in the activation of classical inflammatory pathway (e.g., myeloid differentiation primary response 88 pathway). Besides, intracellularly, the incorporated FFAs are partially diverted into ceramides, which are thought to be involved in the activation of caspase-1, thereby contributing to the proinflammatory phenotype in obesity [74].

In adipose tissue, macrophages can have multiple phenotypes, which are designated as classically activated M1 and alternatively activated M2 macrophages. M1 macrophages have a proinflammatory phenotype and are activated by proinflammatory factors such as interferon-gamma (IFN-γ) and LPS, which bind IFN-γ-receptor and TLR4, respectively [75]. Once activated, ATMs secrete several proinflammatory factors, including tumor necrosis factor (TNF)-α, IL-1β, IL-6, IL-12, and IL-23, which in turn provoke a T-helper type 1 (Th1) response [76]. The Th1 response results in recruitment and activation of CD4+ T-cells, which primarily secrete IFN-γ, subsequently stimulating polarization of monocytes toward M1 macrophages [77]. Besides secreting proinflammatory cytokines, M1 macrophages can attract monocytes and other immune cells by releasing chemoattractant cytokines such as C-C motif ligand (CCL)-2 and C-X-C motif ligand (CXCL)-9 and -10. The secretion of these proinflammatory factors is associated with the development of obesity-induced insulin resistance, as these factors promote signaling cascades involved in cellular-stress responses such as c-Jun N-terminal kinases 1/2 (JNK1/2), IκB kinase (IKKβ)-nuclear factor kappa-light-chain-enhancer of activated B-cells (NF-κβ), extracellular signal-regulated kinase 1 and 2 (ERK1/2), and p38 mitogen-activated protein kinase (p38-MAPK) [78]. The activation of these signaling cascades in turn interferes with insulin signaling by decreasing phosphorylation of insulin receptor substrate (IRS)-1 and -2 and phosphatidylinositol-3,4-bisphosphate 3-kinase (PI-3 K) activity, leading to insulin resistance [79].

On the other hand, M2 macrophages play a role in tissue remodeling and wound healing, and have antiinflammatory properties [80, 81]. During the development of obesity, dynamic changes occur in the immune cell population in adipose tissue. Specifically, a greater number of M1 macrophages infiltrate adipose tissue. Surprisingly, the number of M2 macrophages do not necessarily diminish in obesity but may even increase [82]. Clearly, adipose tissue in obesity is characterized by a major shift in the M1/M2 ratio, which results in a greater percentage of proinflammatory M1 macrophages [49, 83]. In contrast, in adipose tissue in lean individuals, the M1/2 ratio is lower than in obesity, with relatively fewer M1 macrophages [84]. M2 macrophages secrete antiinflammatory cytokines, including IL-10, IL-4, and transforming growth factor-β (TGF-β), and M2 macrophages are also known to suppress immune responses [85]. Furthermore, M2 ATMs play a key role in tissue remodeling because they are capable of synthesizing and secreting ECM components such as collagen and fibronectin [86].

Several cytokines secreted by adipocytes can influence various immune cells in a paracrine manner, thereby contributing to both pro- and antiinflammatory processes within the adipose tissue. For example, previous work by Vieira-Potter et al. [87] demonstrated that adiponectin is able to promote differentiation of ATMs to M2 macrophages. Given that expression of adiponectin is decreased in obese individuals [88], this could prevent polarization of ATMs to M2 macrophages, which may ultimately contribute to decreased insulin sensitivity. Indeed, adiponectin is widely recognized for its antiinflammatory features [89]. In obesity, the reduction in plasma adiponectin levels is associated with increased systemic levels of C-reactive protein (CRP), IL-6, and TNF-α [89], thereby playing a pivotal role in modulating the inflammatory state in obesity.

In addition to macrophages, other innate and adaptive immune cells contribute to an inflammatory phenotype of adipose tissue in obesity. Rodent studies have demonstrated that mast cells, neutrophils, B-lymphocytes, and T-lymphocytes all increase in abundance during the progressive development of obesity [77]. Interestingly, data demonstrate that all these immune cells exert negative effects on insulin sensitivity [90]. Conversely, eosinophils and regulatory T cells (Tregs), which counterbalance other immune cells such as macrophages, are also reduced during the development of obesity in rodents [91]. Moreover, the number of Tregs is positively associated with insulin resistance, hence reduced numbers of Tregs have a negative effect on insulin sensitivity [49].

Interestingly, proinflammatory immune cells can also have beneficial effects. During fasting and weight loss, the increase in lipolysis will attract macrophages to adipose tissue, which will phagocytize newly available lipids, and thereby, prevent lipotoxicity in the local microenvironment of adipose tissue [92]. Indeed, increased release of FFA, induced by lipolysis, is a key regulator in the recruitment of ATMs. ATMs phagocytize lipids and become lipid-laden macrophages, which serve as a buffer to suppress the continued increase in extracellular lipids in adipose tissue. Nevertheless, when this process is dysregulated and an excessive influx of macrophages occurs, a proinflammatory phenotype will prevail and exert detrimental effects on tissue homeostasis.

EXTRACELLULAR MATRIX REMODELING

Adipocytes express a wide variety of ECM proteins and enzymes required to break down various components of the ECM. The expression of these genes is highly regulated by changes in nutrient availability [93]. The ECM of adipose tissue is continuously adapting to the nutritional and hormonal cues that adipocytes receive during adipose tissue remodeling. In general, the ECM of adipose tissue consists of components similar to other tissues in the body [94], including fibrillar proteins (e.g., collagens), glycoproteins (e.g., fibronectin) and proteoglycans (e.g., hyaluronan). The adipose tissue ECM primarily serves mainly as a transducer of mechanical stress and thereby prevents the rupture of adipocytes and the subsequent release of intracellular TAG [94]. Reduced stiffness of the ECM may allow healthy expansion of adipose tissue, whereas an ECM that is too rigid may reduce the expansion capacity of adipocytes, thereby limiting excess nutrient storage [95, 96]. Moreover, it could be hypothesized that this may evoke stress-related pathways, inflammation, and ectopic fat deposition.

In obesity, the drastic expansion of adipose tissue results in tissue fibrosis in both humans and rodents [97, 98]. Adipose tissue fibrosis is associated with insulin resistance in obese individuals, although the underlying mechanisms remain to be elucidated. The formation of fibrotic adipose tissue is thought to be the result of increased ECM deposition [99]. This ultimately results in adipocyte death, which will activate macrophages via the classical pathway, resulting in an inflammatory response followed by metabolic complications, as previously discussed. Interestingly, Khan et al. [100] have demonstrated that depleting collagen VI in the ECM enhances adipocyte survival and improves metabolic homeostasis. In contrast, increased deposition of collagen in adipose tissue is associated with insulin resistance [101].

In addition to composition of the ECM, the expression pattern of ECM-cell transmembrane receptors, the so-called integrins, seems to contribute to the pathophysiology of insulin resistance in obesity. In adipose tissue, the expression of β-2-integrin is associated with the development of high-fat diet-induced insulin resistance in obesity [99]. Indeed, murine β-2-integrin knock-in mice show an increased inflammatory profile within adipose tissue, as well as increased insulin resistance [102]. In humans, the expression of certain subsets of β-2 integrins (CD11a-d) is increased in both adipose tissue and circulating monocytes of obese individuals [99], yet the specific role of integrins in the pathophysiology of obesity in humans remains to be elucidated. In addition to integrins, CD44, a cell-surface adhesion glycoprotein, appears to be positively correlated with both the inflammatory and glycemic profile of type 2 diabetics. In particular, when CD44 pairs with one of its ligands, osteopontin, this provokes localized inflammation and insulin resistance [103].

Besides disproportionate deposition of ECM components and expression of cellular transmembrane receptors, increased activity of matrix metalloproteinases (MMPs) are thought to contribute to fibrosis and insulin resistance during adipose tissue remodeling [99]. Generally, MMPs are considered enzymes (endopeptidases) that play a crucial role in ECM composition and are capable of degrading ECM components. In obesity, expression of certain subsets of MMPs is dysregulated. In particular, MMP2 and MMP9 plasma levels are increased, and the latter has been linked to insulin resistance in obese humans [104]. However, the exact underlying mechanisms with respect to the role of certain MMP subsets in the development of insulin resistance in humans remains to be clarified.

To inhibit the activity of MMPs, the human body endogenously expresses tissue inhibitors of metalloproteinases (TIMPs). Four different subsets of TIMPs exist (TIMP1–4). Both plasma TIMP1 and 2 levels are increased in obesity [105]. Strikingly, TIMP1 or TIMP2 knock-out mice are hyperphagic, obese, and insulin resistant [106, 107]. These experimental data suggest that enhanced TIMP activity, or reduction of MMP activity, is potentially associated with beneficial metabolic effects by improved regulation of adipose tissue remodeling in obesity.

ANGIOGENESIS

Adipose tissue is one of the most vascularized tissues in the human body and is characterized by a well-organized microcirculation. This large capillary network is unlike other vascular systems, as it is capable of growing throughout life. In the context of adipose tissue remodeling, angiogenesis plays a key role [108]. The vasculature network provides the adipose tissue with oxygen, nutrients, growth factors, and other cytokines, which are all essential in the development of mature adipocytes from stem cells [109].

During adipose tissue remodeling, adipocytes tightly control adipokine secretion. In obesity, increased leptin secretion enhances angiogenesis *in vitro* [110]. Besides leptin, other adipokines may induce angiogenesis such as basic fibroblast growth factor (bFGF) and vascular endothelial growth factor (VEGF), a proangiogenic factor [111, 112] These adipokines stimulate vascular endothelial growth and initiate angiogenesis, respectively.

The expansion of adipose tissue is accompanied by increased vascularization. Thereby, angiogenesis plays a pivotal role in the process of adipogenesis and the progression of obesity. Therefore angiogenesis is considered to be the rate-limiting step of adipose tissue remodeling [96]. Previous investigators demonstrate that adipose tissue expansion is potentially limited by its vascularization, and rapid growth of adipose tissue mass results in inadequate capillarization. Lemoine et al. [113] demonstrated that adipocyte size, in both the subcutaneous and visceral fat depots, is positively correlated with vessel density *per* adipocyte. These authors also showed that vascular endothelial growth factor receptor 2 (VEGFR2) expression, one of three receptors for VEGF, is associated with increased adipose tissue mass. However, previous work by Pasarica et al. [114] and Goossens et al. [48] provides conflicting evidence, because these authors demonstrated that both VEGF expression and capillary density were lower in abdominal subcutaneous white adipose tissue of obese when compared with lean individuals.

ADIPOSE TISSUE OXYGENATION

The underlying cause of adipose tissue dysfunction in obesity is under debate. Recently, adipose tissue oxygenation was hypothesized to be an essential contributor to obesity-induced adipose tissue dysfunction. In particular, an oxygen deficiency (e.g., hypoxia) within adipose tissue was proposed to play a role in the pathogenesis of obesity-related adipose tissue dysfunction by provoking a pro-inflammatory response. The oxygen status, or oxygen tension (pO_2), of a certain tissue depends upon the delicate balance between oxygen supply and oxygen consumption. For oxygen to reach a target tissue, it is transported mainly within red blood cells (RBCs) bound to hemoglobin [115]. The supply of oxygen is therefore dependent on several factors, including [1] the fraction of oxygen in inspired air (F_{IO2}), [2] the capacity of RBCs to carry oxygen throughout the circulation, and [3] local blood flow within target tissues. Numerous reports have demonstrated that adipose tissue blood flow is compromised in both fasting and postprandial obese subjects [48, 116].

The pO_2 within adipose tissue may be a key factor in the regulation of angiogenesis because oxygen levels influence VEGF expression. It has been postulated that, during prolonged intake of excess calories leading to obesity, the adipose tissue vasculature is unable to keep pace with the expanding tissue, resulting in local hypoxia [117, 118]. Indeed, experimental evidence indicates that adipose tissue oxygenation is reduced in obese mice [119], which may be related to a lower secretion of adiponectin and increased secretion of inflammatory adipokines [120].

However, whether adipose tissue hypoxia is a characteristic of obesity remains controversial. Although it has previously been shown that adipose tissue oxygenation is lower in overweight and obese subjects [114], we have demonstrated a greater abdominal subcutaneous adipose tissue pO_2 in obese rather than lean humans, which was associated with insulin resistance [48]. In accordance with the latter, we have recently shown that diet-induced weight loss decreased adipose tissue oxygen tension in humans, which was paralleled by improved insulin sensitivity [121]. Whether altered adipose tissue pO_2 is involved in obesity-related inflammation and insulin resistance is a topic currently under active investigation.

ADIPOSE TISSUE DYSFUNCTION AND LUNG DISEASES

The development of obesity is associated with the development of a variety of respiratory diseases, among which are OSAS, chronic obstructive pulmonary disease, and asthma [122]. In general, obese individuals are at an increased risk for developing respiratory abnormalities even without the presence of any preexisting lung disease. The effect of obesity on respiratory function are primarily dependent upon the dispersion of adipose tissue in the neck, thorax, and abdomen, which can ultimately effect respiratory mechanics, gas exchange, and ventilatory control [123]. Indeed, respiratory function in obesity is characterized by decreased lung compliance, small airway dysfunction, and arterial oxygen desaturation [124]. There is evidence to suggest that adipose tissue expansion, and related systemic low-grade inflammation, is capable of reducing pulmonary function in obese individuals (Figure 1.3) [125].

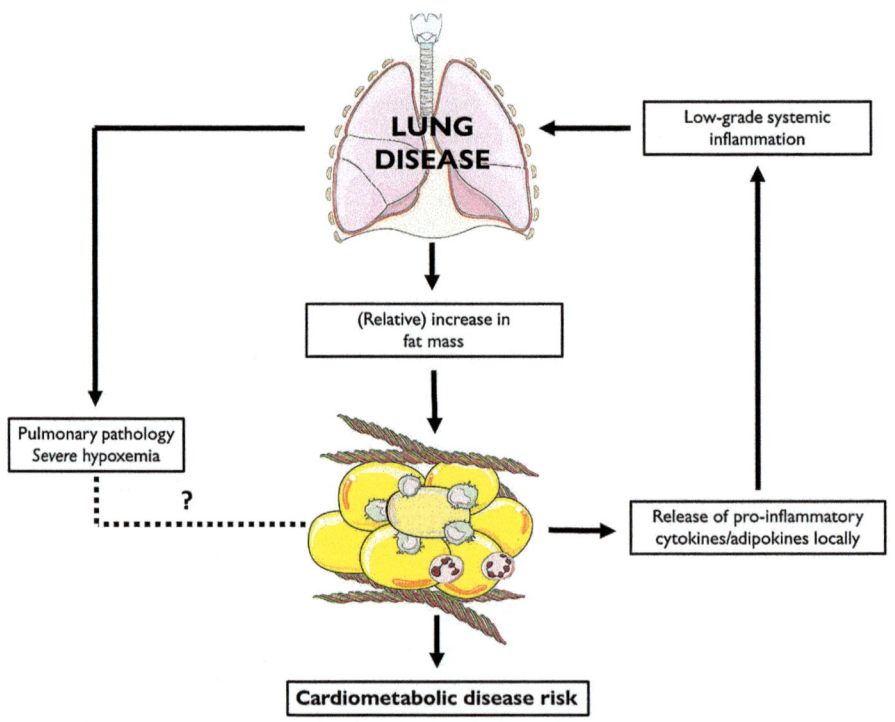

FIGURE 1.3

Adipose tissue dysfunction in patients with lung disease, who are characterized by a relative excess of adipose tissue, may be an important contributor to the increased risk of developing cardiometabolic diseases as well as greater decrements in pulmonary function. On the other hand, systemic hypoxemia in severe COPD might result in adipose tissue hypoxia, thereby affecting adipose tissue function.

Over the last decade, accumulating evidence suggests that the effects of pulmonary diseases reach beyond the lungs, are manifested systemically, and are associated with an increased risk for T2DM and cardiovascular disease. Among other factors, adipose tissue dysfunction may be one of the drivers for the increased cardiometabolic disease risk in individuals with OSAS [126], asthma [127, 128], and COPD and sarcopenia. The latter are characterized by a relative increased adipose tissue mass and a (hidden) loss of fat-free mass (Figure 1.3) [129]. Indeed, impairments in lipid metabolism (i.e., increased lipolysis and reduced fat oxidative capacity), resulting in ectopic fat storage, and low-grade systemic inflammation have been described in patients with lung disease [123, 130]. On the other hand, systemic hypoxemia in severe COPD might result in adipose tissue hypoxia, which in turn may alter adipose tissue function (Figure 1.3).

CONCLUSION

Adipose tissue dysfunction in obesity is associated with a plethora of metabolic and endocrine disturbances, including impairments in lipid and glucose metabolism, local and systemic inflammation, and an altered pattern of adipokine secretion. It is therefore of no surprise that dysfunctional adipose tissue plays a pivotal role in the pathophysiology of obesity-related comorbidities and the development of cardiometabolic diseases. During the progression of obesity, impaired expansion of adipose tissue contributes to lipid accumulation in nonadipose tissues, which is closely associated with metabolic derangements. Moreover, adipose tissue in obesity is characterized by a proinflammatory phenotype. The latter is illustrated by a phenotypic shift toward a higher abundance of proinflammatory macrophages, infiltration of other adaptive and innate immune cells, and the secretion of proinflammatory cytokines. Besides inflammation, ECM remodeling occurs during the progression of obesity characterized by a disproportionate deposition of ECM components and increased MMP activity. These alterations may contribute to the development of adipose tissue fibrosis, and hence, worsen insulin resistance in obesity. Furthermore, adipose tissue oxygen tension, determined by the balance between oxygen supply and consumption, seems to play a role in metabolic regulation. Indeed, animal models have shown a reduced pO_2, referred to as "hypoxia", within the adipose tissue of obese mice. However, human data are conflicting, and recent evidence suggests that adipose tissue pO_2 may be higher in obese insulin-resistant individuals [48]. It remains to be elucidated whether adipose tissue pO_2 exerts a causal role in the progression of obesity-related comorbidities. Moreover, adipose tissue dysfunction may be present in patients with lung diseases such as OSAS, asthma, and COPD, thereby contributing to disease pathology. In fact, adipose tissue dysfunction increases both local and systemic inflammation by secretion of proinflammatory cytokines and acute phase proteins, which might worsen both metabolic health and pulmonary function in patients with lung disease.

Taken together, adipose tissue function is highly important in metabolic regulation, and adipose tissue dysfunction in obesity is associated with various deleterious cardiometabolic consequences. Intervening at the level of the adipose tissue may therefore provide an interesting strategy to improve cardiometabolic health, not only in obese individuals but also in patients with lung disease who are characterized by a relative excess of adipose tissue. Moreover, targeting adipose tissue dysfunction in patients with lung disease may be a valuable approach to alleviate impaired pulmonary function in these individuals.

REFERENCES

[1] Kopelman PG. Obesity as a medical problem. Nature 2000;404(6778):635–43.
[2] Organization WH. Obesity and overweight: fact sheet; World Health Organization Media Centre.
[3] Wang YC, McPherson K, Marsh T, Gortmaker SL, Brown M. Health and economic burden of the projected obesity trends in the USA and the UK. Lancet 2011;378(9793):815–25.
[4] Nguyen JC, Killcross AS, Jenkins TA. Obesity and cognitive decline: role of inflammation and vascular changes. Front Neurosci 2014;8:375.
[5] Luppino FS, de Wit LM, Bouvy PF, Stijnen T, Cuijpers P, Penninx BW, et al. Overweight, obesity, and depression: a systematic review and meta-analysis of longitudinal studies. Arch Gen Psychiatry 2010;67(3):220–9.
[6] De Pergola G, Silvestris F. Obesity as a major risk factor for cancer. J Obes 2013;2013:291546.
[7] Schwartz AR, Patil SP, Laffan AM, Polotsky V, Schneider H, Smith PL. Obesity and obstructive sleep apnea: pathogenic mechanisms and therapeutic approaches. Proc Am Thorac Soc 2008;5(2):185–92.
[8] Yosipovitch G, DeVore A, Dawn A. Obesity and the skin: skin physiology and skin manifestations of obesity. J Am Acad Dermatol 2007;56(6):901–16. [quiz 17-20].
[9] Poirier P, Giles TD, Bray GA, Hong Y, Stern JS, Pi-Sunyer FX, et al. Obesity and cardiovascular disease: pathophysiology, evaluation, and effect of weight loss: an update of the 1997 American Heart Association scientific statement on obesity and heart disease from the obesity Committee of the Council on nutrition, physical activity, and metabolism. Circulation 2006;113(6):898–918.
[10] Goossens GH. The metabolic phenotype in obesity: fat mass, body fat distribution, and adipose tissue function. Obes Facts 2017;10(3):207–15.
[11] Bluher M. Adipose tissue dysfunction contributes to obesity related metabolic diseases. Best Pract Res Clin Endocrinol Metab 2013;27(2):163–77.
[12] Zhou J, Qin G. Adipocyte dysfunction and hypertension. Am J Cardiovasc Dis 2012;2(2):143–9.
[13] Yuan Y, Gao J, Ogawa R. Mechanobiology and mechanotherapy of adipose tissue-effect of mechanical force on fat tissue engineering. Plast Reconstr Surg Glob Open 2015;3(12):e578.
[14] Bohler Jr. H, Mokshagundam S, Winters SJ. Adipose tissue and reproduction in women. Fertil Steril 2010;94(3):795–825.
[15] Fruhbeck G. Overview of adipose tissue and its role in obesity and metabolic disorders. Methods Mol Biol 2008;456:1–22.
[16] Lafontan M, Langin D. Lipolysis and lipid mobilization in human adipose tissue. Prog Lipid Res 2009;48(5):275–97.

[17] Coelho M, Oliveira T, Fernandes R. Biochemistry of adipose tissue: an endocrine organ. Arch Med Sci 2013;9(2):191–200.

[18] Esteve Rafols M. Adipose tissue: cell heterogeneity and functional diversity. Endocrinol Nutr 2014;61(2):100–12.

[19] Pinney SE. Stages of the developing adipocyte: determination and differentiation. In: Rosenfeld CS, editor. The epigenome and developmental origins of health and disease. Academic Press; 2016. p. 267–89.

[20] Ahmadian M, Duncan RE, Jaworski K, Sarkadi-Nagy E, Sul HS. Triacylglycerol metabolism in adipose tissue. Future Lipidol 2007;2(2):229–37.

[21] Nakajima K, Nakano T, Tokita Y, Nagamine T, Inazu A, Kobayashi J, et al. Postprandial lipoprotein metabolism: VLDL vs. chylomicrons. Clin Chim Acta 2011;412(15–16):1306–18.

[22] Camps L, Reina M, Llobera M, Vilaro S, Olivecrona T. Lipoprotein lipase: cellular origin and functional distribution. Am J Physiol 1990;258(4 Pt 1):C673–81.

[23] Stinkens R, Goossens GH, Jocken JW, Blaak EE. Targeting fatty acid metabolism to improve glucose metabolism. Obes Rev 2015;16(9):715–57.

[24] Yost TJ, Jensen DR, Haugen BR, Eckel RH. Effect of dietary macronutrient composition on tissue-specific lipoprotein lipase activity and insulin action in normal-weight subjects. Am J Clin Nutr 1998;68(2):296–302.

[25] Frayn KN, Coppack SW, Fielding BA, Humphreys SM. Coordinated regulation of hormone-sensitive lipase and lipoprotein lipase in human adipose tissue in vivo: implications for the control of fat storage and fat mobilization. Adv Enzyme Regul 1995;35:163–78.

[26] Lass A, Zimmermann R, Oberer M, Zechner R. Lipolysis—a highly regulated multienzyme complex mediates the catabolism of cellular fat stores. Prog Lipid Res 2011;50(1):14–27.

[27] Verboven K, Hansen D, Jocken JWE, Blaak EE. Natriuretic peptides in the control of lipid metabolism and insulin sensitivity. Obes Rev 2017;18(11):1243–59.

[28] Duncan RE, Ahmadian M, Jaworski K, Sarkadi-Nagy E, Sul HS. Regulation of lipolysis in adipocytes. Annu Rev Nutr 2007;27:79–101.

[29] Sethi JK, Vidal-Puig AJ. Thematic review series: adipocyte biology. Adipose tissue function and plasticity orchestrate nutritional adaptation. J Lipid Res 2007;48(6):1253–62.

[30] Klok MD, Jakobsdottir S, Drent ML. The role of leptin and ghrelin in the regulation of food intake and body weight in humans: a review. Obes Rev 2007;8(1):21–34.

[31] Karastergiou K, Mohamed-Ali V. The autocrine and paracrine roles of adipokines. Mol Cell Endocrinol 2010;318(1–2):69–78.

[32] Fedorenko A, Lishko PV, Kirichok Y. Mechanism of fatty-acid-dependent UCP1 uncoupling in brown fat mitochondria. Cell 2012;151(2):400–13.

[33] Dawkins MJ, Scopes JW. Non-shivering thermogenesis and brown adipose tissue in the human new-born infant. Nature 1965;206(980):201–2.

[34] Iyer RB, Guo CC, Perrier N. Adrenal pheochromocytoma with surrounding brown fat stimulation. AJR Am J Roentgenol 2009;192(1):300–1.

[35] Ologun GO, Patel ZM, Rana NK, Trecartin A, Shen A, Trostle D, et al. Large unilateral adrenal mass with surrounding Brown fat: a case report. Cureus 2017;9(8):e1552.

[36] Loh RKC, Kingwell BA, Carey AL. Human brown adipose tissue as a target for obesity management; beyond cold-induced thermogenesis. Obes Rev 2017;18(11):1227–42.

[37] Park A, Kim WK, Bae KH. Distinction of white, beige and brown adipocytes derived from mesenchymal stem cells. World J Stem Cells 2014;6(1):33–42.

[38] van der Lans AA, Hoeks J, Brans B, Vijgen GH, Visser MG, Vosselman MJ, et al. Cold acclimation recruits human brown fat and increases nonshivering thermogenesis. J Clin Invest 2013;123(8):3395–403.

[39] Bordicchia M, Liu D, Amri EZ, Ailhaud G, Dessi-Fulgheri P, Zhang C, et al. Cardiac natriuretic peptides act via p38 MAPK to induce the brown fat thermogenic program in mouse and human adipocytes. J Clin Invest 2012;122(3):1022–36.

[40] Broeders EP, Vijgen GH, Havekes B, Bouvy ND, Mottaghy FM, Kars M, et al. Thyroid hormone activates Brown adipose tissue and increases non-shivering thermogenesis—a cohort study in a Group of Thyroid Carcinoma Patients. PLoS One 2016;11(1):e0145049.

[41] Broeders EP, Nascimento EB, Havekes B, Brans B, Roumans KH, Tailleux A, et al. The bile acid chenodeoxycholic acid increases human Brown adipose tissue activity. Cell Metab 2015;22(3):418–26.

[42] Cinti S. Between brown and white: novel aspects of adipocyte differentiation. Ann Med 2011;43(2):104–15.

[43] Jensen MD. Role of body fat distribution and the metabolic complications of obesity. J Clin Endocrinol Metab 2008;93(11 Suppl 1):S57–63.

[44] Tchkonia T, Thomou T, Zhu Y, Karagiannides I, Pothoulakis C, Jensen MD, et al. Mechanisms and metabolic implications of regional differences among fat depots. Cell Metab 2013;17(5):644–56.

[45] Macotela Y, Emanuelli B, Mori MA, Gesta S, Schulz TJ, Tseng YH, et al. Intrinsic differences in adipocyte precursor cells from different white fat depots. Diabetes 2012;61(7):1691–9.

[46] Fox CS, Massaro JM, Hoffmann U, Pou KM, Maurovich-Horvat P, Liu CY, et al. Abdominal visceral and subcutaneous adipose tissue compartments: association with metabolic risk factors in the Framingham Heart Study. Circulation 2007;116(1):39–48.

[47] Lettner A, Roden M. Ectopic fat and insulin resistance. Curr Diab Rep 2008;8(3):185–91.

[48] Goossens GH, Bizzarri A, Venteclef N, Essers Y, Cleutjens JP, Konings E, et al. Increased adipose tissue oxygen tension in obese compared with lean men is accompanied by insulin resistance, impaired adipose tissue capillarization, and inflammation. Circulation 2011;124(1):67–76.

[49] Rosen ED, Spiegelman BM. What we talk about when we talk about fat. Cell 2014;156(1–2):20–44.

[50] Tchoukalova YD, Votruba SB, Tchkonia T, Giorgadze N, Kirkland JL, Jensen MD. Regional differences in cellular mechanisms of adipose tissue gain with overfeeding. Proc Natl Acad Sci U S A 2010;107(42):18226–31.

[51] Spalding KL, Arner E, Westermark PO, Bernard S, Buchholz BA, Bergmann O, et al. Dynamics of fat cell turnover in humans. Nature 2008;453(7196):783–7.

[52] Unger RH. Lipid overload and overflow: metabolic trauma and the metabolic syndrome. Trends Endocrinol Metab 2003;14(9):398–403.

[53] Virtue S, Vidal-Puig A. Adipose tissue expandability, lipotoxicity and the metabolic syndrome—an allostatic perspective. Biochim Biophys Acta 2010;1801(3):338–49.

[54] Chavez JA, Summers SA. Lipid oversupply, selective insulin resistance, and lipotoxicity: molecular mechanisms. Biochim Biophys Acta 2010;1801(3):252–65.

[55] Cusi K. Role of obesity and lipotoxicity in the development of nonalcoholic steatohepatitis: pathophysiology and clinical implications. Gastroenterology 2012;142(4). 711–25e6.

[56] Kim JK, Gavrilova O, Chen Y, Reitman ML, Shulman GI. Mechanism of insulin resistance in A-ZIP/F-1 fatless mice. J Biol Chem 2000;275(12):8456–60.

[57] Ganda OP. Lipoatrophy, lipodystrophy, and insulin resistance. Ann Intern Med 2000;133(4):304–6.

[58] Potts JL, Coppack SW, Fisher RM, Humphreys SM, Gibbons GF, Frayn KN. Impaired postprandial clearance of triacylglycerol-rich lipoproteins in adipose tissue in obese subjects. Am J Physiol 1995;268(4 Pt 1):E588–94.

[59] Karpe F, Dickmann JR, Frayn KN. Fatty acids, obesity, and insulin resistance: time for a reevaluation. Diabetes 2011;60(10):2441–9.

[60] Jung UJ, Choi MS. Obesity and its metabolic complications: the role of adipokines and the relationship between obesity, inflammation, insulin resistance, dyslipidemia and nonalcoholic fatty liver disease. Int J Mol Sci 2014;15(4):6184–223.

[61] Zhang HH, Souza SC, Muliro KV, Kraemer FB, Obin MS, Greenberg AS. Lipase-selective functional domains of perilipin A differentially regulate constitutive and protein kinase A-stimulated lipolysis. J Biol Chem 2003;278(51):51535–42.

[62] Wang Y, Sullivan S, Trujillo M, Lee MJ, Schneider SH, Brolin RE, et al. Perilipin expression in human adipose tissues: effects of severe obesity, gender, and depot. Obes Res 2003;11(8):930–6.

[63] Boden G. Free fatty acids (FFA), a link between obesity and insulin resistance. Front Biosci 1998;3:d169–75.

[64] Cinti S, Mitchell G, Barbatelli G, Murano I, Ceresi E, Faloia E, et al. Adipocyte death defines macrophage localization and function in adipose tissue of obese mice and humans. J Lipid Res 2005;46(11):2347–55.

[65] Sun S, Ji Y, Kersten S, Qi L. Mechanisms of inflammatory responses in obese adipose tissue. Annu Rev Nutr 2012;32:261–86.

[66] Vandanmagsar B, Youm YH, Ravussin A, Galgani JE, Stadler K, Mynatt RL, et al. The NLRP3 inflammasome instigates obesity-induced inflammation and insulin resistance. Nat Med 2011;17(2):179–88.

[67] Sharma D, Kanneganti TD. The cell biology of inflammasomes: mechanisms of inflammasome activation and regulation. J Cell Biol 2016;213(6):617–29.

[68] Stienstra R, Joosten LA, Koenen T, van Tits B, van Diepen JA, van den Berg SA, et al. The inflammasome-mediated caspase-1 activation controls adipocyte differentiation and insulin sensitivity. Cell Metab 2010;12(6):593–605.

[69] Weisberg SP, McCann D, Desai M, Rosenbaum M, Leibel RL, Ferrante Jr. AW. Obesity is associated with macrophage accumulation in adipose tissue. J Clin Invest 2003;112(12):1796–808.

[70] Caricilli AM, Nascimento PH, Pauli JR, Tsukumo DM, Velloso LA, Carvalheira JB, et al. Inhibition of toll-like receptor 2 expression improves insulin sensitivity and signaling in muscle and white adipose tissue of mice fed a high-fat diet. J Endocrinol 2008;199(3):399–406.

[71] Saberi M, Woods NB, de Luca C, Schenk S, Lu JC, Bandyopadhyay G, et al. Hematopoietic cell-specific deletion of toll-like receptor 4 ameliorates hepatic and adipose tissue insulin resistance in high-fat-fed mice. Cell Metab 2009;10(5):419–29.

[72] Boutagy NE, McMillan RP, Frisard MI, Hulver MW. Metabolic endotoxemia with obesity: is it real and is it relevant? Biochimie 2016;124:11–20.

[73] Konner AC, Bruning JC. Toll-like receptors: linking inflammation to metabolism. Trends Endocrinol Metab 2011;22(1):16–23.

[74] Watanabe Y, Nagai Y, Takatsu K. Activation and regulation of the pattern recognition receptors in obesity-induced adipose tissue inflammation and insulin resistance. Nutrients 2013;5(9):3757–78.

[75] Martinez FO, Gordon S. The M1 and M2 paradigm of macrophage activation: time for reassessment. F1000Prime Rep 2014;6:13.

[76] Arango Duque G, Descoteaux A. Macrophage cytokines: involvement in immunity and infectious diseases. Front Immunol 2014;5:491.

[77] Huh JY, Park YJ, Ham M, Kim JB. Crosstalk between adipocytes and immune cells in adipose tissue inflammation and metabolic dysregulation in obesity. Mol Cells 2014;37(5):365–71.

[78] Chen L, Chen R, Wang H, Liang F. Mechanisms linking inflammation to insulin resistance. Int J Endocrinol 2015;2015:508409.

[79] Fujishiro M, Gotoh Y, Katagiri H, Sakoda H, Ogihara T, Anai M, et al. Three mitogen-activated protein kinases inhibit insulin signaling by different mechanisms in 3T3-L1 adipocytes. Mol Endocrinol 2003;17(3):487–97.

[80] Koh TJ, DiPietro LA. Inflammation and wound healing: the role of the macrophage. Expert Rev Mol Med 2011;13:e23.

[81] Mills CD. M1 and M2 macrophages: oracles of health and disease. Crit Rev Immunol 2012;32(6):463–88.

[82] Fujisaka S, Usui I, Bukhari A, Ikutani M, Oya T, Kanatani Y, et al. Regulatory mechanisms for adipose tissue M1 and M2 macrophages in diet-induced obese mice. Diabetes 2009;58(11):2574–82.

[83] Lumeng CN, Bodzin JL, Saltiel AR. Obesity induces a phenotypic switch in adipose tissue macrophage polarization. J Clin Invest 2007;117(1):175–84.

[84] Lumeng CN, DelProposto JB, Westcott DJ, Saltiel AR. Phenotypic switching of adipose tissue macrophages with obesity is generated by spatiotemporal differences in macrophage subtypes. Diabetes 2008;57(12):3239–46.

[85] Castoldi A, Naffah de Souza C, Camara NO, Moraes-Vieira PM. The macrophage switch in obesity development. Front Immunol 2015;6:637.

[86] Novak ML, Koh TJ. Macrophage phenotypes during tissue repair. J Leukoc Biol 2013;93(6):875–81.

[87] Vieira-Potter VJ. Inflammation and macrophage modulation in adipose tissues. Cell Microbiol 2014;16(10):1484–92.

[88] Weiss R, Dufour S, Groszmann A, Petersen K, Dziura J, Taksali SE, et al. Low adiponectin levels in adolescent obesity: a marker of increased intramyocellular lipid accumulation. J Clin Endocrinol Metab 2003;88(5):2014–8.

[89] Ouchi N, Walsh K. Adiponectin as an anti-inflammatory factor. Clin Chim Acta 2007;380(1–2):24–30.

[90] Mathis D. Immunological goings-on in visceral adipose tissue. Cell Metab 2013;17(6):851–9.

[91] Feuerer M, Herrero L, Cipolletta D, Naaz A, Wong J, Nayer A, et al. Lean, but not obese, fat is enriched for a unique population of regulatory T cells that affect metabolic parameters. Nat Med 2009;15(8):930–9.

[92] Kosteli A, Sugaru E, Haemmerle G, Martin JF, Lei J, Zechner R, et al. Weight loss and lipolysis promote a dynamic immune response in murine adipose tissue. J Clin Invest 2010;120(10):3466–79.

[93] Maquoi E, Munaut C, Colige A, Collen D, Lijnen HR. Modulation of adipose tissue expression of murine matrix metalloproteinases and their tissue inhibitors with obesity. Diabetes 2002;51(4):1093–101.

[94] Mariman EC, Wang P. Adipocyte extracellular matrix composition, dynamics and role in obesity. Cell Mol Life Sci 2010;67(8):1277–92.

[95] Alkhouli N, Mansfield J, Green E, Bell J, Knight B, Liversedge N, et al. The mechanical properties of human adipose tissues and their relationships to the structure and composition of the extracellular matrix. Am J Physiol Endocrinol Metab 2013;305(12):E1427–35.

[96] Sun K, Kusminski CM, Scherer PE. Adipose tissue remodeling and obesity. J Clin Invest 2011;121(6):2094–101.

[97] Divoux A, Tordjman J, Lacasa D, Veyrie N, Hugol D, Aissat A, et al. Fibrosis in human adipose tissue: composition, distribution, and link with lipid metabolism and fat mass loss. Diabetes 2010;59(11):2817–25.

[98] Marcelin G, Ferreira A, Liu Y, Atlan M, Aron-Wisnewsky J, Pelloux V, et al. A PDGFRalpha-mediated switch toward CD9(high) adipocyte progenitors controls obesity-induced adipose tissue fibrosis. Cell Metab 2017;25(3):673–85.

[99] Lin CTH, Kang L. Adipose extracellular matrix remodelling in obesity and insulin resistance. Biochem Pharmacol 2016;119:8–16.

[100] Khan T, Muise ES, Iyengar P, Wang ZV, Chandalia M, Abate N, et al. Metabolic dysregulation and adipose tissue fibrosis: role of collagen VI. Mol Cell Biol 2009;29(6):1575–91.

[101] Berria R, Wang L, Richardson DK, Finlayson J, Belfort R, Pratipanawatr T, et al. Increased collagen content in insulin-resistant skeletal muscle. Am J Physiol Endocrinol Metab 2006;290(3):E560–5.

[102] Meakin PJ, Morrison VL, Sneddon CC, Savinko T, Uotila L, Jalicy SM, et al. Mice lacking beta2-integrin function remain glucose tolerant in spite of insulin resistance, neutrophil infiltration and inflammation. PLoS One 2015;10(9):e0138872.

[103] Liu LF, Kodama K, Wei K, Tolentino LL, Choi O, Engleman EG, et al. The receptor CD44 is associated with systemic insulin resistance and proinflammatory macrophages in human adipose tissue. Diabetologia 2015;58(7):1579–86.

[104] Derosa G, Ferrari I, D'Angelo A, Tinelli C, Salvadeo SA, Ciccarelli L, et al. Matrix metalloproteinase-2 and -9 levels in obese patients. Endothelium 2008;15(4):219–24.

[105] Hopps E, Caimi G. Protein oxidation in metabolic syndrome. Clin Invest Med 2013;36(1):E1–8.

[106] Gerin I, Louis GW, Zhang X, Prestwich TC, Kumar TR, Myers Jr. MG, et al. Hyperphagia and obesity in female mice lacking tissue inhibitor of metalloproteinase-1. Endocrinology 2009;150(4):1697–704.

[107] Jaworski DM, Sideleva O, Stradecki HM, Langlois GD, Habibovic A, Satish B, et al. Sexually dimorphic diet-induced insulin resistance in obese tissue inhibitor of metalloproteinase-2 (TIMP-2)-deficient mice. Endocrinology 2011;152(4):1300–13.

[108] Cao Y. Angiogenesis modulates adipogenesis and obesity. J Clin Invest 2007;117(9):2362–8.

[109] Lemoine AY, Ledoux S, Larger E. Adipose tissue angiogenesis in obesity. Thromb Haemost 2013;110(4):661–8.

[110] Park HY, Kwon HM, Lim HJ, Hong BK, Lee JY, Park BE, et al. Potential role of leptin in angiogenesis: leptin induces endothelial cell proliferation and expression of matrix metalloproteinases in vivo and in vitro. Exp Mol Med 2001;33(2):95–102.

[111] Hausman GJ, Richardson RL. Adipose tissue angiogenesis. J Anim Sci 2004;82(3):925–34.

[112] Cross MJ, Claesson-Welsh L. FGF and VEGF function in angiogenesis: signalling pathways, biological responses and therapeutic inhibition. Trends Pharmacol Sci 2001;22(4):201–7.

[113] Lemoine AY, Ledoux S, Queguiner I, Calderari S, Mechler C, Msika S, et al. Link between adipose tissue angiogenesis and fat accumulation in severely obese subjects. J Clin Endocrinol Metab 2012;97(5):E775–80.

[114] Pasarica M, Sereda OR, Redman LM, Albarado DC, Hymel DT, Roan LE, et al. Reduced adipose tissue oxygenation in human obesity: evidence for rarefaction, macrophage chemotaxis, and inflammation without an angiogenic response. Diabetes 2009;58(3):718–25.

[115] Hodson L. Adipose tissue oxygenation: effects on metabolic function. Adipocyte 2014;3(1):75–80.

[116] Virtanen KA, Lonnroth P, Parkkola R, Peltoniemi P, Asola M, Viljanen T, et al. Glucose uptake and perfusion in subcutaneous and visceral adipose tissue during insulin stimulation in nonobese and obese humans. J Clin Endocrinol Metab 2002;87(8):3902–10.

[117] Trayhurn P, Wood IS. Adipokines: inflammation and the pleiotropic role of white adipose tissue. Br J Nutr 2004;92(3):347–55.

[118] Trayhurn P. Hypoxia and adipose tissue function and dysfunction in obesity. Physiol Rev 2013;93(1):1–21.

[119] Goossens GH, Blaak EE. Adipose tissue dysfunction and impaired metabolic health in human obesity: a matter of oxygen? Front Endocrinol (Lausanne) 2015;6:55.

[120] Ye J. Emerging role of adipose tissue hypoxia in obesity and insulin resistance. Int J Obes (Lond) 2009;33(1):54–66.

[121] Vink RG, Roumans NJ, Cajlakovic M, Cleutjens JPM, Boekschoten MV, Fazelzadeh P, et al. Diet-induced weight loss decreases adipose tissue oxygen tension with parallel changes in adipose tissue phenotype and insulin sensitivity in overweight humans. Int J Obes (Lond) 2017;41(5):722–8.

[122] Koenig SM. Pulmonary complications of obesity. Am J Med Sci 2001;321(4):249–79.

[123] Franssen FM, O'Donnell DE, Goossens GH, Blaak EE, Schols AM. Obesity and the lung: 5. Obesity and COPD. Thorax 2008;63(12):1110–7.

[124] Salome CM, King GG, Berend N. Physiology of obesity and effects on lung function. J Appl Physiol (1985) 2010;108(1):206–11.

[125] Mancuso P. Obesity and lung inflammation. J Appl Physiol (1985) 2010;108(3):722–8.

[126] Thorn CE, Knight B, Pastel E, McCulloch LJ, Patel B, Shore AC, et al. Adipose tissue is influenced by hypoxia of obstructive sleep apnea syndrome independent of obesity. Diabetes Metab 2017;43(3):240–7.

[127] Fenger RV, Gonzalez-Quintela A, Linneberg A, Husemoen LL, Thuesen BH, Aadahl M, et al. The relationship of serum triglycerides, serum HDL, and obesity to the risk of wheezing in 85,555 adults. Respir Med 2013;107(6):816–24.

[128] Sideleva O, Suratt BT, Black KE, Tharp WG, Pratley RE, Forgione P, et al. Obesity and asthma: an inflammatory disease of adipose tissue not the airway. Am J Respir Crit Care Med 2012;186(7):598–605.

[129] Song WJ, Kim SH, Lim S, Park YJ, Kim MH, Lee SM, et al. Association between obesity and asthma in the elderly population: potential roles of abdominal subcutaneous adiposity and sarcopenia. Ann Allergy Asthma Immunol 2012;109(4):243–8.

[130] Wouters EF, Reynaert NL, Dentener MA, Vernooy JH. Systemic and local inflammation in asthma and chronic obstructive pulmonary disease: is there a connection? Proc Am Thorac Soc 2009;6(8):638–47.

Complex interface between immunity and metabolism: The lung as a target organ

Claudio Procaccini*, **Veronica De Rosa***, **Francesco Perna**†, **Giuseppe Matarese***,‡

Istituto per l'Endocrinologia e l'Oncologia Sperimentale, Consiglio Nazionale delle Ricerche (IEOS-CNR), Napoli, Italy* Dipartimento di Medicina Clinica e Chirurgia, Università degli Studi di Napoli "Federico II", Napoli, Italy† Treg Cell Lab, Dipartimento di Medicina Molecolare e Biotecnologie Mediche, Università degli Studi di Napoli "Federico II", Napoli, Italy‡

INTRODUCTION

The intimate connection between immunity and nutritional status has been considered for centuries because of the threats of scarce food supplies that frequently occurred in the past and that still occur in poor countries [1]. A series of studies and practical evidence have confirmed that undernutrition, in the early stages of life, can strongly associate with an impaired immune function against infectious agents [2]. Many of those aspects have radically changed with the advent of the industrial revolution and the technological advancements of the 20th century, together with improved and increased production in the agriculture and farming industries, particularly in the western world. Nowadays, malnourished people are commonly targeted by pathogens, with undernutrition representing the most common cause of secondary immunodeficiency worldwide [1, 2]. However, the long-standing concept that nutrients can support immune responses against pathogens still has inadequate scientific support due to the limited studies and literature on the topic. A basic principle tied to this concept is that immune responses require consistent energy to occur. In times of famine and/or reduced energy intake, particularly in very young subjects (where metabolism is higher), immune responses could fail for two main reasons: the reduced amount of energy (mainly stored as fat reduced by lack of food) cannot adequately fuel immune cell proliferation/expansion, or the production of antibodies and cytokines; the reduction of adipose-tissue-derived adipocytokines (which signal the brain and periphery that sufficient food is stored as fat) cannot support proper immune reactivity. As a result, starvation-induced immunosuppression still represents the most frequent cause of secondary immunodeficiency worldwide [3]. In affluent counties, where nutritional overload frequently occurs, the availability of food and the improvement of hygienic conditions are associated with a reduced prevalence of infections [4, 5]. However, the daily exposure to an excess of calories (and subsequent overnutrition and increased storage of body fat) has facilitated hyperactivation of intracellular nutrient energy-sensing pathways and metabolic overload in peripheral tissues and organs, including in the immune system [6, 7]. To this aim, in this chapter we highlight how overweight and metabolic pressure affect lung immunometabolism and pathophysiological conditions, suggesting the lung as an organ at the interface between immune and metabolic regulation.

NEURAL CONTROL OF FOOD INTAKE

Scientific evidence has shown that obesity has become a worldwide epidemic in the last 20 years, and this is particularly true in the more affluent countries of the western world [8, 9]. Obesity is generally defined as a pathological condition in which the total amount of triglycerides stored in adipose tissue is abnormally increased and is associated with a wide variety of adverse outcomes, such as type 2 diabetes (T2D), insulin resistance, inflammation, and cardiovascular diseases. These complications associated with obesity clearly reduce life expectancy and, as a whole, have enormous economic consequences and a strong social impact. This dysmetabolic condition comes from an imbalance between energy intake and energy expenditure, and it usually derives from the interaction between different genetic factors with an abundance of caloric intake [8, 9].

Human obesity can have a genetic cause, and indeed today about 20 genes are known whose mutation gives rise to an autosomal form of obesity; however, the vast majority of obesity forms are not due to genetic cause but instead derive from the intricate and complex interaction between genes and environment [10]. In this context, leptin and ghrelin are the main peripheral signals, which regulate food intake. They both act at the hypothalamic level, and the former activates proopiomelanocortin (POMC) neurons, by inhibiting food intake (anorexigenic effect), and the genetic deficit of leptin or its receptor causes obesity [10]. On the contrary, ghrelin activates agouti-related protein/ neuropeptide Y (Agrp/NPY) neurons, inducing food intake (orexigenic action) [11]. Leptin is a hormone structurally similar to other cytokines mainly produced by adipose tissue in proportion to body fat mass. As a hormone, leptin regulates food intake and basal metabolism, whereas more recent data have shown that, as a cytokine, it is able to influence thymus homeostasis, and it also promotes T-cell differentiation into T helper (Th1/Th17)/proinflammatory phenotype (Figure 2.1). Also, leptin plays a role in the control of immunological self-tolerance and in the pathogenesis of chronic inflammatory diseases (i.e., autoimmune disorders), representing a key factor in the crosstalk between immunity and metabolism [12].

THE LINK BETWEEN OBESITY AND CHRONIC INFLAMMATION

As previously mentioned, obesity has been associated with a series of negative consequences for human health, such as the increased risk of cardiovascular disorders, including atherosclerosis, diabetes, fatty liver disease, and also cancer. All these

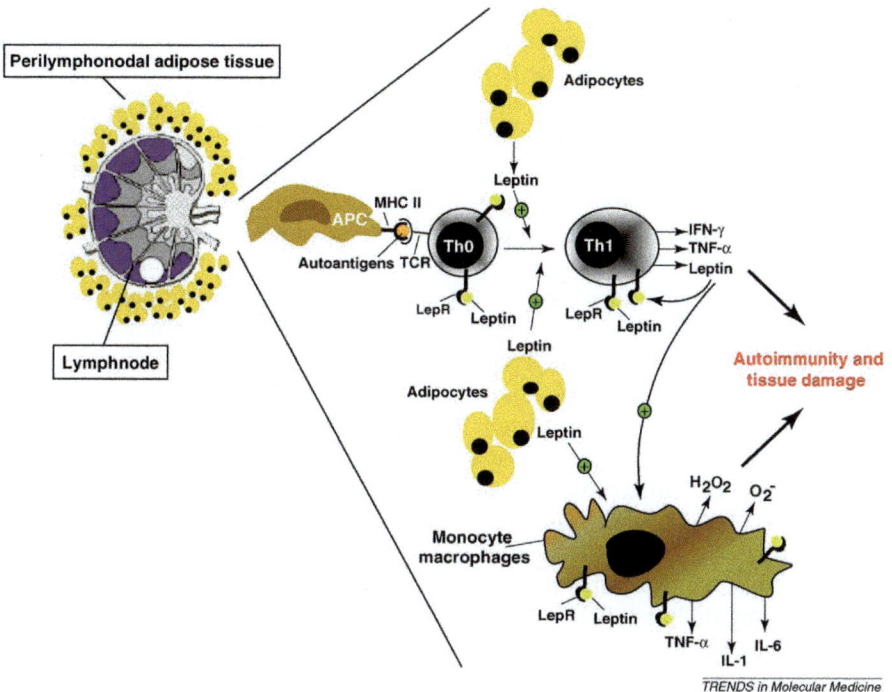

FIGURE 2.1

Schematic model of the role of leptin in promoting Th1 immunity and breaking self-immune tolerance. Adipocyte-derived leptin modulates the activity and function of antigen-presenting cells by increasing antigen presentation by major histocompatibility complex II molecules and priming toward Th1 responses. This increases IFN-g and TNF-a secretion by Th1 cells, which also secrete leptin that amplifies Th1 priming in an autocrine loop. IFN-g, TNF-a, and leptin produced by Th1 cells promote the activation of monocyte/macrophages and innate immunity, inducing the secretion of oxygen radicals and tissue damage. Taken together, these mediators can facilitate protection from infections on the one side, and promotion of autoimmunity on the other.

From Trends Molecular Medicine.

pathological conditions recognize a common basis in their pathogenesis, which is chronic inflammation, characterized by an abnormal production of proinflammatory cytokines, able to activate a network of inflammatory signaling pathways, amplifying and supporting the inflammatory process itself. These conditions are due to chronic "low-grade" inflammation, which is a typical feature of obesity [13]. On the contrary, a reduction in body weight is associated with a marked improvement of the inflammatory processes and a normalization of the different biological parameters, generally altered during inflammation [14].

In recent years, the scientific community has focused its interest on the study of the relationship between immune response and nutrition/metabolism (more recently defined as immunometabolism). In this context, it has been demonstrated that some genetic alterations (i.e., mutation/loss of function of leptin (Lep), leptin receptor (LepR), POMC, proprotein convertase 1 (PCSK1), and the melanocortin-4 receptor (MC4-R)) can cause obesity and, at the same time, can also significantly influence immune responses [15–18]. These studies have clarified the role of the immune system and of its alterations in the pathogenesis of obesity; therefore the immune function in obesity has become a factor of particular interest and relevance, becoming over the years a potential therapeutic target for different immunomediated metabolic disorders.

Adipose tissue is a mix of adipocytes, stromal preadipocytes, immune cells, and endothelium, and it can respond rapidly and dynamically to alterations in nutrient supply, through hypertrophy and hyperplasia affecting adipocytes. With obesity and the progressive enlargement of adipocytes, the blood supply to the adipocytes can be reduced with consequent hypoxia [19]. Hypoxia has been proposed as one of the main etiological factors able to favor the necrosis and immune cells' infiltration in the adipose tissue, with a consequent anomalous and uncontrolled production of proinflammatory factors such as inflammatory chemokines, which in turn recruit other immune cells. All these events cause localized inflammation in the adipose tissue, which propagates an overall systemic inflammation associated with the development of several comorbidities [19].

Recent studies have highlighted the crucial role played by adipocytes and adipose tissue in the generation of inflammatory responses and mediators. The first link between obesity and the immune system has been highlighted thanks to the discovery that tumor necrosis factor-α (TNF-α) and interleukin-6 (IL-6) are strongly expressed in the adipose tissue of obese mice and humans [20]. Exogenous administration of such molecules is able to induce insulin resistance, suggesting a clear link among obesity, diabetes, and chronic inflammation [21]. In addition, adipocytes share several characteristics with immune cells (including T cells, macrophages, and dendritic cells (DCs)): among them, the activation of the complement, the production of mediators of inflammation to the pathogen, and the phagocytic properties [22]. Moreover, confirming the close relationship between adipocytes and the immune system, there is the evidence that many genes that encode for transcription factors, cytokines, inflammatory-signaling molecules, and fatty acid transporters are essential for the genesis of adipocytes, and are also strongly expressed in macrophages or regulatory T cells (Treg) [23, 24], where they are able to modulate their function, similarly to what occurs in the adipocytes.

Adipose tissue during obesity is infiltrated by a large number of macrophages, and this recruitment is linked to systemic inflammation and insulin resistance [25], whereas weight loss results in a reduction in the number macrophages in the adipose tissue with a consequent reduction of proinflammatory profiles of obese individuals [26]. These findings were supported by several studies that demonstrated that the expression of genes encoding for proteins involved in inflammatory processes are strongly altered in white adipose tissue (WAT) in murine models of obesity [27]. Furthermore, Wiesberg and colleagues have also shown that these variations in gene expression in WAT were essentially related to the macrophages' infiltration in obese mice [25]. The macrophages resident in the adipose tissue are those mainly responsible for the production of TNF-α, IL-6, and inducible nitric oxide synthase (iNOS). In line with the previously discussed data, it is interesting to note that a reduction in body weight is accompanied not only by an improvement in the inflammatory process, but also by a decrease in the expression of those genes that encode for inflammation-related proteins, generally altered during obesity [28].

It is now evident that the recruitment of macrophages into adipose tissue is the initial event in inflammation, able to trigger the onset of insulin resistance during obesity. Overnutrition induces adipocytes to secrete chemokines such as monocyte chemoattractin-1 protein (MCP-1), leukotriene B4 (LTB4), and others, creating a chemotactic gradient that attracts monocytes into adipose tissue, where they undergo activation. Once proinflammatory macrophages migrate into adipose tissue, they also secrete their chemokines, attracting other macrophages and creating an inflammatory feed-forward process of a self-sustaining loop (Figure 2.2).

Recently it has been shown that the classification of obese individuals correlated with the presence of crown-like structures, which are the main histological features of obesity and represent an accumulation of macrophages around dead adipocytes in the inflamed adipose tissue [19]. Because the main function of macrophages is essentially to remove the apoptotic cells, the investigators themselves hypothesized that the presence of these crown-like structures in the adipose tissue reflected a pro-inflammatory state that it is due, in part, to a compromise of the phagocytic process mediated by macrophages in the obese individuals [19].

Several subtypes of macrophages are involved in the inflammation of adipose tissue induced by obesity. Schematically, the macrophages that accumulate in the adipose tissue of obese mice mainly express genes associated with an M1 or "classically activated" macrophage phenotype, whereas macrophages present in the adipose tissue of lean mice express genes associated with an M2 or "alternatively activated" macrophage phenotype [29]. Both these populations express F4/80 and CD11b, and M1-type macrophages also express CD11c [29]. In the obese state, M1-type macrophages can accumulate lipids, assuming a foamy characteristic appearance in adipose tissue [30]. Tissue macrophages respond to changes in the local environment by changing their polarization state in an extremely dynamic manner, depending on the conditions. Therefore in the last few years, the classification into M1 and M2 macrophages has only been considered a simplification of the polarization states of the macrophages that can occur *in vivo*.

In general, stimulation with cytokines of Th1, including interferon-γ (IFN-γ), or with bacterial products leads to the generation of M1 macrophages, which produce proinflammatory cytokines (including TNF and IL-6), express iNOS, and produce reactive oxygen species (ROS) and nitrogen intermediates [30]. In contrast, Th2-type cytokines, such as IL-4 and IL-13, polarize macrophages toward the acquisition of an M2 phenotype, which is associated with an increased production of IL-10 and with the inhibition of proinflammatory cytokine secretion. In this specific macrophage subset, the expression of arginase-1, macrophage mannose receptor 1, genes

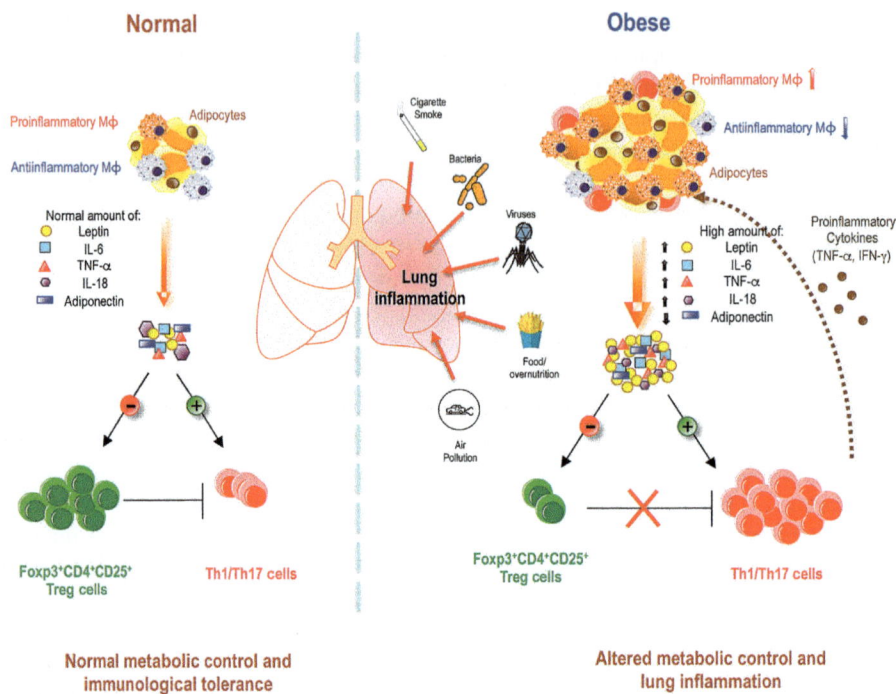

FIGURE 2.2

Model of immunometabolic effects of adipose tissue in normal and obese subjects during lung inflammation. In normal subjects, secretion of adipocyte-derived leptin and other cytokines (such as IL-6, IL-18, TNF-α, and adiponectin) associates with a normal control of metabolic functions and with a balance between the number of Th1/Th17 cells and Treg cells, which are functionally able to suppress immune and autoimmune responses. Conversely, during obesity, the high amount of leptin secreted by adipocytes, together with other proinflammatory cytokines, accounts for an altered control of metabolic functions, leading to expansion of proinflammatory Th1/Th17 cells on one side, and a low proportion of Treg cells infiltrating adipose tissue on the other. Lung inflammation is a complex biological process that occurs in response to harmful stimuli and whose function is to eliminate the cause of cell injury and initiate the repair process. The causes of chronic pulmonary inflammation are manifold. This may include response to bacterial and viral pathogens, inhalation of environmental pollutants (including cigarette smoke, mycotoxins), or substances in the workplace. In this context, overnutrition and obesity-induced perturbation in the balance between Th1-, Th17, and Th2-type signals may influence the recruitment and activation of M1-type (proinflammatory) macrophages in adipose tissues, as well as in the lung, thereby generating a pathogenic environment. Tissue macrophages respond to changes in the local environment by changing their polarization status, and they recruit other inflammatory cells into a self-sustaining loop. Along with the increased number and function of Th1 cells and macrophages in adipose tissue and in the lung, a higher number of CD8[+] T cells and mast cells have been reported during lung inflammation.

encoding for IL-1 receptor antagonist and other genes known to play an immuno-regulatory role on inflammatory cells, is also strongly increased (compared with M1 macrophages) [30]. Therefore any element that alters the balance between Th1 and Th2 signals (a condition that generally occurs during obesity) can affect the recruitment and activation of macrophages toward one phenotype or the other, thereby generating either a pathogenic and inflammatory environment or a noninflammatory and protective environment (Figure 2.2).

OBESITY AS AN AUTOIMMUNE DISORDER?

The presence of immune cells infiltrating the adipose tissue of obese subjects is now considered one of the characteristic, pathological lesions occurring during obesity. The cells of innate immunity (i.e., macrophages, neutrophils, NK) that secrete cytokines and chemokines are recruited by adipocytes. In addition to the cells of innate immunity, however, recent experimental evidence has shown that T cells in adipose tissue show specific rearrangements of the T-cell-receptor (TCR), thus also suggesting the presence of clonal T cells infiltrating the adipose tissue [31].

These data led scientists to hypothesize that obesity can recognize an autoimmune basis in its pathogenesis. In general, the criteria for considering a pathology as "autoimmune" are: (1) the presence of immune cell infiltrates in the target organ; (2) the presence of circulating auto-antibodies with consequent activation of the complement; (3) TCRs of a clonal nature; (4) secretion of pro-inflammatory cytokines; (5) alterations of the regulatory mechanisms controlling immune response (i.e., Treg cells); and (6) association with other autoimmune diseases. With regard to Treg cells, they have been identified in adipose tissue of normal individuals, yet massively reduced in obesity, further suggesting their possible role in the control of an autoimmune attack against adipose tissue [32].

Many of the aforementioned features of autoimmune diseases are also present in obesity, but the hypothetical auto-antigen present in adipose tissue, able of triggering an abnormal autoimmune response is still unknown and is the subject of intense studies in the field. Recently, it has been hypothesized that different metabolic mediators may induce dysregulation and inflammation in the hypothalamus, which results in an uncontrolled food intake. In this context, for example saturated fatty acids (such as palmitate) have been shown to induce hypothalamic inflammation, through the production of ROS [33], or endoplasmic reticulum stress [34], which induce leptin-resistance with consequent increase in body weight.

KEY ADIPOCYTOKINES

It is well established that adipose tissue is not only involved in energy storage but also functions as an immunometabolic organ that secretes various bioactive substances, which are collectively referred to as adipokines. The dysregulated expression of these factors, caused by excess adiposity and adipocyte dysfunction, has been linked to the pathogenesis of various disease processes through altered immune responses.

LEPTIN AND ADIPONECTIN

Leptin is the product of the *obese* (*ob*) gene. It is involved in the regulation of energy homeostasis [35] and is almost exclusively expressed and produced by WAT. Circulating levels and adipose tissue mRNA expression of leptin are strongly associated with body mass index (BMI) and fat mass in obesity, whereas reduction of leptin secretion occurs after fasting [36].

Although leptin acts mainly at the level of the central nervous system to regulate food intake and energy expenditure, there is a relationship between leptin and the low-grade inflammatory state in obesity, suggesting that leptin could exert peripheral biological effects as a function of its cytokine-like structure [37]. Indeed, leptin receptors belong to the cytokine class I receptor family, and recent evidence has shown that leptin has a role as proinflammatory cytokine. In particular, leptin stimulates, in innate immune responses, the production of several proinflammatory mediators such as IL-1, IL-6, IL-12, and TNF, and moreover it is able to activate neutrophil chemotaxis and stimulate production of ROS [37]. Leptin promotes activation and phagocytosis by monocytes/macrophages and their secretion of leukotriene B_4 (LTB_4), cyclo-oxygenase 2 (COX2), and nitric oxide, whereas on NK cells, this hormone sustains their cytotoxicity through activation of signal transducer and activator of transcription 3 (STAT3) and IL-2 [12], as testified by the evidence that *db/db* mice have a deficit in NK cell development (Figure 2.1).

Leptin has different effects on proliferation and cytokine production by human naive (CD45RA$^+$) and memory (CD45RO$^+$) CD4$^+$ T cells (both of which express LepRb). Leptin promotes proliferation and IL-2 secretion by naive T cells, whereas it minimally affects the proliferation of memory cells (on which it promotes a bias toward Th1-cell responses). Another important role of leptin in adaptive immunity is highlighted by the observation that leptin deficiency in *ob/ob* mice is associated with immunosuppression and thymic atrophy [3]. On the other hand, recently it has been reported that leptin can act as a negative signal for the expansion of human naturally occurring fork-head box P3 (Foxp3$^+$) CD4$^+$CD25^{high} Treg cells [38], a cellular subset involved in the prevention of autoimmune diseases. De Rosa et al. showed that freshly isolated human Treg cells produce leptin and express high levels of leptin receptor (LepR). *In vitro* neutralization with anti-leptin monoclonal antibody (mAb) following anti-CD3/CD28 stimulation resulted in Treg cell proliferation [38]. Leptin has been shown to inhibit rapamycin-induced proliferation of Treg cells by increasing activation of the mammalian target of rapamycin (mTOR), a 289-kDa serine/threonine protein kinase inhibited by rapamycin. In addition, under normal conditions, Treg cells secreted leptin, which activated mTOR in an autocrine manner to maintain their state of hyporesponsiveness. Finally, it has been shown that Treg cells from *db/db* mice exhibited decreased mTOR activity and increased proliferation compared with that of wild-type cells [7, 39]. More recently, it has been shown that leptin-induced mTOR pathway activation defines a specific molecular and transcriptional signature, which controls CD4[+] T effector (CD4[+]CD25[−], Teff) responses [40]. Together, these data suggest that the leptin-mTOR axis sets the threshold for the responsiveness of Treg cells, and this pathway might integrate cellular energy status with metabolic-related signaling in both Treg cells and Teff that use this information to control immune tolerance.

Leptin-deficient mice are resistant to induction of active and adoptively transferred experimental autoimmune encephalomyelitis (EAE) [12]. Importantly, a surge of serum leptin anticipates the onset of clinical manifestations of EAE, and this event correlates with inflammatory anorexia, weight loss, and the development of pathogenic T-cell responses against myelin. Interestingly, in relapsing-remitting multiple sclerosis (RRMS) patients, an inverse correlation between serum leptin and percentage of circulating Treg cells was also observed. Moreover, treatment of WT mice with soluble LepR fusion protein (LepR:Fc) increased the percentage of Tregs and ameliorated the clinical course and progression of disease in relapsing-experimental autoimmune encephalomyelitis (R-EAE), an animal model of RRMS [41]. Two recent reports have shown that adipose tissue in normal individuals is a preferential site of accumulation of Treg cells. Their precise role in this tissue is still object of extensive investigation, but what is clear is that in obese, insulin-resistant mice, their number is reduced, and their supplementation with adoptive transfer experiments was able to dampen the inflammatory state and insulin resistance associated with obesity by increasing IL-10 and Th2/regulatory-type cytokines [32, 42]. All these data indicate that leptin could be the molecular link between obesity and reduced number, and probably impaired function of Treg cell observed in this condition, and these alterations could determine a further amplification of the inflammatory process through the recruitment of other cell types of adaptive and innate immunity such as CD8[+], Th1, mast cells, and macrophages [43, 44].

Adiponectin is mainly produced in WAT by mature adipocytes, with increasing expression and secretion during adipocyte differentiation, and by nonfat cells, but it also can be found in skeletal muscle cells, cardiac myocytes, and endothelial cells. Adiponectin levels inversely correlate with visceral obesity and insulin resistance, and weight loss is a potent inducer of adiponectin synthesis, thus suggesting that adiponectin may play a protective role against atherosclerosis and insulin resistance. TNF suppresses adiponectin secretion in adipocyte, and its production is also regulated by other proinflammatory cytokines such as IL-6 [45]. Moreover, adiponectin may modulate the TNF-α-induced inflammatory response, as it has been shown that adiponectin reduces TNF-α secretion of macrophages [45].

In contrast to leptin, early studies have indicated that adiponectin has an antiinflammatory effect on endothelial cells, inhibiting NF-kB activation and TNF-induced adhesion-molecule expression (vascular cell adhesion molecule-1 (VCAM-1), endothelial-leukocyte adhesion molecule-1 (*E*-selectin), and intracellular adhesion molecule-1 (ICAM-1)). Adiponectin induces the secretion of some antiinflammatory cytokines, such as IL-10 and IL-1RA (receptor antagonist), by human monocytes, macrophages, and DCs, and suppresses the production of INF-γ [46].

Interestingly, it was observed that the addition of adiponectin results in a significant diminution of antigen-specific T-cell proliferation and cytokine production. A paper by Tsang et al. suggests that the immunomodulatory effect of adiponectin on immune response could be at least partially mediated by its ability to alter DC functions [47]. Indeed, adiponectin-treated DCs show a lower production of IL-12p40 and a lower expression of CD80, CD86, and histocompatibility complex class II (MHCII). Moreover, in coculture experiments of T cells and adiponectin-treated DCs, a reduction in T-cell proliferation and IL-2 production, and an higher percentage of Treg cells was observed, suggesting that adiponectin could also control regulatory T-cell homeostasis.

THE LUNG AS AN IMMUNOMETABOLIC ORGAN

The increased morbidity and mortality associated with obesity has been recently linked to its role in several chronic pathological conditions, including cardiovascular and metabolic diseases, lower back pain, osteoarthritis and cancer. Obesity has also been associated with respiratory symptoms and diseases, including asthma, chronic obstructive pulmonary disease (COPD), exertional dyspnea, obstructive sleep apnea syndrome (OSAS), pulmonary embolism, and aspiration pneumonia [48, 49]. More specifically, weight gain and rising BMI have been shown to be associated with decreased lung volumes, which are reflected by a more restrictive ventilatory pattern on spirometry. Several cross-sectional and longitudinal studies in the last few years have demonstrated that a rise in BMI reduces different parameters of pulmonary function, such as forced expiratory volume in 1 second (FEV1), forced vital capacity (FVC), functional residual capacity (FRC), and the expiratory reserve volume (ERV) [50, 51]. In association with these functional changes, lungs also "experience" the metabolic pressure exerted by the excessive adipocyte mass typical of overweight and obese subjects, in terms of overproduction of inflammatory mediators and adipocytokines, all able to affect lung interstitial space and function. For all these reasons, the lung could be considered an "immunometabolic" target organ, in which the crosstalk between adipose tissue and immune system plays a key role in the control of the inflammatory state and respiratory functions.

IMMUNOMETABOLISM OF ASTHMA

Asthma is a chronic inflammatory disease of the airway wall. The very first studies on asthma found that CD4[+] T lymphocytes were present in asthma biopsies. The immune response to inhaled allergens (such as house dust mite (HDM), cockroach, pollen grains, or fungal spores) is mainly characterized by an aberrant Th2 lymphocyte response. Th2-type cytokines cause airway eosinophilia (IL-5), goblet cell metaplasia (GCM; IL-4 and IL-13), and bronchial hyperreactivity (BHR; IL-4 and IL-13), all important features of asthma [52]. Animal models of asthma, in which these Th2-type cytokines have been neutralized, have clearly demonstrated the importance of such cytokines in promoting allergic-type airway inflammation. Indeed, IL-4-deficient mice display deficient IgE synthesis and are protected from developing asthma because of several defects in eosinophil recruitment [53]. Also IL-5-deficient mice do not develop airway or bone marrow eosinophilia, and eosinophil-deficient mice show defects in airway wall remodeling, which is another feature of persistent asthma [54]. Over the last few years, the view that asthma is an exclusively Th2-mediated disease has been partially modified by the finding that other cytokines such as IL-9, IL-17, and IL-22 are highly expressed in the airways of mouse models of asthma or in human subjects with asthma. Indeed Wakashin et al. have recently shown that cotransfer of antigen-specific Th17 cells with Th2 cells boosted eosinophilic airway inflammation in mice, and this effect was also observed by overexpression of IL-23, which increased the number of Th17 cells [55]. Moreover IL-17 has been shown to counteract the immunoregulatory and antiinflammatory effects of Treg cells, thus increasing inflammation and BHR [56].

Over the past few years, IL-9-producing $CD4^+$ T (Th9) cells have been identified as a distinct subset, different from the classical Th2 cells, as they require other transcription factors for their development. Th9 cells promote T-cell proliferation, IgE, and IgG production, sustain survival and maturation of eosinophils, and mastocytosis [57–59], and lung-specific IL-9 overexpression increased airway inflammation, goblet cells metaplasia, and BHR, which on the contrary were reduced by blocking of IL-9 activity [60]. With regard to Treg cells, Lewkowich and colleagues have shown that these cells suppress features of asthma by inhibiting the activation of airway DCs (through IL-10 and TGF-β production) [61], and asthmatic children have been shown to display a reduced number of pulmonary Foxp3+ Treg cells. In this context, the cytokines TNF-α and IL-6, which are overproduced in asthmatic airway subjects, could be responsible for the inhibition of Treg-cell functions [62].

As previously mentioned, cross-sectional and longitudinal studies have linked obesity with asthma. More specifically, the frequency of symptoms of breathlessness and wheezing increases with BMI in patients with asthma. A recent study [63] has demonstrated that the prevalence of asthma is higher by 38% in overweight patients and by 92% in obese patients. On the contrary, obese subjects who showed weight loss by diet or bariatric surgery displayed improvements in their lung function and asthma symptoms. [64]. It has also been suggested that systemic inflammation associated with obesity may contribute to glucocorticoid insensitivity observed in a few subjects with asthma. Indeed, Peters-Golden et al. [65] demonstrated that the clinical response to beclomethasone declined with increasing BMI. Obese subjects have elevated inflammatory mediators, such as IL-6, IL-8, and TNF-α, which potentially could affect glucocorticoid sensitivity [66]. In this context, increased leptin levels in obesity have also been implicated in the pathogenesis of asthma; indeed exogenous leptin administration in mice strongly promotes airway hyperresponsiveness (AHR) and increases IgE production [67]. Overall these data suggest a clear link between immunometabolism and lung functions in the context of allergy and asthma, and further studies are needed to dissect this important link more in depth.

IMMUNOMETABOLISM OF COPD

COPD is a chronic inflammatory lung disease that causes airflow obstruction. The main cause of COPD in developed countries is tobacco smoking, however genetic and environmental factors have been also recently implicated in the pathogenesis of this disease [68]. Cigarette smoke and other inhaled irritants can active epithelial cells to release local and systemic inflammatory molecules, such as TNF-α, IL-6, IL-8, granulocyte-macrophage colony-stimulating factor (GM-CSF), and monocyte chemoattractant protein-1 (MCP-1/CCL2), which in turn activates and attracts inflammatory cells including neutrophils and macrophages into the lungs [69]. Moreover, quantitative and qualitative alterations of B cells, CD4$^+$, and CD8$^+$ T cells have also been observed both in airways and blood of COPD subjects compared with healthy individuals [70–72]. Recent data sustain the hypothesis that tobacco smoking can generate new self and foreign epitopes, and by altering the presentation of self-molecules, it may lead to the activation of an aberrant autoreactive adaptive immune response [73, 74]. Indeed, Th1 and Th17 CD4$^+$ T cells have been shown to accumulate in the lungs of subjects affected by stable COPD; Th1 cells secrete huge amounts of IFN-γ, whereas Th17 cells regulate tissue inflammation by producing IL-17A and IL-17F [75, 76].

As for asthma, COPD has also been associated with metabolic disorders, such as obesity. COPD and obesity share a series of complex features, with both conditions associated with deterioration in lung function, hypoxia, and a low-grade systemic inflammation, which predispose to increasing morbidity and mortality [77]. Specifically, COPD has been shown to worsen airflow obstruction and hypoxia, and in addition to smoking, to increase pulmonary and systemic oxidative stress. Systemic inflammation is one of the major features of COPD [78], and increased serum levels of different proinflammatory molecules have been reported, together with augmented levels of oxidative stress [79]. Several studies have suggested that systemic inflammation associates with an excess in fat mass in COPD subjects. In particular, TNF-α, IL-6, and leptin plasma levels are significantly increased in overweight/obese COPD subjects compared with normal-weight patients, [80] and the likelihood of having elevated C-reactive protein (CRP) was about three times higher in obese compared with normal-weight patients [81]. In addition, abdominal fat mass is positively associated with CRP levels in patients with COPD [82]. In contrast, there are more recent data supporting a protective role of overweight and obesity on mortality in COPD, and this evidence seems to be particularly strong in patients with more severe airway obstruction. However, several factors may partly have altered these results. The influence of obesity *per se* on both respiratory mechanisms and pulmonary volumes, the extent of emphysema and the role of muscle mass rather than BMI may all be important in partly justifying the paradox. Future studies, which will focus on body composition, weight changes, and COPD phenotypes, are needed to shed light on this complex phenomenon.

IMMUNOMETABOLISM OF PULMONARY FIBROSIS

Pulmonary fibrosis is a chronic progressive interstitial lung disease of unknown etiology, associated with a poor prognosis and often culminates in lung transplantation or death from respiratory failure [83]. Familial pulmonary fibrosis (FPF) and Hermansky-Pudlak syndrome (HPS) are two genetic disorders that may serve as models for the study of pulmonary fibrosis. Lung inflammatory cells have been found phenotypically altered in the alveolar milieu in pulmonary fibrosis and may contribute to the pathogenesis of this disease.

In this context, $CD11b^+$ alveolar macrophages isolated from patients with idiopathic pulmonary fibrosis (IPF) can transdifferentiate to lymphatic endothelial cells *in vitro*, differently from healthy subjects [84]. In addition, alveolar lymphocytosis, characterized by activated $CD4^+$ T cells, is considered a preclinical feature of FPF, and alveolar inflammation precedes the development of fibrotic lung disease in patients with a telomerase mutation [85, 86]. Also, HPS is associated with high concentrations of bronchoalveolar lavage cells, alveolar macrophage activation, and fibrogenic mast cells [87, 88]. All these data clearly indicate that immune cell dysfunction and influx into lung tissue are features of patients at high risk of inherited pulmonary fibrosis and may lead to alveolar injury and the development of lung fibrosis.

Recently, El-Chemaly and colleagues have reported high concentrations of activated T and B cells associated with increased cytokine levels in pulmonary fibrosis subjects. Consistent with these findings, gene expression and serum proteomic analyses in such patients revealed an upregulation of genes associated with mitosis and cell cycle control in circulating mononuclear cells, as well as altered levels of several analytes, including leptin, cytokines, and growth factors [89]. Moreover, Kang and colleagues recently performed a metabolic profiling of IPF patients and discovered 25 metabolite signatures of IPF. These metabolite signatures clearly indicated alteration in different metabolic pathways: adenosine triphosphate degradation pathway, glycolysis, and glutathione biosynthesis pathway, thus providing additional insight into understanding the disease and the potential for developing novel biomarkers [90]. In line with these findings, recent data have shown the central role of glycolytic reprogramming in myofibroblast differentiation in the pathogenesis of lung fibrosis, thus suggesting that targeting this pathway may represent an effective therapeutic tool for pulmonary fibrosis [91]. Once again, these data linking pulmonary fibrosis, inflammation, and metabolism suggest the lung as an important immunometabolic organ, whose study will help to dissect the intimate link between immunity and metabolism.

IMMUNOMETABOLISM OF TUBERCULOSIS

Tuberculosis (TB) is an infectious disease that represents one of the major health concerns worldwide for mankind. *Mycobacterium tuberculosis* (MTB), the etiological agent of TB, is a highly successful intracellular pathogen that primarily infects lungs. On invasion, MTB activates both innate and adaptive immune responses in the host. The innate immune response elicited by MTB involves professional phagocytes including macrophages, monocytes, DCs, neutrophils [92], natural killer cells (NK) [93], and complement system [94]. These innate immune components, upon stimulation, induce the expression of proinflammatory cytokines/chemokines that contribute to local and systemic immune cell activation and proliferation. In the absence of innate immunity to eliminate MTB, adaptive immunity comes into play, providing an effective protection against infection. The pathogen activates the immune system and, in this scenario, $CD4^+$ and $CD8^+$ T cells induce a Th1-type immunity [95], which, following the production of IFN-γ, activates the microbicidal mechanism of macrophages to eliminate MTB [96]. Once infected, a majority of the individuals develop latent TB due to persistence of MTB for years or decades, and a relatively small proportion (5%–10%) of infected people develop an active form of TB [97].

Only a small percentage of infected individuals develop active TB after infection [98, 99]. A delicate balance presumably exists between the pathogen and the human host. Increasing evidence suggests that leptin differentially regulates metabolic, neuroendocrine, and immune functions in humans [100, 101], and leptin-deficient *ob/ob* mice were shown to be highly susceptible to pulmonary infection by MTB [102]. Further linking metabolism and TB, recent evidence has suggested that low body weight is associated with risk of TB [103–105], severity of disease [106, 107], unfavorable response to treatment [108], and relapse [109]. Moreover, a cholesterol-rich diet accelerated the sterilization rate of sputum cultures in pulmonary TB patients, suggesting that cholesterol should be used as a complementary measure in antitubercular treatment [110].

Two cohort studies showed that obesity and overweight were associated with lower risk of active TB in adults [111] and elderly adults [112], respectively. Two prospective studies found that obese/overweight HIV-infected subjects had a lower risk of TB development than did those of normal weight, thus indicating that obesity is a protective factor in TB development [113, 114]. Interestingly, a recent paper by Neyrolles and colleagues revealed that mycobacteria can persist without replication in adipose tissue [115], suggesting that this organ might constitute a vast reservoir where the tubercle bacillus could persist for long periods of time and avoid both killing by antimicrobials and recognition by the host immune system. This evidence must be taken into account against elderly malnutrition to reduce the risk of reactivation of TB; therefore nutritional supplementation might represent an important adjunct to chemotherapy in this regard. Future immunometabolic studies of TB should investigate the underlying mechanisms and clinical and epidemiological consequences of the relation between obesity/overweight and active TB.

CONCLUDING REMARKS

The link between immunometabolism and lung pathophysiology is fascinating. It involves a global pressure on peripheral tissues and immune cells that fail to maintain their highly tuned homeostatic mechanisms because of the continuous metabolic pressure induced by the excess of caloric intake and nutrients (Figure 2.2) [116]. Genetic predisposition, epigenetics, immune regulation, and the environment have long been considered as the key pathogenetic players in lung inflammation and function. Among the environmental factors, metabolic overload should also be included as a major contributor to lung inflammation. It is interesting to note that metabolic overload could be reduced via immunometabolic manipulation, through pharmacological and nonpharmacological strategies, for an early reduction of metabolic pressure typically associated with excessive caloric intake [116]. For example, drugs acting as signals of pseudo-starvation (i.e., metformin or rapamycin) or caloric restriction (i.e., fasting, chronic caloric restriction) could represent novel means for the control of lung inflammation and its pathophysiology in a series of major lung disorders.

ACKNOWLEDGMENTS

G.M. is supported by grants from the EU Ideas Programme, ERC-Starting Independent Grant "menTORingTregs" n. 310496, Telethon Grant n. GGP17086, FISM Grant n. 2016/R/18; VDR is supported by FISM Grant 2014/R/21. The authors wish to thank Dr. Deriggio Faicchia for the artwork and critically reading of the manuscript. This work is dedicated to the memory of Eugenia Papa and Serafino Zappacosta.

REFERENCES

[1] McFarlane H. Cell-mediated immunity in protein-calorie malnutrition. Lancet 1971;2:1146–7.
[2] Bhargava A. Undernutrition, nutritionally acquired immunodeficiency, and tuberculosis control. Br Med J 2016;355:i5407.
[3] Lord GM, et al. Leptin modulates the T-cell immune response and reverses starvation induced immunosuppression. Nature 1998;394:897–901.
[4] Bach JF. The effect of infections on susceptibility to autoimmune and allergic diseases. N Engl J Med 2002;347:911–20.
[5] Ehlers S, Kaufmann SH, Participants of the 99(th) Dahlem Conference. Infection, inflammation, and chronic diseases: consequences of a modern lifestyle. Trends Immunol 2010;31:184–90.
[6] Matarese G, La Cava A. The intricate interface between immune system and metabolism. Trends Immunol 2004;25:193–200.
[7] Procaccini C, Galgani M, De Rosa V, Matarese G. Intracellular metabolic pathways control immune tolerance. Trends Immunol 2012;33:1–7.
[8] Friedman JM. Obesity: causes and control of excess body fat. Nature 2009;21:340–2.
[9] Hotamisligil GS. Inflammation and metabolic disorders. Nature 2006;14:860–7.
[10] O'Rahilly S. Human genetics illuminates the paths to metabolic disease. Nature 2009;19:307–14.
[11] Horvath TL, Bruning JC. Developmental programming of the hypothalamus: a matter of fat. Nat Med 2006;12:52–3.
[12] La Cava A, Matarese G. The weight of leptin in immunity. Nat Rev Immunol 2004;4:371–9.
[13] Symonds ME, Sebert SP, Hyatt MA, Budge H. Nat Rev Endocrinol 2009;5:604–10.

[14] Van Dielen FM, Buurman WA, Hadfoune M, Nijhuis J, Greven JW. Macrophage inhibitory factor, plasminogen activator inhibitor-1, other acute phase proteins, and inflammatory mediators normalize as a result of weight loss in morbidly obese subjects treated with gastric restrictive surgery. J Clin Endocrinol Metab 2004;89:4062.

[15] Montague CT, Farooqi IS, Whitehead JP, Soos MA, Rau H, Wareham NJ, et al. Congenital leptin deficiency is associated with severe early-onset obesity in humans. Nature 1997;387:903–8.

[16] Clément K, Vaisse C, Lahlou N, Cabrol S, Pelloux V, Cassuto D, et al. A mutation in the human leptin receptor gene causes obesity and pituitary dysfunction. Nature 1998;392:398–401.

[17] Krude H, Biebermann H, Luck W, Horn R, Brabant G, Grüters A. Severe early-onset obesity, adrenal insufficiency and red hair pigmentation caused by POMC mutations in humans. Nat Genet 1998;19:155–7.

[18] Vaisse C, Clement K, Guy-Grand B, Froguel P. A frameshift mutation in human MC4R is associated with a dominant form of obesity. Nat Genet 1998;20:113–4.

[19] Cinti S, Mitchell G, Barbatelli G, Murano I, Ceresi E, Faloia E, et al. Adipocyte death defines macrophage localization and function in adipose tissue of obese mice and humans. J Lipid Res 2005;46:2347–55.

[20] Galic S, Oakhill JS, Steinberg GR. Adipose tissue as an endocrine organ. Mol Cell Endocrinol 2010;316:129–39.

[21] Hotamisligil GS, Spiegelman BM. Tumor necrosis factor alpha: a key component of the obesity-diabetes link. Diabetes 1994;43:1271–8.

[22] Dixit VD. Adipose-immune interactions during obesity and caloric restriction: reciprocal mechanisms regulating immunity and health span. J Leukoc Biol 2008;84:882–92.

[23] Totonoz P, Nagy L, Alvarez JG, Thomazy VA, Evans RM. PPARgamma promotes monocyte/macrophage differentiation and uptake of oxidized LDL. Cell 1998;93:241.

[24] Cipolletta D, Feuerer M, Li A, Kamei N, Lee J, Shoelson SE. PPAR-γ is a major driver of the accumulation and phenotype of adipose tissue Treg cells. Nature 2012;486(7404):549–53.

[25] Weisberg SP, McCann D, Desai M, Rosenbaum M, Leibel RL, Ferrante Jr AW. Obesity is associated with macrophage accumulation in adipose tissue. J Clin Investig 2003;112:1796–808.

[26] Cancello R, Henegar C, Viguerie N, Taleb S, Poitou C, Rouault C. Reduction of macrophage infiltration and chemoattractant gene expression changes in white adipose tissue of morbidly obese subjects after surgery-induced weight loss. Diabetes 2005;54:2277–86.

[27] Soukas A, Cohen P, Socci ND, Friedman JM. Leptin-specific patterns of gene expression in white adipose tissue. Genes Dev 2000;14:963.

[28] Clement K, Viguerie N, Poitou C, Carette C, Pelloux V, Curat CA. Weight loss regulates inflammation-related genes in white adipose tissue of obese subjects. FASEB J 2004;18:1657.

[29] Lumeng CN, Bodzin JL, Saltiel AR. Obesity induces a phenotypic switch in adipose tissue macrophage polarization. J Clin Investig 2007;117:175–84.

[30] Gordon S. Alternative activation of macrophages. Nat Rev Immunol 2003;3:23–35.

[31] Lumeng CN, Maillard I, Saltiel AR. T-ing up inflammation in fat. Nat Med 2009;15:846–7.

[32] Feuerer M, Herrero L, Cipolletta D, Naaz A, Wong J, Nayer A, et al. Lean, but not obese, fat is enriched for a unique population of regulatory T cells that affect metabolic parameters. Nat Med 2009;15:930–9.

[33] Horvath TL, Andrews ZB, Diano S. Fuel utilization by hypothalamic neurons: roles for ROS. Trends Endocrinol Metab 2009;20:78–87.

[34] Hotamisligil GS. Inflammation and endoplasmic reticulum stress in obesity and diabetes. Int J Obes 2008;32:52–4.

[35] Zhang Y, Proenca R, Maffei M, Barone M, Leopold L, Friedman JM. Positional cloning of the mouse obese gene and its human homologue. Nature 1994;372:425.

[36] Friedman JM, Halaas JL. Leptin and the regulation of body weight in mammals. Nature 1998;395:763–70.

[37] Procaccini C, Jirillo E, Matarese G. Leptin as an immunomodulator. Mol Asp Med 2012;33:35–45.

[38] De Rosa V, Procaccini C, Calì G, Pirozzi G, Fontana S, Zappacosta S, et al. A key role of leptin in the control of regulatory T cell proliferation. Immunity 2007;26:241–55.

[39] Procaccini C, De Rosa V, Galgani M, Abanni L, Calì G, Porcellini A, et al. An oscillatory switch in mTORkinase activity sets regulatory T cell responsiveness. Immunity 2010;33:929–41.

[40] Procaccini C, De Rosa V, Galgani M, Carbone F, Cassano S, Greco D, et al. Leptin-induced mTOR activation defines a specific molecular and transcriptional signature controlling CD4$^+$ effector T cell responses. J Immunol 2012;189:2941–53.

[41] De Rosa V, Procaccini C, La Cava A, Chieffi P, Nicoletti GF, Fontana S, et al. Leptin neutralization interferes with pathogenic T cell autoreactivity in autoimmune encephalomyelitis. J Clin Investig 2006;116:447–55.

[42] Winer S, Chan Y, Paltser G, Truong D, Tsui H, Bahrami J, et al. Normalization of obesity-associated insulin resistance through immunotherapy. Nat Med 2009;15:921–9.

[43] Nishimura S, Manabe I, Nagasaki M, Eto K, Yamashita H, Ohsugi M, et al. CD8$^+$ effector T cells contribute to macrophage recruitment and adipose tissue inflammation in obesity. Nat Med 2009;15:914–20.

[44] Liu J, Divoux A, Sun J, Zhang J, Clément K, Glickman JN, et al. Genetic deficiency and pharmacological stabilization of mast cells reduce diet-induced obesity and diabetes in mice. Nat Med 2009;15:940–5.

[45] Ouchi N, Walsh K. Adiponectin as an anti-inflammatory factor. Clin Chim Acta 2007;380:24–30.

[46] Wolf AM, Wolf AD, Rumpold H, Enrich B, Tilg H. Adiponectin induces the anti-inflammatory cytokines IL-10 and IL-1RA in human leukocytes. Biochem Biophys Res Commun 2004;323:630–5.

[47] Tsang JY, Li D, Ho D, Peng J, Xu A, Lamb J, et al. Novel immunomodulatory effects of adiponectin on dendritic cell functions. Int Immunopharmacol 2011;11:604–9.

[48] Koenig SM. Pulmonary complications of obesity. Am J Med Sci 2001;321:249–79.

[49] Murugan AT, Sharma G. Obesity and respiratory disease. Chron Respir Dis 2008;5:233–42.

[50] McClean KM, Cardwell CR, Kee F. Longitudinal change in BMI and lung function in middle-aged men in Northern Ireland. Ir J Med Sci 2007;176(Suppl 10):S418.

[51] Jones RL, Nzekwu MU. The effects of body mass index on lung volumes. Chest J 2006;130:827–33.

[52] Lloyd CM, Hessel EM. Functions of T cells in asthma: more than just T(H)2 cells. Nat Rev Immunol 2010;10:838–48.

[53] Brusselle GG, Kips JC, Tavernier J, Van Der Heyden JG, Cuvelier CA, Pauwels RA, et al. Attenuation of allergic airway inflammation in IL-4 deficient mice. Clin Exp Allergy 1994;24:73–80.

[54] Humbles AA, Lloyd CM, McMillan SJ, Friend DS, Xanthou G, McKenna EE, et al. A critical role for eosinophils in allergic airways remodeling. Science 2004;305:1776–9.

[55] Wakashin H, Hirose K, Maezawa Y, Kagami S, Suto A, Watanabe N, et al. IL-23 and Th17 cells enhance Th2-cell-mediated eosinophilic airway inflammation in mice. Am J Respir Crit Care Med 2008;178:1023–32.

[56] Zhao J, Lloyd CM, Noble A. Th17 responses in chronic allergic airway inflammation abrogate regulatory T-cell-mediated tolerance and contribute to airway remodeling. Mucosal Immunol 2012;6:335–46.

[57] Veldhoen M, Uyttenhove C, Van Snick J, Helmby H, Westendorf A, Buer J, et al. Transforming growth factor-beta 'reprograms' the differentiation of T helper 2 cells and promotes an interleukin 9- producing subset. Nat Immunol 2008;9:1341–6.

[58] Staudt V, Bothur E, Klein M, Lingnau K, Reuter S, Grebe N, et al. Interferon-regulatory factor 4 is essential for the developmental program of T helper 9 cells. Immunity 2010;33:192–202.

[59] Kearley J, Erjefalt JS, Andersson C, Benjamin E, Jones CP, Robichaud A, et al. IL-9 governs allergen-induced mast cell numbers in the lung and chronic remodeling of the airways. Am J Respir Crit Care Med 2011;183:865–75.

[60] Temann UA, Ray P, Flavell RA. Pulmonary overexpression of IL-9 induces Th2 cytokine expression, leading to immune pathology. J Clin Investig 2002;109:29–39.

[61] Lewkowich IP, Herman NS, Schleifer KW, Dance MP, Chen BL, Dienger KM, et al. CD4+CD25+ T cells protect against experimentally induced asthma and alter pulmonary dendritic cell phenotype and function. J Exp Med 2005;202:1549–61.

[62] Nguyen KD, Vanichsarn C, Nadeau KC. TSLP directly impairs pulmonary Treg function: association with aberrant tolerogenic immunity in asthmatic airway. Allergy, Asthma Clin Immunol 2010;6:4.

[63] Beuther DA, Sutherland ER. Overweight, obesity and incident asthma: a meta-analysis of prospective epidemiologic studies. Am J Respir Crit Care Med 2007;175:661–6.

[64] Aaron SD, Fergusson D, Dent R, et al. Effect of weight reduction on respiratory function and airway reactivity in obese women. Chest J 2004;125:2046–52.

[65] Peters-Golden M, Swern A, Bird SS, et al. Influence of body mass index on the response to asthma controller agents. Eur Respir J 2006;27:495–503.

[66] Sin DD, Sutherland ER. Obesity and the lung. 4. Obesity and asthma. Thorax 2008;63:1018–23.

[67] Shore SA, Schwartzman IN, Mellema MS, et al. Effect of leptin on allergic airway responses in mice. J Allergy Clin Immunol 2005;115:103–9.

[68] Barnes PJ. Chronic obstructive pulmonary disease. N Engl J Med 2000;343(4):269–80.

[69] Yamasaki K, Eeden SFV. Lung macrophage phenotypes and functional responses: role in the pathogenesis of COPD. Int J Mol Sci 2018;19(2).

[70] Roberts MEP, Higgs BW, Brohawn P, Pilataxi F, Guo X, Kuziora M, et al. CD4+ T-cell profiles and peripheral blood ex vivo responses to T-cell directed stimulation delineate COPD phenotypes. Chron Obstruct Pulmon Dis 2015;2(4):268–80.

[71] Rovina N, Koutsoukou A, Koulouris NG. Inflammation and immune response in COPD: where do we stand? Mediat Inflamm J 2013;2013:413735.

[72] Freeman CM, Martinez FJ, Han MK, Washko Jr GR, McCubbrey AL, Chensue SW, et al. Lung CD8+ T cells in COPD have increased expression of bacterial TLRs. Respir Res 2013;14:13.

[73] Cerami C, Founds H, Nicholl I, Mitsuhashi T, Giordano D, Vanpatten S, et al. Tobacco smoke is a source of toxic reactive glycation products. Proc Natl Acad Sci 1997;94(25):13915–20.

[74] Agustí A, MacNee W, Donaldson K, Cosio M. Hypothesis: does COPD have an auto-immune component? Thorax 2003;58(10):832–4.

[75] Grumelli S, Corry DB, Song LZ, Song L, Green L, Huh J, et al. An immune basis for lung parenchymal destruction in chronic obstructive pulmonary disease and emphysema. PLoS Med 2004;1(1):e8.

[76] Di Stefano A, Caramori G, Gnemmi I, Contoli M, Vicari C, Capelli A, et al. T helper type 17-related cytokine expression is increased in the bronchial mucosa of stable chronic obstructive pulmonary disease patients. Clin Exp Immunol 2009;157(2):316–24.

[77] Ochs-Balcom HM, Grant BJ, Muti P, et al. Pulmonary function and abdominal adiposity in the general population. Chest J 2006;129:853–62.

[78] Gan WQ, Man SF, Senthilselvan A, Sin DD. Association between chronic obstructive pulmonary disease and systemic inflammation: a systematic review and a meta-analysis. Thorax 2004;59(7):574–80.

[79] Vernooy JH, Kucukaycan M, Jacobs JA, et al. Local and systemic inflammation in patients with chronic obstructive pulmonary disease: soluble tumor necrosis factor receptors are increased in sputum. Am J Respir Crit Care Med 2002;166(9):1218–24.

[80] Mancuso P. Obesity and lung inflammation. J Appl Physiol 2010;108:722–8.

[81] Breyer MK, Spruit MA, Celis AP, et al. Highly elevated C-reactive protein levels in obese patients with COPD: a fat chance? Clin Nutr 2009;28(6):642–7.

[82] Rutten EP, Breyer MK, Spruit MA, et al. Abdominal fat mass contributes to the systemic inflammation in chronic obstructive pulmonary disease. Clin Nutr 2010;29(6):756–60.

[83] Martinez FJ, Safrin S, Weycker D, Starko KM, Bradford WZ, King Jr TE, et al. The clinical course of patients with idiopathic pulmonary fibrosis. Ann Intern Med 2005;142:963–7.

[84] El-Chemaly S, Malide D, Zudaire E, Ikeda Y, Weinberg BA, Pacheco-Rodriguez G, et al. Abnormal lymphangiogenesis in idiopathic pulmonary fibrosis with insights into cellular and molecular mechanisms. Proc Natl Acad Sci 2009;106:3958–63.

[85] El-Chemaly S, Ziegler SG, Calado RT, Wilson KA, Wu HP, Haughey M, et al. Natural history of pulmonary fibrosis in two subjects with the same telomerase mutation. Chest J 2011;139:1203–9.

[86] Rosas IO, Ren P, Avila NA, Chow CK, Franks TJ, Travis WD, et al. Early interstitial lung disease in familial pulmonary fibrosis. Am J Respir Crit Care Med 2007;176:698–705.

[87] Kirshenbaum AS, Cruse G, Desai A, Bandara G, Leerkes M, Lee CC, et al. Immunophenotypic and ultrastructural analysis of mast cells in Hermansky-Pudlak syndrome type-1: a possible connection to pulmonary fibrosis. PLoS One 2016;11:e0159177.

[88] Rouhani FN, Brantly ML, Markello TC, Helip-Wooley A, O'Brien K, Hess R, et al. Alveolar macrophage dysregulation in Hermansky-Pudlak syndrome type 1. Am J Respir Crit Care Med 2009;180:1114–21.

[89] El-Chemaly S, Cheung F, Kotliarov Y, O'Brien KJ, Gahl WA, Chen J, et al. The immunome in two inherited forms of pulmonary fibrosis. Front Immunol 2018;31:9–76.

[90] Kang YP, Lee SB, Lee JM, Kim HM, Hong JY, Lee WJ, et al. Metabolic profiling regarding pathogenesis of idiopathic pulmonary fibrosis. J Proteome Res 2016;15(5):1717–24.

[91] Xie N, Tan Z, Banerjee S, Cui H, Ge J, Liu RM, et al. Glycolytic reprogramming in myofibroblast differentiation and lung fibrosis. Am J Respir Crit Care Med 2015;192(12):1462–74.

[92] Van Crevel R, Ottenhoff TH, Van DerMeer JW. Innate immunity to *Mycobacterium tuberculosis*. Clin Microbiol Rev 2002;15(2):294–309.

[93] Morikawa F, Nakano A, Nakano H, Oseko F, Morikawa S. Enhanced natural killer cell activity in patients with pulmonary tuberculosis. Jpn J Med 1989;28(3):316–22.

[94] Schlesinger LS, Bellinger-Kawahara CG, Payne NR, Horwitz MA. Phagocytosis of *Mycobacterium tuberculosis* is mediated by human monocyte complement receptors and complement component C3. J Immunol 1990;144(7):2771–80.

[95] Schaible UE, Collins HL, Kaufmann SH. Confrontation between intracellular bacteria and the immune system. Adv Immunol 1999;71:267–377.

[96] Flynn JL, Chan J, Triebold KJ, Dalton DK, Stewart TA, Bloom BR. An essential role for interferon gamma in resistance to *Mycobacterium tuberculosis* infection. J Exp Med 1993;178(6):2249–54.

[97] Dheda K, Schwander SK, Zhu B, Van Zyl-Smit RN, Zhang Y. The immunology of tuberculosis: from bench to bedside. Respirology 2010;15(3):433–50.

[98] Styblo K. Epidemiology of tuberculosis. In: Royal Nethers Tuberculosis Association selected papers. The Hague: Royal Netherlands Tuberculosis Association; 1991.

[99] American Thoracic Society. Diagnostic standards and classification of tuberculosis in adults and children. Am J Respir Crit Care Med 2000;161:1376–95.

[100] Chan JL, Matarese G, Shetty GK, et al. Differential regulation of metabolic, neuroendocrine, and immune function by leptin in humans. Proc Natl Acad Sci 2006;103:8481–6.

[101] Otero M, Lago R, Gomez R, Lago F, Gomez-Reino JJ, Gualillo O. Leptin: a metabolic hormone that functions like a proinflammatory adipokine. Drug News Perspect 2006;19:21–6.

[102] Wieland CW, Florquin S, Chan ED, et al. Pulmonary *Mycobacterium tuberculosis* infection in leptin-deficient ob/ob mice. Int Immunol 2005;17:1399–408.

[103] Palmer CE, Jablon S, Edwards PQ. Tuberculosis morbidity of young men in relation to tuberculin sensitivity and body build. Am Rev Tuberc 1957;76:517–39.

[104] Edwards LB, Livesay VT, Acquaviva FA, Palmer CE. Height, weight, tuberculous infection, and tuberculous disease. Arch Environ Health 1971;22:106–12.

[105] Tverdal A. Body mass index and incidence of tuberculosis. Eur J Respir Dis 1986;69:355–62.

[106] Van Lettow M, Kumwenda JJ, Harries AD, et al. Malnutrition and the severity of lung disease in adults with pulmonary tuberculosis in Malawi. Int J Tuberc Lung Dis 2004;8:211–7.

[107] Zachariah R, Spielmann MP, Harries AD, Salaniponi FM. Moderate to severe malnutrition in patients with tuberculosis is a risk factor associated with early death. Trans R Soc Trop Med Hyg 2002;96:291–4.

[108] Harries AD, Nkhoma WA, Thompson PJ, Nyangulu DS, Wirima JJ. Nutritional status in Malawian patients with pulmonary tuberculosis and response to chemotherapy. Eur J Clin Nutr 1988;42:445–50.

[109] Khan A, Sterling TR, Reves R, Vernon A, Horsburgh CR. Tuberculosis trials consortium: lack of weight gain and relapse risk in a large tuberculosis treatment trial. Am J Respir Crit Care Med 2006;174:344–8.

[110] Pèrez-Guzmàn C, Vargas MH, Quinonez F, Bazavilvazo N, Aguilar A. A cholesterol rich diet accelerates bacteriologic sterilization in pulmonary tuberculosis. Chest J 2005;127:643–51.

[111] Cegielski JP, Arab L, Cornoni-Huntley J. Nutritional risk factors for tuberculosis among adults in the United States, 1971–1992. Am J Epidemiol 2012;176:409–22.

[112] Leung CC, Lam TH, Chan WM, Yew WW, Ho KS, Leung G, et al. Lower risk of tuberculosis in obesity. Arch Intern Med 2007;167:1297–304.

[113] Hanrahan CF, Golub JE, Mohapi L, Tshabangu N, Modisenyane T, Chaisson RE, et al. Body mass index and risk of tuberculosis and death. AIDS 2010;24:1501–8.

[114] Chang CA, Meloni ST, Eisen G, Chaplin B, Akande P, Okonkwo P, et al. Tuberculosis incidence and risk factors among human immunodeficiency virus (HIV)-infected adults receiving antiretroviral therapy in a large HIV program in Nigeria. Open Forum Infectious Diseases 2015;2:ofv154.

[115] Neyrolles O, Hernandez-Pando R, Pietri-Rouxel F, Fornes P, Tailleux L, Barrios Payan JA, et al. Is adipose tissue a place for *Mycobacterium tuberculosis* persistence? PLoS One 2006;1:e43.

[116] De Rosa V, La Cava A, Matarese G. Metabolic pressure and the breach of immunological self-tolerance. Nat Immunol 2017;18:1190–6.

Obesity and lung function: From childhood to adulthood

3

David Chapman*, Gregory King[†,‡], Erick Forno[§,¶]

Translational Airways Group, School of Life Sciences, The University of Technology Sydney, Sydney, NSW, Australia* The Woolcock Institute of Medical Research, The University of Sydney, Sydney, NSW, Australia[†] Department of Respiratory Medicine, Royal North Shore Hospital, Sydney, NSW, Australia[‡] Department of Pediatrics, University of Pittsburgh School of Medicine, Pittsburgh, PA, United States[§] Division of Pulmonary Medicine, UPMC Children's Hospital of Pittsburgh, Pittsburgh, PA, United States[¶]

ABBREVIATIONS

AHR	airway hyperresponsiveness
BMI	body mass index
CAMP	Childhood Asthma Management Program
CC	closing capacity
CO	carbon monoxide
COPD	chronic obstructive pulmonary disease
DEXA	dual-energy X-ray absorptiometry
DLCO	diffusing capacity of the lungs for carbon monoxide
EMG	electromyography
ERV	expiratory reserve volume
FEF$_{25-75}$	mean forced expiratory flow between 25% and 75% of FVC
FeNO	fractional exhaled nitric oxide
FEV$_1$	forced expiratory volume in 1 second
FRC	functional residual capacity
FVC	forced vital capacity
Grs	respiratory system conductance
IC	inspiratory capacity
MEF$_{50}$	maximal forced expiratory flow with 50% of FVC remaining
MEF$_{75}$	maximal forced expiratory flow with 75% of FVC remaining
MRI	magnetic resonance imaging
P0.1	airway occlusion pressure at 0.1 second
PaO2	partial pressure of O_2 in arterial blood
PEEPi	intrinsic positive end-expiratory pressure
PEF	peak expiratory flow rate
Raw	airway resistance
Rrs	respiratory system resistance
RV	residual volume

Mechanisms and Manifestations of Obesity in Lung Disease. https://doi.org/10.1016/B978-0-12-813553-2.00003-8

TLC	total lung capacity
Xrs	respiratory system reactance
Zrs	respiratory system impedance

INTRODUCTION

Obesity affects lung function in otherwise healthy individuals from childhood to old age. The changes are highly variable but are clinically relevant because obesity, in conjunction with changes associated with aging, cause respiratory symptoms. We will describe the changes in lung function and mechanics associated with obesity and describe how these effects potentially change with increasing age and potentially cause respiratory symptoms.

Obesity is a well-known risk factor for metabolic syndrome, diabetes, and cardiovascular disease, among other comorbidities and complications. Some of these processes start as early as adolescence or even the (pre-)school years. Research in the last few decades has shown that obesity and its metabolic complications also have important effects on the respiratory system. There is ample and growing evidence that obesity affects lung function in otherwise healthy adults and children. It is important for clinicians to understand these changes, as well as researchers who work in the field of obesity. For clinicians, this knowledge is relevant to interpretation of lung function measurements, which are routinely done in clinical practice, and to understand how obesity may cause or modify respiratory symptoms in obesity alone, or in obesity that coexists with disease. For researchers, knowledge of how obesity affects lung function in individuals without disease is important in the interpretation of any *in vitro* or animal models so that they relate accurately to human physiology. In this chapter, we will review clinical, epidemiological, and experimental data on the effects of obesity on lung function in adults, children, and adolescents. Given that a preponderance of our understanding of the physiological consequences of obesity on lung function is based on findings in adults, results in pediatric populations are specifically stated, and otherwise can be assumed to come from the adult literature.

LUNG VOLUMES AND RESPIRATORY MECHANICS

The most characteristic effect of obesity on lung function is seen as the reduction in tidal breathing volume [e.g., functional residual capacity (FRC) (Figure 3.1)]. FRC is determined by the balance between outward recoil of the chest wall [1] and the inward recoil of the lung's elastic parenchymal tissue. Therefore the reduction in FRC is a due to the external pressure provided by excessive adipose tissue altering the balance between inflationary and deflationary pressures on the lung. There is an exponential relationship between body mass index (BMI) and FRC, with reduced FRC evident even in overweight individuals. The reduction in FRC plateaus at approximately $40 \, kg/m^2$, where FRC begins to encroach on residual volume (RV) [2]. In contrast, there is little effect of mild obesity on total lung capacity (TLC) and RV,

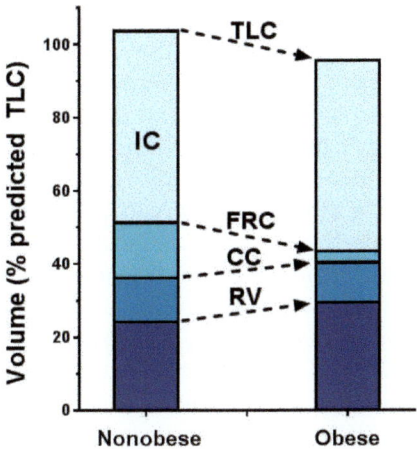

FIGURE 3.1

The effect of obesity on lung volume subdivisions. The major effect of obesity on lung function is a reduction in the volume at which tidal breathing occurs (i.e., reduced function residual capacity (FRC)). There is little effect of mild obesity on total lung capacity (TLC) and residual volume (RV), so that the RV/TLC is normal or slightly increased. Similarly, the inspiratory capacity (IC) is either normal or slightly increased. The volume at which airways begin to close, closing capacity (CC), is unaltered by obesity, so that airway closure will begin to occur during tidal breathing as increased obesity pushes FRC toward and then below CC.

Adapted from Mahadev S, Salome CM, Berend N, King GG. The effect of low lung volume on airway function in obesity. Respir Physiol Neurobiol 2013;188(2):192–9.

and only small decreases in TLC with morbid obesity [2, 3]. The combination of a small reduction in TLC with a relatively preserved RV means that the RV-to-TLC ratio is either normal or slightly increased in obesity [3].

The small reduction in TLC in morbid obesity in adults may be due to a variety of factors, including increased lung stiffness, increased stiffness and mass loading of the chest and abdominal walls, and altered diaphragmatic function. Weight loss after bariatric surgery is associated with increases in expiratory reserve volume (ERV), FRC, and TLC [4, 5]; the mechanisms that explain these improvements are potentially complex and have not yet been extensively explored. Pouwels *et al.* [6] however reported an increase in maximal inspiratory muscle pressures 3 and 6 months after bariatric surgery. Normal maximal inspiratory pressures, as well as reduced pressures, have been reported in obesity, and so the contribution of inspiratory muscle weakness to lung volume changes is probably variable between individuals [3, 7–9]. In contrast to adults, there is some evidence that maximal inspiratory pressure is reduced in overweight and obese children [10]. There have also been highly conflicting results on whether obesity affects lung or chest wall stiffness in adults [11–15]. This may be due to differences in the populations studied, awake versus under anesthesia and paralysis, and technical difficulties obtaining reliable esophageal pressure measurements in obesity. Furthermore, these studies have been confined to adult populations, so we can only speculate upon the effects of obesity on lung and chest wall stiffness in pediatric populations. However, overall it is likely that total respiratory system stiffness is increased due either to increases in chest wall and/or lung stiffness, and with probably a high degree of variability between obese individuals.

There are only a handful of pediatric studies evaluating the effect of obesity on lung volumes. However, the effects of obesity on lung volumes are different to that seen in adults. In a retrospective review of medical records from 327 nonasthmatic children and adolescents, Davidson and colleagues reported that increased BMI z-score was associated with lower FRC, ERV, and RV [16]. Similarly, the extent of obesity, measured by dual-energy X-ray absorptiometry (DEXA), has also been negatively correlated with FRC [17]. In a recent, prospective study of 168 adolescents (86 of normal weight and 82 with obesity), Rastogi et al. [18] reported reduced RV, ERV, and FRC in obesity, with increased inspiratory capacity (IC) suggesting little to no effect on TLC. Consistent with a lack of effect of obesity on TLC in childhood, two other studies have reported reduced FRC/end-expiratory lung volume in obese adolescents despite similar TLC volumes when compared with normal weight adolescents [19, 20]. However, other studies have yielded conflicting results, with some reporting that obese adolescents have higher inspiratory lung volumes and TLC than adolescents of normal weight [20], yet others have reported no significant differences [19]. The reduction in RV seen with obesity in children is likely because, in adults, RV is determined by chest wall stiffness, expiratory muscle strength, and airway closure [21, 22], whereas in children older than around 10 years of age [23], airway closure is minimal or nonexistent, so RV may be determined by the mechanics of the chest wall and lung, and the strength of the expiratory muscles [24].

Obesity likely impairs the downward movement of the diaphragm due to increased abdominal stiffness and mass loading, and could therefore limit lung expansion at full inspiration and also affect the tidal respiratory cycle. Furthermore, there may also be additional effects due to impairment of diaphragmatic function, although published data on diaphragmatic function in obesity is scarce. Watson et al. [25] reported that truncal fat in obese individuals measured by magnetic resonance imaging (MRI), which would impair diaphragmatic descent, did not explain the reduction in TLC in obese subjects in supine posture. However, the maximal BMI was only 44 kg/m^2, they did not measure transdiaphragmatic pressures, and TLC was measured by helium dilution method in supine posture, which may be inaccurate in obesity due to increased airway closure during tidal breathing [26, 27]. Steier et al. [3] reported no difference in transdiaphragmatic pressures in obese subjects compared with nonobese, healthy controls who were matched for age and height. Therefore, diaphragmatic strength appears to be normal in obese subjects, and the reduction in maximal inspiratory muscle pressures may be due to the obesity-related changes in the chest wall and nondiaphragmatic respiratory muscles. Nevertheless, end-expiratory gastric and esophageal pressures, as well as pressure swings during breathing, were greater in obese subjects, confirming an increase in the work of breathing.

Ventilatory drive, as measured by airway occlusion pressure at 0.1 second (P0.1) and diaphragm electromyography (EMG) activity, is increased in obese subjects without obesity hypoventilation syndrome, presumably to compensate for increased ventilatory loads [8, 28–30]. The ventilatory loads in obese subjects with obesity-hypoventilation syndrome are similar to obese subjects without hypoventilation, and hypercapnia appears to be due to a failure to increase respiratory drive centrally, in response to additional ventilatory loading [29]. Furthermore, diaphragm activity may persist into expiration in obese individuals, with one report that increased adiposity correlates with increased expiratory time with diaphragm activity [30].

Consistent with the lack of effect of obesity on RV, there is also no effect on the volume at which airway closure occurs, as measured by closing capacity (CC) [27]. Therefore, airway closure will begin to occur during tidal breathing as increased obesity forces FRC to and below CC (Figure 3.1) [27, 31]. Importantly, the difference between FRC and CC, which becomes negative in morbid obesity, correlates with reduced partial pressure of oxygen in arterial blood (PaO_2) in obesity [31]. These findings suggest that not only is the reduction in FRC the most consistent effect of obesity on lung volumes, but that it likely contributes greatly to the presence of symptoms and functional limitation.

SPIROMETRY

In adults, there is little effect of obesity on conventional spirometric variables. The forced expiratory volume in 1 second (FEV_1) and forced vital capacity (FVC) decrease slightly with increasing BMI [32]. Both remain within the normal range even in severe obesity and, as both change to the same extent, the FEV_1/FVC ratio remains normal or slightly increased [32, 33].

Interestingly, the effect of obesity on spirometric outcomes is almost opposite in children. The recognition that excess weight affected lung function in children likely started in the 1970s. Schroenberg *et al.* [34] reported increased spirometry with increasing weight up to a certain point, but then a decrease with a further increase in body weight. They suggested this was a "muscularity-obesity" effect. Further studies from that time had very small sample sizes or focused on severely overweight children. Another truly epidemiological study was published in 1997, when Lazarus and Speizer [35] analyzed a nationally representative cross-sectional sample of 2464 Australian children and reported that height-adjusted FEV_1 and FVC increased with higher subject weight, whereas height- and weight-adjusted FEV_1 and FVC decreased with increasing percent body fat. In 2003, Tantisira *et al.* [36] reported that a higher BMI was associated with higher FEV_1 and FVC among children who participated in the Childhood Asthma Management Program (CAMP), a multicenter clinical trial of long-term asthma controller medications that followed 1041 children over 4 years. The authors reported that the BMI-related increase was larger for FVC than for FEV_1, and thus BMI was also associated with lower FEV_1/FVC. Of interest, the increases in FEV_1 and FVC were opposite to the initial hypothesis of the authors, who proposed that obesity would be associated with lower lung function and a restrictive deficit, as had been reported previously in adults.

Since the early 2000s, dozens of studies have evaluated the association between obesity and spirometric measurements of lung function, and many have found obesity-related decreases in FEV_1/FVC [37–43]. Although one would initially assume this was related to airway obstruction, several studies have reported increases in FEV_1 with obesity, which is not consistent with an obstructive deficit. Moreover, BMI may be associated with more pronounced increases in FVC than in FEV_1, which may at least in part contribute to the low FEV_1/FVC seen in these children. This has led to the speculation that obesity in childhood leads to "airway dysanapsis," a term used to describe an incongruence or asymmetry between the growth of the lung parenchyma and that of the caliber of the airways [44, 45]. This asymmetric growth can be reflected by normal or supranormal FEV_1 and FVC, with larger effects on FVC, which lead to a low FEV_1/FVC ratio (in contrast to a genuinely obstructive deficit, in which the low FEV_1/FVC ratio is due to decreases in FEV_1). Dysanapsis has been described in breath-hold divers, for instance, who force their lung volumes above TLC when diving to provide extra oxygen and protect against increased pressures while diving [46]. Analyzing data from six independent cohorts of children with and without asthma, we recently reported that overweight/obese children and adolescents have 2–4 times higher risk of airway dysanapsis than those of normal weight [47]. Dysanapsis was associated with larger lung volumes (VC, TLC), confirming that the asymmetry likely stems from larger lung volumes rather than narrow airways. Dysanapsis in these obese children was associated with low mean forced expiratory flow between 25% and 75% of FVC (FEF_{25-75}), maximal forced expiratory flow with 50% FVC remaining (MEF_{50}), and MEF_{75}, suggesting that the airway caliber relative to lung size is reduced uniformly – at least throughout small- and medium-sized airways. The risk of dysanapsis with obesity was higher among boys than girls, with no significant differences by race/ethnicity. These findings may explain results from prior studies that reported obesity-related increases in FEV_1 and FVC, with larger increases in FVC leading to low FEV_1/FVC ratios [38].

The age of onset of obesity may affect lung function in children, as some obesity-related changes may start very early in life. Den Dekker *et al.* [48] recently published a metaanalysis of 24 birth cohorts in which greater birthweight and infant weight gain were associated with higher FEV_1 and FVC but lower FEV_1/FVC at school age. These results are more consistent with our report of airway dysanapsis in obese children, suggesting that these changes may happen quite early in life. However, it is likely that the changes seen with obesity may differ along the lifespan. Strunk and colleagues [49] performed a *post hoc* analysis of CAMP using data from asthmatic participants followed up to ~26–30 years of age and reported that children who were not obese during the trial but became obese as adults had significant decreases in FEV_1 and FEV_1/FVC (compared with those who were never obese), but obesity was not associated with significant changes in FVC over an average of 16 years of follow-up. Although these findings cannot necessarily be extrapolated to nonasthmatic children, one could speculate that this behaviour may be an "intermediate stage" between obesity effects earlier in childhood (airway dysanapsis, with normal/high FEV_1 and FVC but a low FEV_1/FVC) and later effects in adulthood (restriction, with low FEV_1 and FVC and a normal FEV_1/FVC). Future studies therefore will be needed to determine whether there is a "progression" of lung changes with obesity throughout an individual's life or whether there are separate, distinct phenotypes that may depend on the age of onset, duration, or severity of obesity.

We recently performed a large metaanalysis on obesity and lung function in adults and children, including 44 studies ($n=23,460$ subjects) for FEV_1, 30 studies ($n=16,913$) for FVC, 34 studies ($n=28,494$) for FEV_1/FVC, and 16 studies ($n=13,627$) for FEF_{25-75} [50]. This confirmed the differences in effects of obesity on spirometry between children and adults. In children, overweight/obesity was not associated with significant changes in FEV_1 percent of predicted. However, when raw FEV_1 values were adjusted for age, sex, and height (but not weight), overweight/obese children had a higher mean FEV_1 of 410 mL (95% confidence interval=280 to 540 mL). Similarly, overweight/obesity was associated with no changes in FVC as percent of predicted but with an increase of 250 mL (95% confidence interval=190 to 300 mL) when reported in adjusted absolute values. Obesity was associated with lower FEV_1/FVC (-2.5%, 95% confidence interval=-1.8% to -3.0%) and lower FEF_{25-75} (-4.7%, 95% confidence interval=-2.6% to -6.9%). This analysis further supports the explanation of dysanapsis in obesity at least partially explaining the spirometric differences. Given our large sample size, we were able to perform a metaregression analysis to evaluate potential effect modifiers and found that age and sex may modify the association between obesity and certain measures of lung function. For FEV_1, the effect of obesity was more pronounced in males. For FVC, the effect of obesity was more pronounced with increasing age, with the opposite effect in FEV_1/FVC. For example, using the beta coefficients from the metaregression, we can estimate that, for an obese 45- to 55-year-old, FVC would be reduced by approximately 6%–8% without any significant change in FEV_1/FVC. In contrast, for an obese 10- to 12-year-old, we would estimate that FEV_1/FVC would be reduced by approximately 4%–5% without any change in FVC [50].

DIFFUSING CAPACITY

In adults, obesity does not alter lung carbon monoxide (CO) diffusing capacity, with either normal [51] or supranormal [52, 53] values reported. An increase in CO diffusing capacity in obesity is assumed to be a result of the increased blood volume [54]. Consistent with this theory, one study measured the components of lung CO diffusing capacity in obesity, reporting that membrane diffusion was mildly reduced, but capillary blood volume was increased [51]. In children, there is minimal literature, and the data is conflicting. Davidson *et al.* [16] reported a positive correlation between BMI z-score and percent predicted diffusing capacity of CO. In contrast, Li *et al.* [17] report that 21/64 obese children had impaired CO diffusing capacity. Future studies with larger population samples will be able to determine whether the effect of obesity on lung diffusing capacity in children differs from that in adults.

OSCILLATORY AIRWAY MECHANICS

The forced oscillation technique has been used to measure respiratory system impedance in children and adults. Pressure oscillations (usually continuous sinusoidal waves made up of multiple frequencies commonly between 5 and ~35 Hz) are applied to the airway opening. Complex analysis of the relationship between pressure changes and flow changes allows the calculation of respiratory system impedance (Zrs), which includes the impedance of the airways, lungs, and chest wall [55]. Impedance can then be further split into its components of respiratory system resistance (Rrs) and reactance (Xrs). Rrs reflects airway caliber whereas reactance reflects the elastic (elastance) and inertive (inertance) properties of the respiratory system. A *lower* Xrs reflects increased respiratory system stiffness. From an interpretative point of view, Rrs and Xrs are also sensitive to heterogeneous narrowing and airway closure [56, 57] and therefore are sensitive measures of airway function. Increased Rrs and a lower Xrs (usually negative in value at lower frequencies around 5 Hz) indicate greater airway narrowing, heterogeneity of airway caliber, and airway closure. These processes make the respiratory system stiffer (increased elastance), from an oscillatory point of view, hence Xrs is reduced. Rrs and Xrs are highly lung volume dependent [58].

Because the most prominent effect of obesity in adults is reduced FRC, one would expect that Rrs and Xrs, which are measured at FRC, would be affected in obesity. As lung volume is reduced from TLC to RV, there is a nonlinear decrease in respiratory system conductance (Grs - reciprocal of Rrs) [58], reflecting reductions in airway caliber (Figure 3.2). Therefore, airway caliber will be reduced in obesity simply due to the reduction in FRC [59]. However, in an epidemiological study of young adults with airway caliber measured by plethysmographic airway resistance, airway resistance was still increased after adjustment for lung volume and age in obese males compared with nonobese males [60], which suggested an additional effect of obesity on airway function other than just reduced operating lung volume. This was confirmed by Oppenheimer *et al.* [61] who reported that, when obese subjects voluntarily increased their FRC to their predicted levels, airway caliber increased but did not normalize. This residual abnormality in airway caliber could be due to increased airway smooth muscle tone because obese subjects have an enhanced improvement in small airway function following bronchodilator administration [62]. Weight loss following bariatric surgery improves Rrs at FRC [5], however no predicted values nor comparison with matched controls were used in that study, and it remains uncertain whether weight loss normalizes airway function, as measured by respiratory system or lung impedance.

Respiratory system elastance is also dependent upon lung volume, although the relationship is different to that of Grs. In normal healthy subjects, Xrs changes minimally between TLC and FRC [58]. However, just below FRC, there is a critical point below which Xrs dramatically falls (worsens) as breathing approaches RV. This critical point likely reflects the onset of airway closure. Again, reduced FRC as seen in obesity should therefore worsen Xrs during tidal breathing. Indeed, a recent study

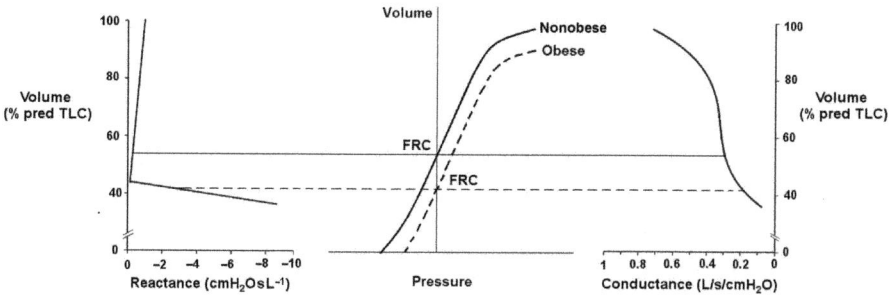

FIGURE 3.2

The effect of obesity on respiratory and airway mechanics. Excessive adipose tissue alters the balance between inflationary and deflationary pressures on the lung, altering the position of the pressure-volume curve. The resulting reduction in functional residual capacity (FRC) has consequences on volume-dependent measurements, such as oscillatory airway mechanics measured during tidal breathing. A reduction in FRC would lead to more negative respiratory system reactance, suggesting more airway closure, and reduced respiratory system conductance, reflecting reduced airway caliber. As discussed in the text, the effect of obesity on airway mechanics may be further exaggerated if there are pathological changes to airway structure or function. Nonobese data is represented by the bold line and obese data by the dashed line.

Respiratory system reactance and conductance curves have been adapted from Kelly VJ, Brown NJ, Sands SA, Borg BM, King GG, Thompson BR. Effect of airway smooth muscle tone on airway distensibility measured by the forced oscillation technique in adults with asthma. J Appl Physiol (1985) 2012;112(9):1494–503.

reported that Xrs was decreased in obese individuals when tidal breathing occurred below the volume at which airway closure begins (i.e., FRC is at or below CC) [27]. The presence of intrinsic positive end-expiratory pressure (PEEPi) [63] and positive intrapleural pressures [64] also confirm the presence of airway closure in obesity, particularly in morbidly obese individuals with BMI around 45 kg/m^2 or greater. The occurrence of airway closure during tidal breathing could cause cyclic reopening and closure of airways causing small airway damage [65], which would be expected to contribute to permanent small airways dysfunction. Additionally, abnormalities in Xrs are correlated with increased capillary blood volume, as derived from the diffusing capacity of the lungs for CO (DLCO), suggesting that pulmonary vascular congestion due to obesity may also contribute to small airway dysfunction [61]. Weight loss after bariatric surgery also improves Xrs [5].

A few studies have evaluated the effect of obesity on airway mechanics in children. In a study of 518 preschool healthy children, Kalhoff *et al.* [66] reported no association between BMI and either Rrs or Xrs. In contrast, a significant but small correlation was found between BMI and Rrs in a cohort of children aged 5–7 years with a history of bronchiolitis in infancy [67]. Similarly, a Japanese study developing pediatric reference values noted that 21/37 of children excluded due to a BMI > 25 kg/m^2 had Rrs more than one standard deviation above the predicted values [68]. Airway resistance (Raw) measured by the interrupter technique has been shown to be greatly increased in severely obese children, with Raw measured over 200% of predicted in 29/35 children who were able to perform the test [69]. Given the small number of studies and the somewhat conflicting data, future studies are needed to determine the effect of obesity on airway mechanics in children and whether these changes are related to reduced FRC or to permanent airways disease. However, the increased resistance suggests airways are narrower in obese children, which appears to be consistent with the spirometric findings.

VENTILATION AND VENTILATION DISTRIBUTION

There is a complex interaction between the severity of obesity and ventilation distribution. Holley *et al.* [26] measured ventilation using ^{133}Xenon scintigraphy in a small population of obese participants and found that ventilation was normal in obese patients with an ERV > 400 mL. However, in those with an ERV < 400 mL, basal airway closure led to a greater proportion of ventilation occurring in the upper lung zones, and because perfusion remained predominantly basal (normal), these patients had abnormalities in ventilation/perfusion ratios.

The multiple breath nitrogen washout allows measurement of ventilation distribution. The washout analysis provides measurements of heterogeneity of ventilation (i.e., variability of the magnitude of regional ventilation) in diffusion-dependent and convection-dependent airways. Diffusion dependent ventilation occurs more peripherally and in healthy subjects, modeling suggests that this occurs around the acinar entrance. Ventilation in diffusion—dependent but not convection—dependent airways was increased in obesity, suggesting small airway dysfunction, which is consistent with the observed increase in airway closure [27, 63, 64].

Increased ventilation heterogeneity in obese subjects has been shown to alter particle deposition [70] and likely contributes to reduced arterial oxygen tension [26, 31]. One study has investigated the effect of weight loss on ventilation, reporting reduced pulmonary shunt with no effect on perfusion; however, there was a paradoxical worsening of ventilation heterogeneity [71]. Because participants remained overweight/obese after the intervention, the apparent paradox could be due to two mechanisms; firstly, submaximal weight loss may recruit (i.e., reopen) some, but not all airways, thereby worsening ventilation distribution. Subsequent further weight loss toward normal might therefore reduce ventilation heterogeneity. Secondly, obesity could have led to permanent small airways disease that becomes apparent following weight loss.

AIRWAY RESPONSIVENESS

An effect of obesity on airway responsiveness has long been speculated, although there is limited evidence of increased airway responsiveness in obese subjects. In adults, cross-sectional studies have reported that obesity is associated with increased airway responsiveness [72] and that obesity is associated with an increased risk for the subsequent development of airway hyperresponsiveness (AHR) [73]. In contrast, one study suggests that obese subjects are at no greater risk of AHR compared with normal weight controls [32].The literature in children is no less conflicting, with studies reporting that obesity is associated with increased AHR among girls only [74–76], among boys only [77], and yet others in which there were no significant differences in AHR by obesity status [19, 78]. Karampatakis and coauthors [79] recently reported that prepubertal obese children with asthma have more prominent AHR than their nonobese counterparts, and that children who were obese and had insulin resistance had the highest mean levels of AHR of all groups. The evaluation of AHR in obese children may thus be confounded by several factors, including sex, age, and asthma status. To further add to the complexity, hormones may modify the effect of obesity on AHR as Sposato *et al.* [76] reported that obese prepubertal children showed lower mean levels of AHR than their normal-weight peers, whereas obese postpubertal adolescents showed higher AHR.

Part of this conflicting data may be due to the fact that airway responsiveness is traditionally measured by FEV_1, and this measure may not be sensitive to differences in small airway function. Obese subjects have increased airway closure, measured as changes in FVC, during methacholine compared with normal weight subjects. Two studies have used the forced oscillation technique to measure the change in small airway function during methacholine challenge in obese and normal weight subjects. Salome *et al.* [80] reported that, although changes in FEV_1 and respiratory system resistance did not differ between groups, obese subjects had exaggerated changes in respiratory system reactance, reflecting increased elastance (or stiffness). This increased stiffness was associated with increased dyspnea in the obese subjects. Similarly, Torchio *et al.* [81] reported that the association between BMI and airway responsiveness was considerably stronger when measured by respiratory system reactance than by respiratory system resistance.

As discussed earlier, the most prominent lung volume change with obesity is a reduction in FRC, and this in itself would be expected to reduce transpulmonary pressure and the forces opposing bronchoconstriction. Increased airway responsiveness due to low lung volume breathing has been demonstrated experimentally in normal weight subjects breathing below FRC either voluntarily [82] or with chest wall strapping [83]. Computational modeling suggests that natural variations in properties of the airway wall in the general population would result in a certain proportion of the population being vulnerable to the effects of low lung volume breathing on increased airway responsiveness [84]. This suggests that obesity would only be expected to lead to AHR in some individuals and may begin to explain the many conflicting studies as to the role of obesity on AHR.

RESPONSE TO SUPINE POSTURE

It has long been recognized that postural change from the seated to the supine position leads to a reduction in FRC in normal weight subjects [85]. However, the effect of supine posture in obese subjects is likely dependent upon the severity of obesity

(Figure 3.3). In mild-moderate obesity, FRC is reduced from seated to supine, but the extent is dampened compared to normal weight subjects [85]. However, there is little to no effect of supine posture on FRC in morbid obesity [86] due to the fact that ERV is virtually nonexistent and therefore FRC is unable to fall further. Indeed, the effect of supine posture on FRC can be regained in morbid obese subjects following weight loss [86]. The severity of obesity (waist-to-hip ratio) correlates with the ratio of CC to FRC in obese subjects while supine, suggesting that increased adiposity promotes airway closure during tidal breathing while supine [87].

Esophageal pressure (reflecting pleural pressure) is increased when nonobese subjects move from seated to supine [88]. Despite increased oesophageal pressure in obese participants while seated, the change in oesophageal pressure was not different between obese and nonobese. At first glance, this appears surprising given that obese subjects have a dampened fall in lung volume when supine. However, the change

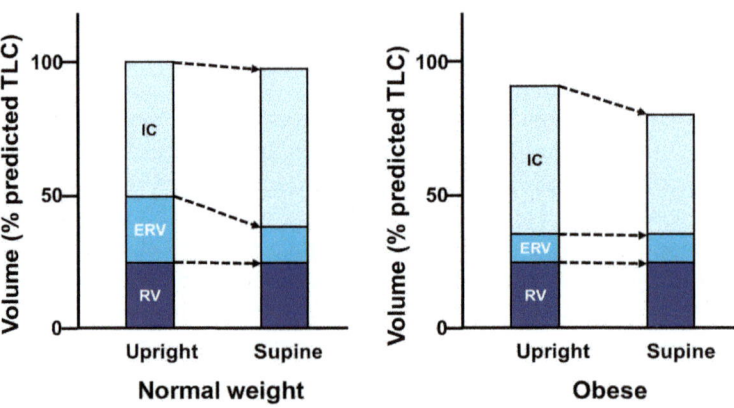

FIGURE 3.3

The effect of supine posture on lung volume subdivisions in normal weight and obese participants. In normal weight participants, postural change from the seated to the supine position leads to a reduction in functional residual capacity (FRC) with little change in total lung capacity (TLC) or residual volume (RV). However, the extent of the reduction in FRC is diminished in mild obesity and completely absent in severe/morbid obesity. The lack of effect of supine posture in severe obesity is due to the fact that ERV is virtually nonexistent, and therefore FRC is unable to fall further. There is also a small reduction in TLC in obesity when changing to supine posture. Therefore there are opposite effects of supine posture on inspiratory capacity (IC) in nonobese and obese individuals; IC is increased in nonobese individuals but is reduced in obesity.

This illustrative representation has been based on data reported by Steier J, Lunt A, Hart N, Polkey MI, Moxham J. Observational study of the effect of obesity on lung volumes. Thorax 2014;69(8):752–9; Tucker DH, Sieker HO. The effect of change in body position on lung volumes and intrapulmonary gas mixing in patients with obesity, heart failure, and emphysema. Am Rev Respir Dis 1960;82:787–91; and Watson RA, Pride NB. Postural changes in lung volumes and respiratory resistance in subjects with obesity. J Appl Physiol (1985) 2005;98(2):512–7.

in oesophageal pressure when supine is due to several factors including the change in lung volume and airway closure. Therefore in nonobese subjects, the increase in pleural pressure may be due to decreased lung volume, whereas in obese subjects, it may be due to an increase in airway closure/expiratory flow limitation, resulting in similar increases in oesophageal pressure between obese and nonobese individuals [63, 89]. This would also explain why obese subjects could have an increase in Rrs and Xrs when supine despite no change in lung volume [7].

The literature on the effects of supine posture in children is extremely limited and we are only aware of one study from the Dental literature [90]. In normal weight children, changing from the seated to supine position led to a decrease in FEV_1, FVC, peak expiratory flow rate (PEF) and FEF_{25-75}. Although obese children had increased spirometric values in the seated position, consistent with the literature, the effect of supine posture was the same as in normal weight children. No measurement of static lung volumes was performed. As such, future research is required to determine whether obesity in childhood, and its effect on airway dysanapsis, may alter the response to supine posture.

EXHALED NITRIC OXIDE

Fractional exhaled nitric oxide (FeNO), a marker of eosinophilic airway inflammation, has been used to assess the potential effects of obesity on airway inflammation. The limited studies in adults to date have produced conflicting results in obese participants with normal lung function, with reports of slightly increased [91], slightly decreased [92], and no change in FeNO [93]. Singleton *et al.* [94] analyzed data from 10,817 Americans who participated in the National Health and Nutrition Examination Survey and reported no correlation between BMI and FeNO in nonasthmatic or in asthmatic participants, suggesting any effect of obesity on FeNO is trivial. However, Maniscalco *et al.* [92] reported a negative correlation between FeNO and FRC in severe obese patients, suggesting that FeNO may only be appreciably reduced in severe disease. Consistent with reduced FeNO in severe obesity only, weight loss increased FeNO to levels equivalent to healthy controls when BMI improved from 45 to $30 \, kg/m^2$ [92]. However, it remains to be determined whether a reduction in FeNO in severe obesity is due to airway closure/expiratory flow limitation caused by the reduced FRC, increased metabolic activity of excess adipose tissue, or other yet-to-be-determined mechanisms.

There is also a lack of information on the effect of obesity on FeNO in pediatric populations. In a small, cross-sectional study, the presence of metabolic syndrome in children without asthma was associated with higher FeNO [95]. In contrast, a recent population-based study of 1717 children aged 5–18 years showed an inverse association between BMI and FeNO [96]. However, when the analysis was stratified by atopy, the association was significant among atopic children but not among those without atopy. Theses conflicting results may be in part due to the nonspecific nature of BMI as a marker of obesity. In a population-based cohort of 6178 children, den Dekker *et al.* [97] recently reported that a high android/gynoid fat mass was associated with lower FeNO, but that a larger preperitoneal fat mass was associated with higher FeNO instead. Interestingly, the relationship between obesity and FeNO is likely confounded by the duration of obesity and the presence of asthma. Chen *et al.* reported that long-term adiposity in Taiwanese children was associated with risk of asthma and low FeNO, whereas short-term increases in adiposity were associated with asthma and high FeNO [98]. We have reported that, in children, the association between obesity and asthma was significant only among those with normal/low fractional exhaled nitric oxide [99], whereas there was no obesity-asthma association among participants with high FeNO.

INTERACTION BETWEEN OBESITY AND LUNG DISEASE

Although the interaction between obesity and lung diseases such as asthma and chronic obstructive pulmonary disease (COPD) will be dealt with in subsequent chapters, we thought it important to highlight the interaction of obesity with lung function in disease. The most studied effects are those in asthma. As can be seen in Table 3.1, the effect of obesity on FEV_1 is greater in adults without asthma than in adults with asthma. Similarly, although there is an effect of obesity on FRC in nonasthmatics, this effect is reduced or nonexistent in asthma. Although this initially appears strange, Nikolacakis *et al.* [100] report that, in obese patients with asthma, lung volumes are a net result of the combined effects of obesity and asthma. In our example of FRC, the almost normal values in obese asthmatics is due to the low lung volume breathing due to obesity being offset by the hyperinflation due to asthma. A similar effect has been suggested in COPD, where obese patients have less hyperinflation (FRC, RV, and TLC) and greater IC than their nonobese counterparts [101]. Indeed, measurement of lung volumes before and after weight loss suggests that obesity may "mask" mild hyperinflation in smokers [4]. Therefore, caution should be taken when interpreting lung volume data from obese patients.

CONCLUSION

Obesity exerts an additional load upon the respiratory system, changing the balance of pressures between the inward recoil of the lung and the outward pressure of the chest. The major effect is a reduction in the volume at which tidal breathing occurs, leading to reduced airway caliber, increased respiratory system resistance, and increased respiratory system stiffness. Importantly, as the severity of obesity increases, tidal breathing begins to occur at or below CC, which predisposes to expiratory flow limitation and airway closure during tidal breathing. Many of these changes in lung function due to obesity are likely to contribute to respiratory symptoms and may induce changes in the sensitivity to bronchoconstriction. Interestingly, the effect of obesity on spirometry is dependent upon age, with reductions in FEV_1 and FVC in adults, whereas in children the most predominant deficit is in FEV_1/FVC, potentially reflecting asymmetric growth between the parenchyma and the airways. Future, ideally longitudinal,

Table 3.1 Pooled Estimates of Overweight or Obesity and Lung Function by Age Group and Asthma

	FEV_1 (%pred)	FVC (%pred)	$FEV_1/$FVC (%)	FEF_{25-75} (%pred)	TLC (%pred)	RV (%pred)	FRC (%pred)
By age and asthma							
Adult, asthma	**−1.7 (−2.3, −1.2)**[a]	**−3.2 (−6.0, −0.4)**	**−1.2 (−1.8, −0.6)**	−3.3 (−10.1, +3.6)	**−3.3 (−5.2, −1.4)**	**−4.1 (−8.2, −0.1)**	−0.7 (−12.5, +11.0)[a]
Adult, no asthma	**−6.9 (−11.1, −2.8)**[a]	**−7.5 (−11.4, −3.7)**	−0.9 (−1.9, +0.1)	–	**−5.3 (−8.2, −2.4)**	**−6.5 (−10.2, −2.7)**	**−23.6 (−31.1, −16.1)**[a]
Child, asthma	0.4 (−1.2, +2.0)	0.4 (−0.6, +1.5)	**−1.8 (−2.4, −1.3)**	−2.5 (−6.6, +1.5)	−3.6 (−8.0, +0.8)	–	–
Child, no asthma	−0.9 (−5.4, +3.6)	2.8 (−6.8, +12.5)	**−2.8 (−3.9, −1.8)**	−4.4 (−10.0, +1.2)	**−3.7 (−6.2, −1.3)**	–	–

[a]WMDs are significantly different between the groups (e.g., asthma vs. no asthma within age group).

Adapted from Forno E, Han YY, Mullen J, Celedon JC. Overweight, obesity, and lung function in children and adults—a meta-analysis. J Allergy Clin Immunol Pract 2018;6(2):570–81.e10.

Shown are weighted mean differences (WMDs) for overweight/obese versus normal weight. Significant WMDs are in bold.

studies will be needed to determine whether the effect of obesity on lung function changes throughout the lifespan (i.e., obese children have reduced FEV_1/FVC and as they become obese adults this slowly transitions to reductions in both FEV_1 and FVC with normal FEV_1/FVC), or whether they represent different phenotypes that depend on obesity onset, duration, and severity. In particular for pediatrics, further research is needed on the effect of obesity on lung volumes and AHR. Similarly, large randomized controlled clinical trials are needed to evaluate whether weight management interventions lead to improvements in lung function.

REFERENCES

[1] West JB, Luks AM. West's respiratory physiology. 10th ed. Philadephia, United States: Lippincott Williams and Wilkins; 2015.

[2] Jones RL, Nzekwu MM. The effects of body mass index on lung volumes. Chest 2006;130(3):827–33.

[3] Steier J, Lunt A, Hart N, Polkey MI, Moxham J. Observational study of the effect of obesity on lung volumes. Thorax 2014;69(8):752–9.

[4] Thomas PS, Cowen ER, Hulands G, Milledge JS. Respiratory function in the morbidly obese before and after weight loss. Thorax 1989;44(5):382–6.

[5] Al-Alwan A, Bates JH, Chapman DG, Kaminsky DA, DeSarno MJ, Irvin CG, et al. The nonallergic asthma of obesity. A matter of distal lung compliance. Am J Respir Crit Care Med 2014;189(12):1494–502.

[6] Pouwels S, Buise MP, Smeenk FW, Teijink JA, Nienhuijs SW. Comparative analysis of respiratory muscle strength before and after bariatric surgery using 5 different predictive equations. J Clin Anesth 2016;32:172–80.

[7] Yap JC, Watson RA, Gilbey S, Pride NB. Effects of posture on respiratory mechanics in obesity. J Appl Physiol (1985) 1995;79(4):1199–205.

[8] Steier J, Jolley CJ, Seymour J, Roughton M, Polkey MI, Moxham J. Neural respiratory drive in obesity. Thorax 2009;64(8):719–25.

[9] Faria AG, Ribeiro MA, Marson FA, Schivinski CI, Severino SD, Ribeiro JD, et al. Effect of exercise test on pulmonary function of obese adolescents. J Pediatr (Rio J) 2014;90(3):242–9.

[10] da Rosa GJ, Schivinski CI. Assessment of respiratory muscle strength in children according to the classification of body mass index. Rev Paul Pediatr 2014;32(2):250–5.

[11] Hedenstierna G, Santesson J. Breathing mechanics, dead space and gas exchange in the extremely obese, breathing spontaneously and during anaesthesia with intermittent positive pressure ventilation. Acta Anaesthesiol Scand 1976;20(3):248–54.

[12] Naimark A, Cherniack RM. Compliance of the respiratory system and its components in health and obesity. J Appl Physiol 1960;15:377–82.

[13] Pelosi P, Croci M, Ravagnan I, Cerisara M, Vicardi P, Lissoni A, et al. Respiratory system mechanics in sedated, paralyzed, morbidly obese patients. J Appl Physiol (1985) 1997;82(3):811–8.

[14] Sharp JT, Henry JP, Sweany SK, Meadows WR, Pietras RJ. The total work of breathing in normal and obese men. J Clin Invest 1964;43:728–39.

[15] Suratt PM, Wilhoit SC, Hsiao HS, Atkinson RL, Rochester DF. Compliance of chest wall in obese subjects. J Appl Physiol Respir Environ Exerc Physiol 1984;57(2):403–7.

[16] Davidson WJ, Mackenzie-Rife KA, Witmans MB, Montgomery MD, Ball GD, Egbogah S, et al. Obesity negatively impacts lung function in children and adolescents. Pediatr Pulmonol 2014;49(10):1003–10.

[17] Li AM, Chan D, Wong E, Yin J, Nelson EA, Fok TF. The effects of obesity on pulmonary function. Arch Dis Child 2003;88(4):361–3.

[18] Rastogi D, Bhalani K, Hall CB, Isasi CR. Association of pulmonary function with adiposity and metabolic abnormalities in urban minority adolescents. Ann Am Thorac Soc 2014;11(5):744–52.

[19] Mansell AL, Walders N, Wamboldt MZ, Carter R, Steele DW, Devin JA, et al. Effect of body mass index on response to methacholine bronchial provocation in healthy and asthmatic adolescents. Pediatr Pulmonol 2006;41(5):434–40.

[20] Mendelson M, Michallet AS, Perrin C, Levy P, Wuyam B, Flore P. Exercise training improves breathing strategy and performance during the six-minute walk test in obese adolescents. Respir Physiol Neurobiol 2014;200:18–24.

[21] Cohn JE, Donoso HD. Mechanical properties of lung in normal men over 60 years old. J Clin Invest 1963;42:1406–10.

[22] Pedersen OF, Thiessen B, Naeraa N, Lyager S, Hilberg C. Factors determining residual volume in normal and asthmatic subjects. Eur J Respir Dis 1984;65(2):99–105.

[23] Mansell A, Bryan C, Levison H. Airway closure in children. J Appl Physiol 1972;33(6):711–4.

[24] Schrader PC, Quanjer PH, Olievier IC. Respiratory muscle force and ventilatory function in adolescents. Eur Respir J 1988;1(4):368–75.

[25] Watson RA, Pride NB, Thomas EL, Ind PW, Bell JD. Relation between trunk fat volume and reduction of total lung capacity in obese men. J Appl Physiol (1985) 2012;112(1):118–26.

[26] Holley HS, Milic-Emili J, Becklake MR, Bates DV. Regional distribution of pulmonary ventilation and perfusion in obesity. J Clin Invest 1967;46(4):475–81.

[27] Mahadev S, Salome CM, Berend N, King GG. The effect of low lung volume on airway function in obesity. Respir Physiol Neurobiol 2013;188(2):192–9.

[28] Lourenco RV. Diaphragm activity in obesity. J Clin Invest 1969;48(9):1609–14.

[29] Sampson MG, Grassino A. Neuromechanical properties in obese patients during carbon dioxide rebreathing. Am J Med 1983;75(1):81–90.

[30] Sampson MG, Grassino AE. Load compensation in obese patients during quiet tidal breathing. J Appl Physiol Respir Environ Exerc Physiol 1983;55(4):1269–76.

[31] Farebrother MJ, McHardy GJ, Munro JF. Relation between pulmonary gas exchange and closing volume before and after substantial weight loss in obese subjects. Br Med J 1974;3(5927):391–3.

[32] Schachter LM, Salome CM, Peat JK, Woolcock AJ. Obesity is a risk for asthma and wheeze but not airway hyperresponsiveness. Thorax 2001;56(1):4–8.

[33] Sin DD, Jones RL, Man SF. Obesity is a risk factor for dyspnea but not for airflow obstruction. Arch Intern Med 2002;162(13):1477–81.

[34] Schoenberg JB, Beck GJ, Bouhuys A. Growth and decay of pulmonary function in healthy blacks and whites. Respir Physiol 1978;33(3):367–93.

[35] Lazarus R, Colditz G, Berkey CS, Speizer FE. Effects of body fat on ventilatory function in children and adolescents: cross-sectional findings from a random population sample of school children. Pediatr Pulmonol 1997;24(3):187–94.

[36] Tantisira KG, Litonjua AA, Weiss ST, Fuhlbrigge AL, Childhood Asthma Management Program Research G. Association of body mass with pulmonary function in the Childhood Asthma Management Program (CAMP). Thorax 2003;58(12):1036–41.

[37] Weinmayr G, Forastiere F, Buchele G, Jaensch A, Strachan DP, Nagel G, et al. Overweight/obesity and respiratory and allergic disease in children: International Study of Asthma and Allergies in Childhood (ISAAC) phase two. PLoS One. 2014;9(12):e113996.

[38] Cibella F, Bruno A, Cuttitta G, Bucchieri S, Melis MR, De Cantis S, et al. An elevated body mass index increases lung volume but reduces airflow in Italian schoolchildren. PLoS One 2015;10(5):e0127154.

[39] Chen YC, Huang YL, Ho WC, Wang YC, Yu YH. Gender differences in effects of obesity and asthma on adolescent lung function: results from a population-based study. J Asthma 2016;1–7.

[40] Huang F, Del-Rio-Navarro BE, Torres-Alcantara S, Perez-Ontiveros JA, Ruiz-Bedolla E, Saucedo-Ramirez OJ, et al. Adipokines, asymmetrical dimethylarginine, and pulmonary function in adolescents with asthma and obesity. J Asthma 2017;54(2):153–61.

[41] Baek HS, Kim YD, Shin JH, Kim JH, Oh JW, Lee HB. Serum leptin and adiponectin levels correlate with exercise-induced bronchoconstriction in children with asthma. Ann Allergy Asthma Immunol 2011;107(1):14–21.

[42] Lang JE, Holbrook JT, Wise RA, Dixon AE, Teague WG, Wei CY, et al. Obesity in children with poorly controlled asthma: Sex differences. Pediatr Pulmonol 2013;48(9):847–56.

[43] Spathopoulos D, Paraskakis E, Trypsianis G, Tsalkidis A, Arvanitidou V, Emporiadou M, et al. The effect of obesity on pulmonary lung function of school aged children in Greece. Pediatr Pulmonol 2009;44(3):273–80.

[44] Green M, Mead J, Turner JM. Variability of maximum expiratory flow-volume curves. J Appl Physiol 1974;37(1):67–74.

[45] Mead J. Dysanapsis in normal lungs assessed by the relationship between maximal flow, static recoil, and vital capacity. Am Rev Respir Dis 1980;121(2):339–42.

[46] Lemaitre F, Clua E, Andreani B, Castres I, Chollet D. Ventilatory function in breath-hold divers: effect of glossopharyngeal insufflation. Eur J Appl Physiol 2010;108(4):741–7.

[47] Forno E, Weiner DJ, Mullen J, Sawicki G, Kurland G, Han YY, et al. Obesity and airway dysanapsis in children with and without asthma. Am J Respir Crit Care Med 2017;195(3):314–23.

[48] den Dekker HT, Sonnenschein-van der Voort AM, de Jongste JC, Anessi-Maesano I, Arshad SH, Barros H, et al. Early growth characteristics and the risk of reduced lung function and asthma: a meta-analysis of 25,000 children. J Allergy Clin Immunol 2016;137(4):1026–35.

[49] Strunk RC, Colvin R, Bacharier LB, Fuhlbrigge A, Forno E, Arbelaez AM, et al. Airway obstruction worsens in young adults with asthma who become obese. J Allergy Clin Immunol Pract 2015;3(5). 765-71.e2.

[50] Forno E, Han YY, Mullen J, Celedon JC. Overweight, obesity, and lung function in children and adults-a meta-analysis. J Allergy Clin Immunol Pract 2018;6(2). 570–81.e10.

[51] Oppenheimer BW, Berger KI, Ali S, Segal LN, Donnino R, Katz S, et al. Pulmonary vascular congestion: a mechanism for distal lung unit dysfunction in obesity. PLoS One 2016;11(4):e0152769.

[52] Rubinstein I, Zamel N, DuBarry L, Hoffstein V. Airflow limitation in morbidly obese, nonsmoking men. Ann Intern Med 1990;112(11):828–32.

[53] Collard P, Wilputte JY, Aubert G, Rodenstein DO, Frans A. The single-breath diffusing capacity for carbon monoxide in obstructive sleep apnea and obesity. Chest 1996;110(5):1189–93.

[54] Ray CS, Sue DY, Bray G, Hansen JE, Wasserman K. Effects of obesity on respiratory function. Am Rev Respir Dis 1983;128(3):501–6.

[55] Bates JH, Irvin CG, Farre R, Hantos Z. Oscillation mechanics of the respiratory system. Compr Physiol 2011;1(3):1233–72.

[56] Lundblad LK, Thompson-Figueroa J, Allen GB, Rinaldi L, Norton RJ, Irvin CG, et al. Airway hyperresponsiveness in allergically inflamed mice: the role of airway closure. Am J Respir Crit Care Med 2007;175(8):768–74.

[57] Lutchen KR, Jensen A, Atileh H, Kaczka DW, Israel E, Suki B, et al. Airway constriction pattern is a central component of asthma severity: the role of deep inspirations. Am J Respir Crit Care Med 2001;164(2):207–15.

[58] Kelly VJ, Brown NJ, Sands SA, Borg BM, King GG, Thompson BR. Effect of airway smooth muscle tone on airway distensibility measured by the forced oscillation technique in adults with asthma. J Appl Physiol (1985) 2012;112(9):1494–503.

[59] Zerah F, Harf A, Perlemuter L, Lorino H, Lorino AM, Atlan G. Effects of obesity on respiratory resistance. Chest 1993;103(5):1470–6.

[60] King GG, Brown NJ, Diba C, Thorpe CW, Munoz P, Marks GB, et al. The effects of body weight on airway calibre. Eur Respir J 2005;25(5):896–901.

[61] Oppenheimer BW, Berger KI, Segal LN, Stabile A, Coles KD, Parikh M, et al. Airway dysfunction in obesity: response to voluntary restoration of end expiratory lung volume. PLoS One 2014;9(2):e88015.

[62] Desai AG, Togias A, Schechter C, Fisher B, Parow A, Skloot G. Peripheral airways dysfunction in obesity reflects increased bronchomotor tone. J Allergy Clin Immunol 2015;135(3):820–2.

[63] Pankow W, Podszus T, Gutheil T, Penzel T, Peter J, Von Wichert P. Expiratory flow limitation and intrinsic positive end-expiratory pressure in obesity. J Appl Physiol (1985) 1998;85(4):1236–43.

[64] Behazin N, Jones SB, Cohen RI, Loring SH. Respiratory restriction and elevated pleural and esophageal pressures in morbid obesity. J Appl Physiol (1985) 2010;108(1):212–8.

[65] Guerin C, LeMasson S, de Varax R, Milic-Emili J, Fournier G. Small airway closure and positive end-expiratory pressure in mechanically ventilated patients with chronic obstructive pulmonary disease. Am J Respir Crit Care Med 1997;155(6):1949–56.

[66] Kalhoff H, Breidenbach R, Smith HJ, Marek W. Impulse oscillometry in preschool children and association with body mass index. Respirology 2011;16(1):174–9.

[67] Lauhkonen E, Koponen P, Nuolivirta K, Paassilta M, Toikka J, Saari A, et al. Obesity and bronchial obstruction in impulse oscillometry at age 5–7 years in a prospective post-bronchiolitis cohort. Pediatr Pulmonol 2015;50(9):908–14.

[68] Hagiwara S, Mochizuki H, Muramatsu R, Koyama H, Yagi H, Nishida Y, et al. Reference values for Japanese children's respiratory resistance using the LMS method. Allergol Int 2014;63(1):113–9.

[69] Dubern B, Tounian P, Medjadhi N, Maingot L, Girardet JP, Boule M. Pulmonary function and sleep-related breathing disorders in severely obese children. Clin Nutr 2006;25(5):803–9.

[70] Graham DR, Chamberlain MJ, Hutton L, King M, Morgan WK. Inhaled particle deposition and body habitus. Br J Ind Med 1990;47(1):38–43.

[71] Rivas E, Arismendi E, Agusti A, Sanchez M, Delgado S, Gistau C, et al. Ventilation/perfusion distribution abnormalities in morbidly obese subjects before and after bariatric surgery. Chest 2015;147(4):1127–34.

[72] Chinn S, Jarvis D, Burney P. Relation of bronchial responsiveness to body mass index in the ECRHS. European Community Respiratory Health Survey Thorax 2002;57(12):1028–33.

[73] Litonjua AA, Sparrow D, Celedon JC, DeMolles D, Weiss ST. Association of body mass index with the development of methacholine airway hyperresponsiveness in men: the Normative Aging Study. Thorax 2002;57(7):581–5.

[74] Yoo S, Kim HB, Lee SY, Kim BS, Kim JH, Yu JH, et al. Association between obesity and the prevalence of allergic diseases, atopy, and bronchial hyperresponsiveness in Korean adolescents. Int Arch Allergy Immunol 2011;154(1):42–8.

[75] Huang SL, Shiao G, Chou P. Association between body mass index and allergy in teenage girls in Taiwan. Clin Exp Allergy 1999;29(3):323–9.

[76] Sposato B, Scalese M, Migliorini MG, Riccardi MP, Tosti Balducci M, Petruzzelli L, et al. Obesity can influence children's and adolescents' airway hyperresponsiveness differently. Multidiscip Respir Med 2013;8(1):60.

[77] Jang AS, Lee JH, Park SW, Shin MY, Kim DJ, Park CS. Severe airway hyperresponsiveness in school-aged boys with a high body mass index. Korean J Intern Med 2006;21(1):10–4.

[78] Del Rio-Navarro BE, Blandon-Vijil V, Escalante-Dominguez AJ, Berber A, Castro-Rodriguez JA. Effect of obesity on bronchial hyperreactivity among Latino children. Pediatr Pulmonol 2013;48(12):1201–5.

[79] Karampatakis N, Karampatakis T, Galli-Tsinopoulou A, Kotanidou EP, Tsergouli K, Eboriadou-Petikopoulou M, et al. Impaired glucose metabolism and bronchial hyperresponsiveness in obese prepubertal asthmatic children. Pediatr Pulmonol 2016;.

[80] Salome CM, Munoz PA, Berend N, Thorpe CW, Schachter LM, King GG. Effect of obesity on breathlessness and airway responsiveness to methacholine in non-asthmatic subjects. Int J Obes (Lond) 2008;32(3):502–9.

[81] Torchio R, Gobbi A, Gulotta C, Dellaca R, Tinivella M, Hyatt RE, et al. Mechanical effects of obesity on airway responsiveness in otherwise healthy humans. J Appl Physiol (1985) 2009;107(2):408–16.

[82] Ding DJ, Martin JG, Macklem PT. Effects of lung volume on maximal methacholine-induced bronchoconstriction in normal humans. J Appl Physiol (1985) 1987;62(3):1324–30.

[83] Chapman DG, Berend N, Horlyck KR, King GG, Salome CM. Does increased baseline ventilation heterogeneity following chest wall strapping predispose to airway hyperresponsiveness? J Appl Physiol (1985) 2012;113(1):25–30.

[84] Bates JH, Dixon AE. Potential role of the airway wall in the asthma of obesity. J Appl Physiol (1985) 2015;118(1):36–41.

[85] Tucker DH, Sieker HO. The effect of change in body position on lung volumes and intrapulmonary gas mixing in patients with obesity, heart failure, and emphysema. Am Rev Respir Dis 1960;82:787–91.

[86] Sebbane M, El Kamel M, Millot A, Jung B, Lefebvre S, Rubenovitch J, et al. Effect of weight loss on postural changes in pulmonary function in obese subjects: a longitudinal study. Respir Care 2015;60(7):992–9.

[87] Benedik PS, Baun MM, Keus L, Jimenez C, Morice R, Bidani A, et al. Effects of body position on resting lung volume in overweight and mildly to moderately obese subjects. Respir Care 2009;54(3):334–9.

[88] Owens RL, Campana LM, Hess L, Eckert DJ, Loring SH, Malhotra A. Sitting and supine esophageal pressures in overweight and obese subjects. Obesity (Silver Spring) 2012;20(12):2354–60.

[89] Ferretti A, Giampiccolo P, Cavalli A, Milic-Emili J, Tantucci C. Expiratory flow limitation and orthopnea in massively obese subjects. Chest 2001;119(5):1401–8.

[90] Hoge C, Oueis H, Casamassimo PS, Rashid R, Prior S. Physiologic signs during dental treatment in overweight vs normal weight children. Pediatr Dent 2008;30(6):522–9.

[91] Chapman DG, Berend N, King GG, Salome CM. Increased airway closure is a determinant of airway hyperresponsiveness. Eur Respir J 2008;32(6):1563–9.

[92] Maniscalco M, de Laurentiis G, Zedda A, Faraone S, Giardiello C, Cristiano S, et al. Exhaled nitric oxide in severe obesity: effect of weight loss. Respir Physiol Neurobiol 2007;156(3):370–3.

[93] van de Kant KD, Paredi P, Meah S, Kalsi HS, Barnes PJ, Usmani OS. The effect of body weight on distal airway function and airway inflammation. Obes Res Clin Pract 2016;10(5):564–73.

[94] Singleton MD, Sanderson WT, Mannino DM. Body mass index, asthma and exhaled nitric oxide in U.S. adults, 2007–2010. J Asthma 2014;51(7):756–61.

[95] Erkocoglu M, Kaya A, Ozcan C, Akan A, Vezir E, Azkur D, et al. The effect of obesity on the level of fractional exhaled nitric oxide in children with asthma. Int Arch Allergy Immunol 2013;162(2):156–62.

[96] Yao TC, Tsai HJ, Chang SW, Chung RH, Hsu JY, Tsai MH, et al. Obesity disproportionately impacts lung volumes, airflow and exhaled nitric oxide in children. PLoS One 2017;12(4):e0174691.

[97] den Dekker HT, Ros KP, de Jongste JC, Reiss IK, Jaddoe VW, Duijts L. Body fat mass distribution and interrupter resistance, fractional exhaled nitric oxide, and asthma at school-age. J Allergy Clin Immunol 2016;.

[98] Chen YC, Chih AH, Chen JR, Liou TH, Pan WH, Lee YL. Rapid adiposity growth increases risks of new-onset asthma and airway inflammation in children. Int J Obes (Lond) 2017;.

[99] Han YY, Forno E, Celedon JC. Adiposity, fractional exhaled nitric oxide, and asthma in U.S. children. Am J Respir Crit Care Med 2014;190(1):32–9.

[100] Nicolacakis K, Skowronski ME, Coreno AJ, West E, Nader NZ, Smith RL, et al. Observations on the physiological interactions between obesity and asthma. J Appl Physiol (1985) 2008;105(5):1533–41.

[101] Ora J, Laveneziana P, Ofir D, Deesomchok A, Webb KA, O'Donnell DE. Combined effects of obesity and chronic obstructive pulmonary disease on dyspnea and exercise tolerance. Am J Respir Crit Care Med 2009;180(10):964–71.

FURTHER READING

Watson RA, Pride NB. Postural changes in lung volumes and respiratory resistance in subjects with obesity. J Appl Physiol (1985) 2005;98(2):512–7.

Obesity and obstructive lung disease: An epidemiologic review

4

Sapna Bhatia, Akshay Sood

Department of Internal Medicine, University of New Mexico School of Medicine,
Albuquerque, NM, United States

ABBREVIATIONS

ACQ	asthma control questionnaire
BMI	body mass index
COPD	chronic obstructive pulmonary disease
CT	computed tomography
DC	District of Columbia
ECLIPSE	Evaluation of COPD longitudinally to identify predictive surrogate end-points
ER	emergency room
FEV1	forced expiratory volume in 1 second
FVC	forced vital capacity
GOLD	Global Initiative for Obstructive Lung Disease
HIV	human immunodeficiency virus
ICS	inhaled corticosteroids
IgE	immunoglobulin E
LABA	long-acting beta agonist
NHLBI	National Heart, Lung, and Blood Institute
SABA	short-acting beta agonist
PPAR	peroxisome proliferator-activated receptor
U.S.	United States

INTRODUCTION

The prevalence of obesity in American adults and youths has steadily increased from 1999 to 2014 [1]. More than one-third of adults in the U.S. are currently obese, with a higher prevalence in women than men (38.3% vs. 34.3%, respectively, in 2011–14) [1]. Among all racial and ethnic groups, non-Hispanic blacks have the highest age-adjusted rates of obesity in the U.S., with approximately half of this population

Mechanisms and Manifestations of Obesity in Lung Disease. https://doi.org/10.1016/B978-0-12-813553-2.00004-X

affected [1]. The prevalence rates of childhood obesity in the U.S. has more than doubled since 1980 and remains at generally high levels [2], particularly in urban areas with minority populations such as Washington, DC. Approximately 17% of U.S. youth were obese in 2011–14, and these rates are highest in the adolescent population [1]. Although defined by body mass index (BMI) value, obesity represents a heterogeneous collection of adiposity phenotypes, defined by measures of global adiposity, regional adiposity, and metabolically active adiposity.

Obesity is associated with profound changes in lung health and disease. The complex interactions between obesity-related metabolic dysregulation, systemic inflammation, and pulmonary function deficits have impacted the frequency, clinical presentation, and pathophysiology of obstructive lung diseases. The recent epidemic of obesity has been contemporaneous with an epidemic of obstructive lung diseases in the U.S. and the world. The two most common obstructive lung diseases are asthma and chronic obstructive pulmonary disease (COPD), the latter in turn includes two major phenotypes of emphysema and chronic bronchitis. Emphysema, pathologically defined, is clinically diagnosed by the presence of airflow obstruction on postbronchodilator spirometry. Chronic bronchitis is defined by cough and phlegm production for at least 3 months out of the year for at least 2 consecutive years. Based on 2015 data, 7.6% of all adults and 8.4% of all children in the U.S. currently have asthma [3]. In 2015, 3.8% and 1.5% of all adults in the U.S. were diagnosed to have chronic bronchitis and emphysema, respectively [4]. The mechanistic association between obesity and obstructive lung diseases include genetic, epigenetic, mechanical, inflammatory/immunological, metabolic, dietary, and hormonal pathways, as well as shared comorbidities.

ASTHMA
PREVALENCE

Obesity is associated with high rates of prevalent self-reported asthma across a wide range of populations, races, ethnicities, and age ranges, in both developed and developing countries. Data from the 2011–14 U.S. National Health and Nutrition Examination Survey demonstrate that self-reported asthma prevalence is higher among obese adults (11.1%), when compared with overweight (7.8%) and normal weight (7.1%) adults, and this difference is more pronounced among women than men [5]. In a large, multiethnic, population-based, cross-sectional study including 681,122 children aged 6–19 years of age in Southern California, moderate obesity and extreme obesity in children were similarly associated with a high frequency of asthma (37% and 68%, respectively) [6]. An alarming increase in obese asthmatic children was also recently reported in other developed countries, including Netherlands and Portugal [7, 8], as well as developing countries such as China [9]. Several adult and pediatric studies have reported a female gender predilection and dose-response gradient between asthma and BMI value [7, 10]. A limitation in many epidemiological studies examining the asthma-obesity relationship is that the diagnosis of asthma relies on either patient self-report or parent report (in children) of physician diagnosis, use of asthma medication, or hospitalization for asthma exacerbation, and few studies have examined objective tests. However, the use of methacholine challenge test data from large populations of adults confirms that obesity is associated with an increase in bronchial responsiveness [11, 12].

INCIDENCE

Obesity is a risk factor for incident asthma in both adults and children, as reported by a number of prospective studies [8, 13–16]. Moreover, obesity increases the risk for incident asthma in a dose-dependent manner [17, 18]. Sex-related differences in this association have been reported by some but not all studies, and the possible role of female reproductive hormones on the obesity-asthma association is not clear. In a metaanalysis by Chen *et al.* [17], obese boys were at twofold greater risk for incident asthma than obese girls. On the other hand, a metaanalysis published by Egan *et al.* [19] showed that both obese girls and boys were at a similar elevated risk for incident asthma than their normal weight counterparts. There is a need for studies to examine the gender-specific effect of puberty on the obesity-asthma association in children, which may explain the conflicting findings in the literature.

Because BMI includes both fat and muscle mass, some studies have analyzed alternate adiposity phenotypes with respect to incident asthma. For instance, the relationship between abdominal adiposity (as defined by a high waist circumference) and incident asthma has been analyzed [13, 14, 20, 21]. These studies demonstrate that a large waist circumference confers an elevated risk for incident asthma. On the other hand, Assad *et al.* [21], using a longitudinal cohort, demonstrated that BMI was a stronger predictor for incident asthma in women than increased waist circumference or other parameters of the metabolic syndrome. The latter study indicates that incident asthma in women is associated with general adiposity rather than regional adiposity or metabolically active adiposity phenotypes. Among human immunodeficiency virus (HIV)-positive adults in the postantiretroviral therapy era, greater BMI and greater subcutaneous adiposity are both associated with increased risk for incident asthma and airway wall thickness [22]. The literature is therefore unclear whether specific adiposity phenotypes are particularly relevant for asthma.

In the pediatric population, recent literature has focused on the effect of maternal obesity in pregnancy and maternal gestational weight gain on increased risk for incident asthma in the offspring. Each kg/m^2 increase in maternal BMI in pregnancy was associated with a 2%–3% increase in the odds of childhood asthma [23, 24]. This increased risk was evident in children up to the age of 18 years [24]. The causal relationship between maternal obesity and incident asthma in children is however still uncertain, although potential epigenetic mechanisms have been invoked.

DISEASE CONTROL

Obesity may not affect all people with asthma alike. There are at least two major phenotypes of asthma noted in obesity: a later-onset nonallergic type or asthma consequent to obesity, and an earlier-onset type or preexisting asthma complicated by obesity, which is typically associated with increased markers of allergic inflammation (Figure 4.1). It is likely that obesity has distinct effects on these two asthma phenotypes [25]. Obese

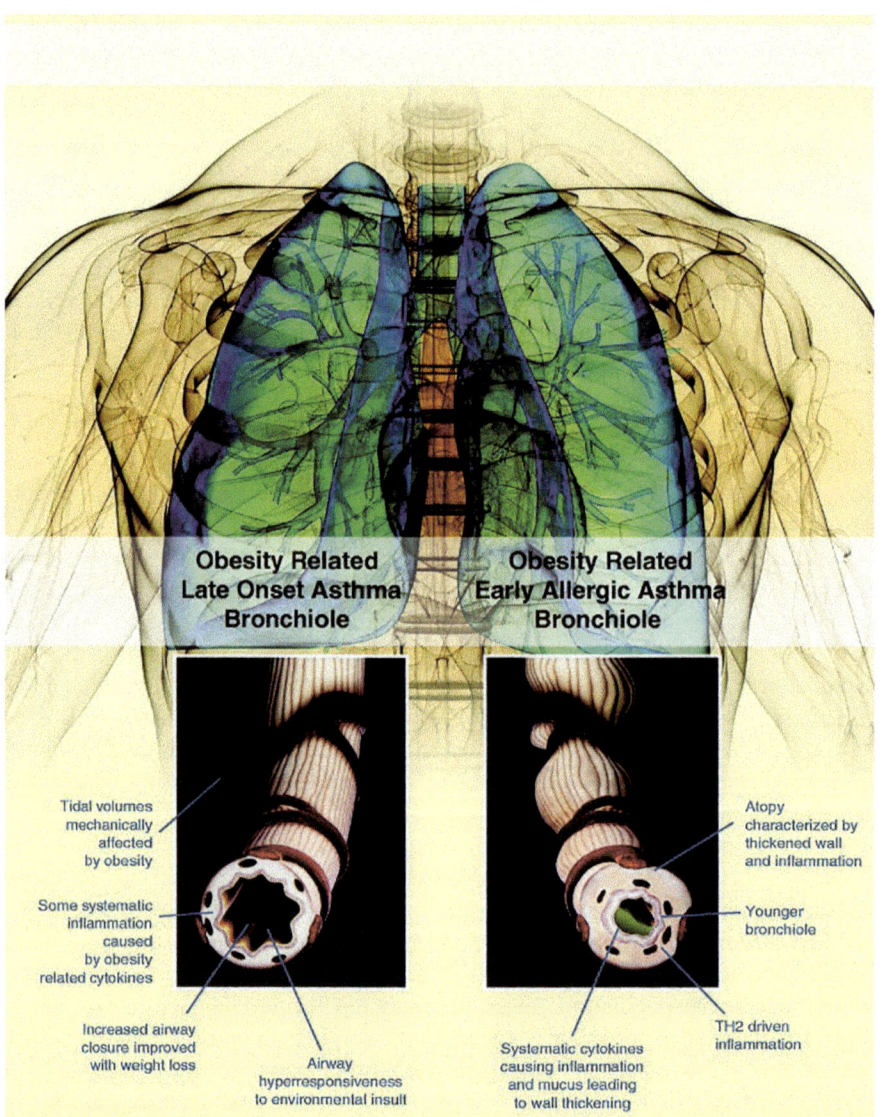

FIGURE 4.1

Obesity may complicate preexisting asthma (asthma complicated by obesity) or lead to *de novo* airway disease (asthma consequent to obesity).

Reproduced with permission from SAGE Publications Limited Mohanan S, Tapp H, McWilliams A, Dulin M.
Obesity and asthma: pathophysiology and implications for diagnosis and management in primary care.
Exp Biol Med (Maywood). 2014;239(11):1531–40.

people with early-onset asthma have more airway obstruction, greater increase in bronchial responsiveness, and greater likelihood of asthma-related medical treatment or admissions, compared with obese people with late-onset asthma [26].

The level of asthma control requires a multifaceted definition, which includes underlying disease severity, responsiveness to treatment, and adequacy of asthma care and management [27]. The relationship between obesity and inadequate control of asthma in adults has been demonstrated in several studies. The largest study to date evaluating >50,000 adults with asthma revealed that a third of those with very poorly controlled asthma were obese [28]. In a recent, large, prospective observational study from Poland, obese subjects with asthma had half the odds of having asthma control, as assessed by the asthma control questionnaire (ACQ), as nonobese subjects. This association was significant in both patient and physician assessment of asthma control, and was independent of the use of asthma controller therapy [29]. Poor control in obese adults with asthma, despite adherence to standard therapies, has also been demonstrated in other studies [30], with some studies showing an inverse linear relationship between BMI/waist circumference/waist-to-height ratio and asthma control [31, 32]. Obese adult subjects with asthma report a worse quality of life than nonobese subjects with asthma. Particularly in the setting of concurrent metabolic syndrome, obese adult subjects with asthma have worse scores than nonobese subjects with asthma on the St. George's Respiratory Questionnaire, Pittsburgh Sleep Quality Index, and Asthma Quality of Life Questionnaire [33].

Obese subjects with asthma consume a greater amount of asthma medications than nonobese subjects with asthma, thus imposing a financial burden on patients and society. In a large study involving the Kaiser Permanente cohort in Southern California, obese subjects with asthma had a greater risk of using frequent short-acting beta agonist (SABA) medications than nonobese subjects with asthma [34]. In addition to SABA, the overall use of maintenance oral steroids and steroid burst therapy was more prevalent among obese asthmatics, compared with nonobese asthmatics [35].

It is reasonable to conclude that obesity in adults results in the development of a difficult-to-control asthma phenotype, characterized by poor clinical control, poor quality of life, and high use of standard and rescue asthma therapies. The mechanism for the difficult-to-control asthma phenotype in obesity is however not known, but concomitant comorbidities of depression, gastroesophageal reflux (GER), and obstructive sleep apnea may be contributory factors.

Unlike adults, the current literature surrounding obesity and control of asthma in the pediatric population is less clear. Although a number of cross-sectional studies support the conclusion that childhood obesity is associated with inadequate asthma control [9, 36–41], several other studies do not demonstrate such association [42–47]. The differences between the findings in these studies may be due to differing sociodemographic data among participants, and varying definitions of asthma control and medication adherence. Apart from asthma control, there is consistency in the literature that obese children and adolescents with asthma experience greater respiratory symptoms such as cough, phlegm production, wheeze, and nocturnal awakenings, when compared with their nonobese counterparts [9, 38, 47].

EXACERBATIONS

Recent cross-sectional and longitudinal cohort studies support that obesity in adults is associated with increased frequency of acute asthma exacerbations [48, 49]. Because the exacerbations in obese subjects are more likely to require emergency room (ER) visits, hospitalizations, mechanical ventilation, higher length of hospital stay, and overall higher hospital charges [34, 50, 51], the data indicate that obese subjects have a greater severity of exacerbation than nonobese subjects. Obesity may have a dose-effect relationship with severity of exacerbation, with the greatest odds for hospitalizations for exacerbation noted in those with severe obesity (BMI $\geq 40\,kg/m^2$) [51]. Similar findings are seen with exacerbations involving obese pregnant women [52]. Obese adults with asthma on omalizumab therapy (a recombinant humanized monoclonal antibody that binds to IgE, indicated in individuals with moderate to severe persistent disease, inadequately controlled on inhaled corticosteroids [ICS]) have twofold greater odds for asthma exacerbations than their nonobese counterparts receiving the same treatment [53]. This suggests that obesity reduces the efficacy of omalizumab, although further studies are needed to understand the mechanism(s) by which this occurs. At 12–24 months following bariatric surgery in obese patients with asthma, the risk for asthma exacerbation requiring ER visits or hospitalization declines to approximately half the original value [54], indicating that surgical weight loss is effective in preventing asthma exacerbations in the obese.

The association of obesity with increased risk for asthma exacerbation is similarly noted among pediatric patients [34, 42], although a study by De Vera and colleagues showed that the severity of the exacerbation was not affected by BMI in Filipino children. The latter study was limited by small sample size and lack of detail on specific asthma triggers [55].

TREATMENT OPTIONS

Currently the National Heart, Lung, and Blood Institute (NHLBI) Expert Panel Report 3 Guidelines for the Diagnosis and Management of Asthma do not differ between obese and lean individuals [27]. The treatment of the obese-asthmatic phenotype is however multidimensional and involves interventions beyond conventional asthma medication use and adherence. Specific interventions such as medical and surgical weight loss, treatment of comorbid conditions, and novel treatment strategies may play an important role in treatment of the obese asthmatic.

Obese asthmatics do not respond as well to standard controller therapy when compared with their nonobese counterparts [29, 30, 56]. Yet ICS and combination ICS/long-acting beta agonists (LABA) appear to be superior to leukotriene receptor antagonist montelukast for the treatment of asthma in the obese, and obese asthmatics do not demonstrate an altered response to rescue therapy [56]. Obese asthmatics are however found to have a decreased response to omalizumab therapy [53].

A recent study by Scott *et al.*, in which patients were randomized into one of three groups (exercise alone, diet alone, or combination of diet and exercise), found that a 12-week intervention that produced weight loss resulted in significant improvement in asthma control. The authors of this study found that a 5%–10% weight loss resulted in clinically important improvements in asthma control in 58% and quality of life in 83% of obese asthmatic subjects [57]. This finding is also supported by Ma *et al.* who demonstrated that a reduction in weight of 10% or more via behavioral weight loss and physical activity intervention resulted in a fourfold increase in achieving asthma control among obese individuals [58]. Medical weight loss achieved through reduced caloric intake, use of sibutramine and orlistat medications, and/or behavioral interventions improved asthma control [59, 60]. Surgical weight loss in obese asthmatics similarly improved asthma control and lung function [61], as well as lowered rates for ER visits or hospitalization for asthma exacerbations [54]. Following bariatric surgery, obese subjects with early-onset allergic asthma showed improved respiratory resistance without any decrease in airway responsiveness, but airway responsiveness decreased in obese subjects with later-onset nonallergic asthma, implying that the two distinct clinical phenotypes of obese patients with asthma may respond differently to surgical weight loss [62].

Obese patients with asthma often have comorbidities, such as sleep-disordered breathing and GER. These comorbidities may, in turn, worsen asthma symptoms. Obese patients with asthma and concomitant symptomatic GER are often treated with proton pump inhibitors. Obese subjects without self-reported symptoms of GER are also often found to have reflux on invasive testing. Empiric treatment of asymptomatic GER, discovered on invasive testing with a proton pump inhibitor, however did not improve asthma symptoms in a randomized controlled trial of subjects with asthma, >50% of whom were also obese [13]. Taken together, these data appear to suggest that further understanding is needed to clarify the GER-asthma relationship, particularly among the obese. A number of older studies suggest that treatment of obstructive sleep apnea improves symptoms and lung function in patients with asthma. Larger studies entailing a more diverse population are however required [56].

There is a need for novel drug therapies targeting noneosinophilic pathways to treat asthma in the obese. For instance, the use of pioglitazone has been considered in the treatment of obese asthmatics. Pioglitazone is an antidiabetic medication that can alter circulating adiponectin and the peroxisome proliferator-activated receptor (PPAR)-gamma pathways and may be useful in the prevention and treatment of asthma in obese subjects. In a pilot study by Dixon *et al.*, the use of pioglitazone in obese adults with asthma, however, raised concern for potential harm, given the weight gain noted in the treatment arm [63]. Intake of omega-3 fatty acids may improve asthma control in obese asthmatics through a number of antiinflammatory pathways, which is currently being studied in an ongoing trial evaluating its use, specifically in obese asthmatics with persistent disease [64].

COPD

PREVALENCE OF EMPHYSEMA PHENOTYPE OF COPD

Obesity is an important comorbidity in patients with COPD. Several cross sectional studies report a frequency of obesity in COPD patients that ranges between 25% and 35% [65–81]. The variation in prevalence of obesity in COPD patients may be explained by different criteria for classifying COPD (Global Initiative for Obstructive Lung Disease [GOLD] versus lower limit of normal value of the forced expiratory volume in 1 second/forced vital capacity or FEV_1/FVC ratio versus self-report of provider diagnosis); different sociodemographic factors; and differing adjustment for covariates between studies. There is consensus however that the prevalence of obesity is higher among subjects with COPD than in the non-COPD population. The prevalence of obesity also seems to vary based upon the severity of COPD, with higher prevalence of obesity noted in less severe stage of disease [66, 71, 75]. The impact of obesity on COPD prevalence is however likely to be underestimated in the literature. A high BMI increases the value of the FEV_1/FVC ratio, potentially impacting the diagnosis of COPD, which is spirometrically defined by a low ratio (i.e., <70% or the statistically defined lower limit of normal value) [82]. The Copenhagen City Heart Study revealed that FEV_1/FVC ratio increased by 0.04 in men and 0.03 in women with every 10-unit increase in BMI. Obesity is therefore associated with a spirometric underdiagnosis of COPD [83], and this should be taken into consideration when examining at risk subjects for COPD [84].

Recent interest has been generated in examining alternative adiposity phenotypes in COPD beyond BMI. Studies report disproportionate accumulation of metabolically active and proinflammatory visceral fat in patients with COPD (Figure 4.2) [85]. In an Evaluation of COPD Longitudinally to Identify Predictive Surrogate Endpoints (ECLIPSE) substudy including 585 participants with COPD, the authors reported that visceral adipose tissue, as measured by computed tomography (CT) images at L2-L3 lumbar vertebral level, was increased in patients with COPD, and muscle attenuation was reduced indicating more muscle fat accumulation [86]. Similarly, a case control study by Lenartova *et al.* demonstrated that visceral fat deposits were highest among men with COPD, when compared with smokers and nonsmokers [66]. Abdominal obesity, defined by android/gynoid percentage fat mass, is also highly prevalent among men and women with COPD [87]. Metabolic syndrome is more frequently present in COPD patients than in healthy subjects [76]. A recent bidirectional longitudinal study noted that healthy adults with a rapid decline in lung function were at risk for developing metabolic syndrome and disproportionate accumulation of intrathoracic visceral fat [88]. It is therefore likely that COPD contributes to greater metabolically active adiposity, indicating a cross-talk between the lung and adipose tissue.

INCIDENCE OF EMPHYSEMA PHENOTYPE OF COPD

Although the prevalence of COPD in obese subjects has been well published in the literature, less is known about the incidence of COPD in obese subjects, partly explained by the confounding effect of BMI on the FEV_1/FVC ratio. In a large

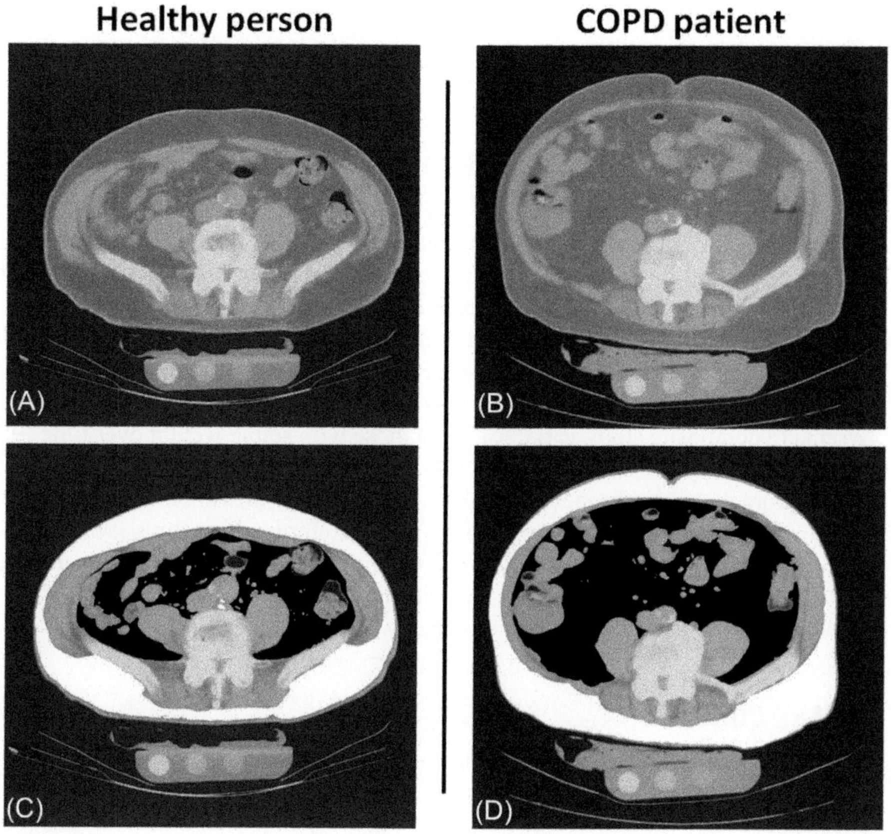

Healthy person   **COPD patient**

FIGURE 4.2

Excessive visceral fat accumulation, but normal subcutaneous fat content in chronic obstructive pulmonary disease (COPD). Abdominal computed tomography (CT) scans from (A) a healthy person and (B) a patient with COPD (FEV_1 53% of predicted) matched for age, sex, and body mass index (29 kg/m^2). The black and white areas in (C) and (D) denote the visceral and subcutaneous fat compartments, respectively.

Reprinted with permission of the American Thoracic Society. Copyright © 2018 American Thoracic Society.
van den Borst B, Gosker HR, Schols AM. Central fat and peripheral muscle: partners in crime in chronic
obstructive pulmonary disease. Am J Respir Crit Care Med 2013;187(1):8–13. The American Journal of
Respiratory and Critical Care Medicine is an official journal of the American Thoracic Society.

prospective cohort study ($n = 113,279$) by Behrens *et al.*, BMI-defined class 2–3 obesity was associated with a greater risk for developing COPD, compared with those with normal BMI. However, after adjustment for waist circumference, the relationship between BMI-defined obesity and incident COPD was attenuated. On the other hand, large waist circumference and higher waist-to-hip ratio was associated with greater risk for incident COPD, even after statistical adjustment for BMI [89]. On the other hand, in a study by Zhou *et al.*, those with high baseline BMI had a lower risk of developing COPD than those with normal BMI [90]. Taken together, the available data suggest that the association between high BMI and incident COPD is unclear, but abdominal obesity appears to increase risk for incident COPD.

PREVALENCE AND INCIDENCE OF CHRONIC BRONCHITIS PHENOTYPE OF COPD

A high prevalence of chronic bronchitis is described in those between 44 and 65 years of age, have ongoing tobacco use, and women [91]; this disease is not seen in children. Several large cross-sectional studies have shown that prevalent chronic bronchitis is more frequent among obese adults, when compared with nonobese adults [92–94]. In a prospective longitudinal study, Canadian adults with the greatest gain in body weight over time were more likely to self-report "chronic bronchitis or emphysema (a composite study variable)" than those with stable weight over time [95]. A large Irish study showed an association between obesity and bronchitis (defined as self-reported chronic bronchitis, COPD, emphysema) in women but not in men, suggesting a sex-obesity interaction on the risk for chronic bronchitis [96]. The potential mechanism related to increased chronic bronchitis in obese adults is not well understood. Larger longitudinal and mechanistic studies are however needed to better understand the association between obesity and chronic bronchitis.

DISEASE CONTROL

Obese COPD patients have worse dyspnea scores; poorer respiratory-specific and general quality of life scores; greater physical activity limitation; lower lung functions; and reduced 6-minute walk test distances, independent of airflow obstruction, compared with nonobese COPD patients [65, 69, 72, 80, 97–105]. Furthermore, obesity may affect these outcomes in a dose-dependent manner [65]. In a cross-sectional study using the U.S.-based Lovelace Smokers' Cohort, Sood *et al.* demonstrated that weight gain in obese smokers was associated with worsening health status and lower lung function [100]. Obesity has been shown to modify the effects of particulate matter exposure in COPD patients, leading to higher prevalence of cough and phlegm production, dyspnea, and rescue medication use, when compared with nonobese counterparts [106, 107]. It has been suggested that the increased dyspnea in obese patients with COPD is partly due to their higher rates of excessive expiratory central airway collapse. This finding occurs either due to the weakening of the tracheobronchial cartilaginous wall (i.e., tracheobronchomalacia) or excessive inward bulging of the posterior membranous wall (i.e., excessive dynamic airway collapse) (Figure 4.3) [104, 108–110].

Interestingly, several large studies have demonstrated that obesity has a protective effect on mortality in patients with COPD, the so-called "obesity paradox" [111–114]. On the other hand, nonobese COPD patients were shown to have higher mortality [111, 112]. A study by Galesanu *et al.* has suggested that confounders such as hypercapnia, exercise capacity, and muscle mass should be considered when interpreting

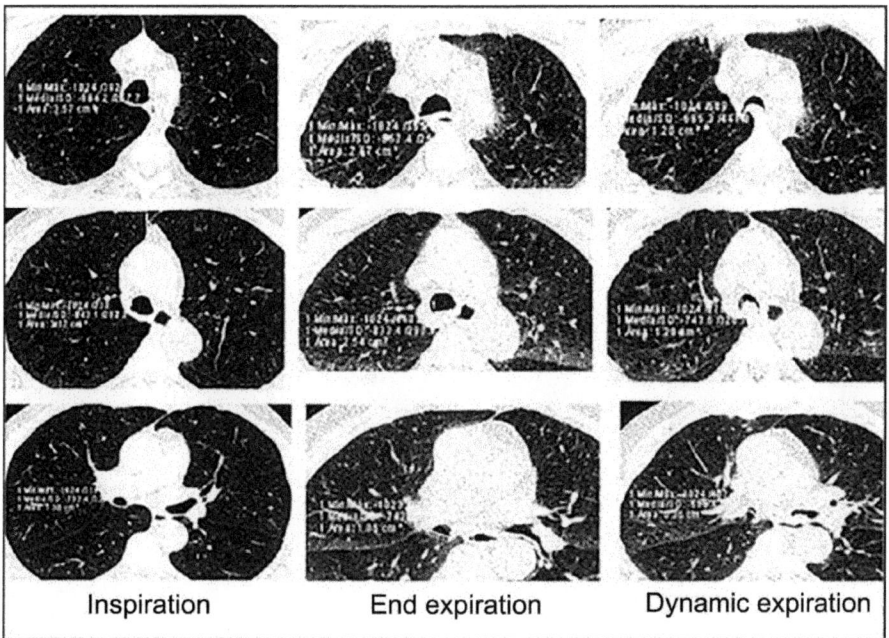

| Inspiration | End expiration | Dynamic expiration |

FIGURE 4.3

Dynamic low-dose computed tomography (CT) images of a patient with excessive dynamic airway collapse in inspiration, end-expiration, and dynamic expiration at three anatomical levels (aortic arch, carina, and bronchus intermedius from top to bottom row in the panel). The panel shows decrease in cross-sectional area from inspiration (left column) to dynamic expiration (right column), indicating excessive dynamic airway collapse due to excessive inward bulging of the posterior membranous wall.

Reproduced with permission from Represas-Represas C, Leiro-Fernandez V, Mallo-Alonso R, Botana-Rial MI, Tilve-Gomez A, Fernandez-Villar A. Excessive dynamic airway collapse in a small cohort of chronic obstructive pulmonary disease patients. Ann Thorac Med 2015;10(2):118–22.

the "paradoxical" association between increased BMI and survival in patients with COPD, as after adjustment for these confounders, the tendency for improved survival in overweight/obese subjects is significantly attenuated (Figure 4.4) [113].

EXACERBATIONS

Data regarding obesity as a risk factor for acute COPD exacerbations appears to be somewhat conflicting in the literature. A large longitudinal cohort study showed that obesity increased the risk of acute COPD exacerbation in a dose-dependent manner [65]. Similarly, a retrospective study by Johannesdottir *et al.* showed that, among hospitalized patients with acute COPD exacerbation, obesity was a risk factor for subsequent hospitalization in the proceeding 12-month period [115]. On the other

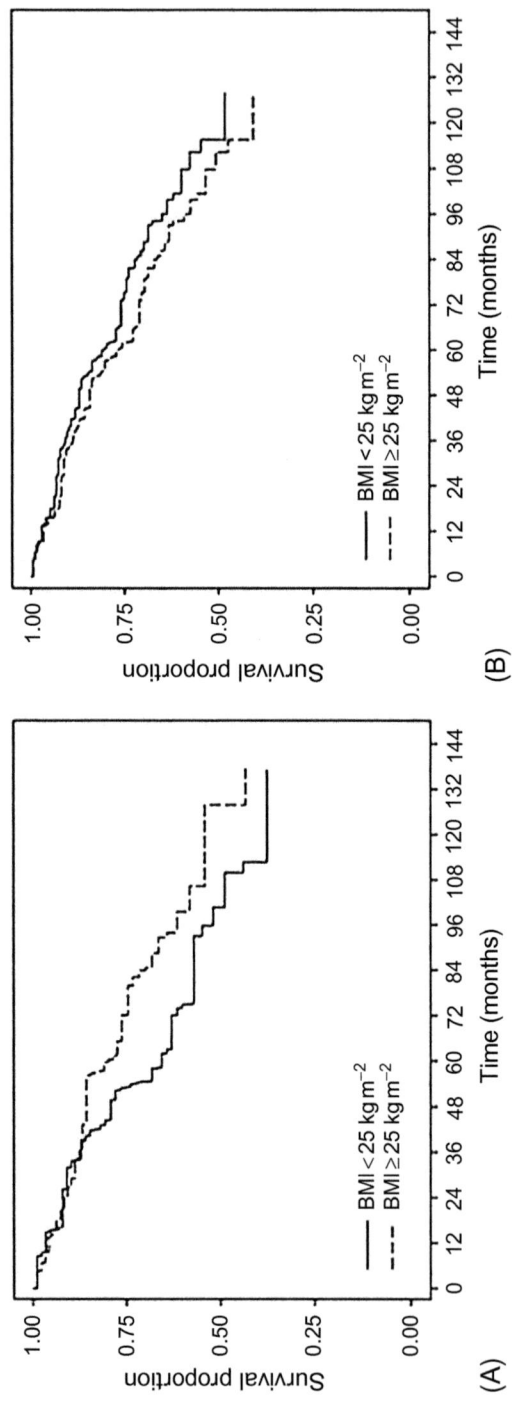

FIGURE 4.4

Excess BMI is described to have a protective effect on mortality in patients with COPD, the so-called "obesity paradox," as shown in Panel A. Overweight/obese subjects tend to have better preserved lung function, muscle mass, and exercise capacity, all important predictors of mortality in COPD. A study by Galesanu et al. has suggested that confounders such as hypercapnia, exercise capacity, and muscle mass should be considered when interpreting the association between BMI and survival in patients with COPD, as after adjustment for these confounders, the tendency for improved survival in overweight subjects vanished, as shown in Panel B [109]. It is possible that these variables rather than fat accumulation may explain why overweight and obese patients with COPD are apparently protected against mortality. The solid lines represent patients with BMI <25 kg/m^2; the dashed lines represent patients with BMI ≥25 kg/m^2.

Reproduced with permission from Hindawi Limited, Galesanu RG, Bernard S, Marquis K, Lacasse Y, Poirier P, Bourbeau J, et al. Obesity in chronic obstructive pulmonary disease: is fatter really better? Can Respir J 2014;21(5):297–301.

hand, utilizing the BREATHE cohort, Koniski *et al.* demonstrated that obesity was not associated with an increased risk for acute COPD exacerbations [70]. A large retrospective chart review study of COPD admittances by Zapatero *et al.* demonstrated that obesity was associated with a lower in-hospital mortality risk and lower early readmittance risk 30 days after discharge, compared with nonobese COPD patients [111]. In addition, a small study by McCormack *et al.* showed that the effect of indoor particulate matter on acute COPD exacerbations tended to be greater among obese versus nonobese participants with COPD, suggesting that obesity may increase susceptibility to particulate air pollution [106]. The mechanism by which obesity directly impacts acute COPD exacerbation or interacts with particulate air pollution on the risk for acute COPD exacerbation is however not known.

TREATMENT OPTIONS

The disease-specific treatment of COPD in obese patients is no different from the treatment of COPD in nonobese patients. Additionally, obese COPD patients may demonstrate disproportionately greater clinical benefit from weight reduction, exercise training, and noninvasive positive pressure ventilation therapy than nonobese COPD patients, but studies in this regard are limited.

Although there appears to be a general consensus that weight loss is beneficial for severely obese COPD patients, the optimal target BMI has yet to be established [116]. Given the "obesity paradox" whereby obesity in moderate-to-severe COPD patients is reported to have protective effects on survival and lung function decline, clinicians may need to consider accepting a "modest" degree of excess body weight in patients with severe COPD as optimal. A study by McDonald *et al.* has shown that a decrease in BMI of $2.4 \, kg/m^2$ from a mean BMI of $36.3 \, kg/m^2$ square at baseline, although maintaining skeletal muscle mass, improved exercise capacity, health status, dyspnea scores, and functional outcomes in obese COPD patients. The participants of this study underwent a 12-week weight reduction program involving meal replacements, dietary counseling, and resistance training [117]. It has also been shown that obese COPD patients who underwent an 8-week, water-based exercise program had improved exercise capacity and health-related quality of life [118]. These intervention studies did not however enroll any nonobese COPD patients as controls. It has also been shown that, during hospitalizations for acute exacerbations of COPD, obese patients who underwent breathing training and limb exercises experienced an improvement in strength, exercise capacity, and psychological distress [119]. Taken together, these data suggest that physical activity and weight loss may be beneficial for the obese COPD patient, but randomized controlled trials comparing obese and nonobese COPD subjects are needed.

The use of noninvasive positive-pressure ventilation therapy has also gained importance in obese COPD patients. In a prospective multicenter cohort study by Borel *et al.*, the use of long-term, noninvasive, positive-pressure ventilation for >5 hours daily was associated with an improved prognosis in obese COPD patients [120]. This study was however limited by lack of control for underlying sleep disordered breathing [120].

CONCLUSIONS

The causal relationship between obesity and asthma is much better defined than that between obesity and COPD. The asthma phenotype in an obese subject is characterized by poor clinical control, and high frequency and severity of acute exacerbations. Obese asthmatics do not respond well to standard controller asthma therapy; weight loss however helps improve asthma control. Decrease in airway hyperresponsiveness is noted with surgical weight loss in the later-onset nonallergic phenotype of obesity-associated asthma but not in the early-onset allergic phenotype. Like asthma, obesity is an important comorbidity in patients with COPD. Although COPD is underdiagnosed in the obese population, obese COPD patients have greater dyspnea and physical activity limitation, as well as lower quality of life and lung function than nonobese COPD patients. Given the "obesity paradox," clinicians may consider accepting a modest degree of excess body weight in patients with severe COPD as optimal. The contribution of more sophisticated phenotypes of obesity, such as visceral or metabolically active adiposity, irrespective of general adiposity, to the pathophysiology of obstructive lung diseases needs to be explored in future studies. Targeted drugs or nutritional interventions that reduce the inflammatory and metabolic activity of fat may open up novel therapies for asthma and COPD in obese subjects.

REFERENCES

[1] Ogden CL, Carroll MD, Fryar CD, Flegal KM. Prevalence of obesity among adults and youth: United States, 2011–2014. NCHS Data Brief No: 219, Hyattsville, MD: National Center for Health Statistics, 2015; 2015.

[2] Aragona E, El-Magbri E, Wang J, Scheckelhoff T, Scheckelhoff T, Hyacinthe A, et al. Impact of obesity on clinical outcomes in urban children hospitalized for status asthmaticus. Hosp Pediatr 2016;6(4):211–8.

[3] Centers for Disease Control and Prevention. FastStats: asthma. National Center for Health Statistics; 2017. Available at: https://www.cdc.gov/nchs/fastats/asthma.htm; (Accessed December 13, 2017).

[4] Centers for Disease Control and Prevention. FastStats: chronic obstructive pulmonary disease. National Center for Health Statistics; 2017. Available at: https://www.cdc.gov/nchs/fastats/copd.htm; (Accessed December 13, 2017).

[5] Akinbami LJ, Fryar CD. Current asthma prevalence by weight status among adults: United States, 2001–2014. NCHS Data Brief 2016;239:1–8.

[6] Black MH, Smith N, Porter AH, Jacobsen SJ, Koebnick C. Higher prevalence of obesity among children with asthma. Obesity (Silver Spring) 2012;20(5):1041–7.

[7] Willeboordse M, van den Bersselaar DL, van de Kant KD, Muris JW, van Schayck OC, Dompeling E. Sex differences in the relationship between asthma and overweight in Dutch children: a survey study. PLoS One 2013;8(10):e77574.

[8] Barros R, Moreira P, Padrao P, Teixeira VH, Carvalho P, Delgado L, et al. Obesity increases the prevalence and the incidence of asthma and worsens asthma severity. Clin Nutr 2016.

[9] Wang D, Qian Z, Wang J, Yang M, Lee YL, Liu F, et al. Gender-specific differences in associations of overweight and obesity with asthma and asthma-related symptoms in 30,056 children: result from 25 districts of Northeastern China. J Asthma 2014;51(5):508–14.

[10] Greenblatt R, Mansour O, Zhao E, Ross M, Himes BE. Gender-specific determinants of asthma among U.S. adults. Asthma Res Pract 2017;3:2.

[11] Eid W, Dawson B, Hopkins-Price P, Vijil JJ, Eagleton L, Henkle J, et al. Association between bronchial hyperreactivity and body mass index in adults. Am J Resp Crit Care Med 2002;165(8):A435 [Abstract].

[12] Celedon JC, Palmer LJ, Litonjua AA, Weiss ST, Wang B, Fang Z, et al. Body mass index and asthma in adults in families of subjects with asthma in Anqing, China. Am J Respir Crit Care Med 2001;164(10 Pt 1):1835–40.

[13] Bhatt NA, Lazarus A. Obesity-related asthma in adults. Postgrad Med 2016;128(6):563–6.

[14] Brumpton BM, Camargo Jr. CA, Romundstad PR, Langhammer A, Chen Y, Mai XM. Metabolic syndrome and incidence of asthma in adults: the HUNT study. Eur Respir J 2013;42(6):1495–502.

[15] Moreira A, Bonini M, Garcia-Larsen V, Bonini S, Del Giacco SR, Agache I, et al. Weight loss interventions in asthma: EAACI evidence-based clinical practice guideline (part I). Allergy 2013;68(4):425–39.

[16] Rzehak P, Wijga AH, Keil T, Eller E, Bindslev-Jensen C, Smit HA, et al. Body mass index trajectory classes and incident asthma in childhood: results from 8 European Birth Cohorts—a Global Allergy and Asthma European Network initiative. J Allergy Clin Immunol 2013;131(6):1528–36.

[17] Chen YC, Dong GH, Lin KC, Lee YL. Gender difference of childhood overweight and obesity in predicting the risk of incident asthma: a systematic review and meta-analysis. Obes Rev 2013;14(3):222–31.

[18] Beuther DA, Sutherland ER. Overweight, obesity, and incident asthma: a meta-analysis of prospective epidemiologic studies. Am J Respir Crit Care Med 2007;175(7):661–6.

[19] Egan KB, Ettinger AS, Bracken MB. Childhood body mass index and subsequent physician-diagnosed asthma: a systematic review and meta-analysis of prospective cohort studies. BMC Pediatr 2013;13:121.

[20] Brumpton B, Langhammer A, Romundstad P, Chen Y, Mai XM. General and abdominal obesity and incident asthma in adults: the HUNT study. Eur Respir J 2013;41(2): 323–9.

[21] Assad N, Qualls C, Smith LJ, Arynchyn A, Thyagarajan B, Schuyler M, et al. Body mass index is a stronger predictor than the metabolic syndrome for future asthma in women. The longitudinal CARDIA study. Am J Respir Crit Care Med 2013;188(3):319–26.

[22] Barton JH, Ireland A, Fitzpatrick M, Kessinger C, Camp D, Weinman R, et al. Adiposity influences airway wall thickness and the asthma phenotype of HIV-associated obstructive lung disease: a cross-sectional study. BMC Pulm Med 2016;16(1):111.

[23] Forno E, Young OM, Kumar R, Simhan H, Celedon JC. Maternal obesity in pregnancy, gestational weight gain, and risk of childhood asthma. Pediatrics 2014;134(2): e535–46.

[24] Rusconi F, Popovic M. Maternal obesity and childhood wheezing and asthma. Paediatr Respir Rev 2016;.

[25] Suratt BT, Ubags NDJ, Rastogi D, Tantisira KG, Marsland BJ, Petrache I, et al. An official American Thoracic Society Workshop Report: obesity and metabolism. An emerging frontier in lung health and disease. Ann Am Thorac Soc 2017;14(6):1050–9.

[26] Holguin F, Bleecker ER, Busse WW, Calhoun WJ, Castro M, Erzurum SC, et al. Obesity and asthma: an association modified by age of asthma onset. J Allergy Clin Immunol 2011;127(6). 1486-93. e2.

[27] National Asthma E, Prevention P. Expert Panel Report 3 (EPR-3): Guidelines for the Diagnosis and Management of Asthma-Summary Report 2007. J Allergy Clin Immunol. 2007;120(5 Suppl):S94–138.

[28] Zahran HS, Bailey CM, Qin X, Moorman JE. Assessing asthma control and associated risk factors among persons with current asthma—findings from the child and adult Asthma Call-back Survey. J Asthma 2015;52(3):318–26.

[29] Rogala B, Majak P, Gluck J, Debowski T. Asthma control in adult patients treated with a combination of inhaled corticosteroids and longacting beta2agonists: a prospective observational study. Pol Arch Intern Med 2017;127(2):100–6.

[30] Scott L, Li M, Thobani S, Nichols B, Morphew T, Kwong KY. Factors affecting ability to achieve asthma control in adult patients with moderate to severe persistent asthma. J Asthma 2016;53(6):644–9.

[31] Singh M, Gupta N, Kumar R. Effect of obesity and metabolic syndrome on severity, quality of life, sleep quality and inflammatory markers in patients of asthma in India. Pneumonol Alergol Pol 2016;84(5):258–64.

[32] Lv N, Xiao L, Camargo Jr. CA, Wilson SR, Buist AS, Strub P, et al. Abdominal and general adiposity and level of asthma control in adults with uncontrolled asthma. Ann Am Thorac Soc 2014;11(8):1218–24.

[33] Kapadia SG, Wei C, Bartlett SJ, Lang J, Wise RA, Dixon AE, et al. Obesity and symptoms of depression contribute independently to the poor asthma control of obesity. Respir Med 2014;108(8):1100–7.

[34] Schatz M, Zeiger RS, Yang SJ, Chen W, Sajjan S, Allen-Ramey F, et al. Prospective study on the relationship of obesity to asthma impairment and risk. J Allergy Clin Immunol Pract 2015;3(4). 560-5.e1.

[35] Gibeon D, Batuwita K, Osmond M, Heaney LG, Brightling CE, Niven R, et al. Obesity-associated severe asthma represents a distinct clinical phenotype: analysis of the British Thoracic Society Difficult Asthma Registry Patient cohort according to BMI. Chest 2013;143(2):406–14.

[36] Hugo MN, Walter PY, Maimouna M, Malea NM, Ubald O, Adeline W, et al. Assessment of asthma control using asthma control test in chest clinics in Cameroon: a cross-sectional study. Pan Afr Med J 2016;23:70.

[37] Lang JE, Hossain J, Holbrook JT, Teague WG, Gold BD, Wise RA, et al. Gastro-oesophageal reflux and worse asthma control in obese children: a case of symptom misattribution? Thorax 2016;71(3):238–46.

[38] Lang JE, Hossain MJ, Lima JJ. Overweight children report qualitatively distinct asthma symptoms: analysis of validated symptom measures. J Allergy Clin Immunol 2015;135(4). 886-93.e3.

[39] McGarry ME, Castellanos E, Thakur N, Oh SS, Eng C, Davis A, et al. Obesity and bronchodilator response in black and Hispanic children and adolescents with asthma. Chest 2015;147(6):1591–8.

[40] Sasaki M, Yoshida K, Adachi Y, Furukawa M, Itazawa T, Odajima H, et al. Factors associated with asthma control in children: findings from a national Web-based survey. Pediatr Allergy Immunol 2014;25(8):804–9.

[41] Borrell LN, Nguyen EA, Roth LA, Oh SS, Tcheurekdjian H, Sen S, et al. Childhood obesity and asthma control in the GALA II and SAGE II studies. Am J Respir Crit Care Med 2013;187(7):697–702.

[42] Ahmadizar F, Vijverberg SJ, Arets HG, de Boer A, Lang JE, Kattan M, et al. Childhood obesity in relation to poor asthma control and exacerbation: a meta-analysis. Eur Respir J 2016;48(4):1063–73.

[43] Lang JE, Holbrook JT, Wise RA, Dixon AE, Teague WG, Wei CY, et al. Obesity in children with poorly controlled asthma: sex differences. Pediatr Pulmonol 2013;48(9):847–56.

[44] Yilmaz O, Sogut A, Bozgul A, Turkeli A, Kader S, Yuksel H. Is obesity related to worse control in children with asthma? Tuberk Toraks 2014;62(1):39–44.

[45] Forte GC, Grutcki DM, Menegotto SM, Pereira RP, Dalcin Pde T. Prevalence of obesity in asthma and its relations with asthma severity and control. Rev Assoc Med Bras (1992) 2013;59(6):594–9.

[46] Giese JK. Pediatric obesity and its effects on asthma control. J Am Assoc Nurse Pract 2014;26(2):102–9.

[47] Sah PK, Gerald Teague W, Demuth KA, Whitlock DR, Brown SD, Fitzpatrick AM. Poor asthma control in obese children may be overestimated because of enhanced perception of dyspnea. J Allergy Clin Immunol Pract 2013;1(1):39–45.

[48] Tay TR, Radhakrishna N, Hore-Lacy F, Smith C, Hoy R, Dabscheck E, et al. Comorbidities in difficult asthma are independent risk factors for frequent exacerbations, poor control and diminished quality of life. Respirology 2016;21(8): 1384–90.

[49] Lang JE. Obesity and asthma in children: current and future therapeutic options. Paediatr Drugs 2014;16(3):179–88.

[50] Schatz M, Zeiger RS, Zhang F, Chen W, Yang SJ, Camargo Jr. CA. Overweight/obesity and risk of seasonal asthma exacerbations. J Allergy Clin Immunol Pract 2013;1(6):618–22.

[51] Hasegawa K, Tsugawa Y, Lopez BL, Smithline HA, Sullivan AF, Camargo Jr. CA. Body mass index and risk of hospitalization among adults presenting with asthma exacerbation to the emergency department. Ann Am Thorac Soc 2014;11(9):1439–44.

[52] Ali Z, Ulrik CS. Incidence and risk factors for exacerbations of asthma during pregnancy. J Asthma Allergy 2013;6:53–60.

[53] Novelli F, Latorre M, Vergura L, Caiaffa MF, Camiciottoli G, Guarnieri G, et al. Asthma control in severe asthmatics under treatment with omalizumab: a cross-sectional observational study in Italy. Pulm Pharmacol Ther 2015;31:123–9.

[54] Hasegawa K, Tsugawa Y, Chang Y, Camargo Jr. CA. Risk of an asthma exacerbation after bariatric surgery in adults. J Allergy Clin Immunol 2015;136(2). 288-94.e8.

[55] De Vera MJ, Gomez MC, Yao CE. Association of obesity and severity of acute asthma exacerbations in Filipino children. Ann Allergy Asthma Immunol 2016;117(1):38–42.

[56] Pradeepan S, Garrison G, Dixon AE. Obesity in asthma: approaches to treatment. Curr Allergy Asthma Rep 2013;13(5):434–42.

[57] Scott HA, Gibson PG, Garg ML, Pretto JJ, Morgan PJ, Callister R, et al. Dietary restriction and exercise improve airway inflammation and clinical outcomes in overweight and obese asthma: a randomized trial. Clin Exp Allergy 2013;43(1):36–49.

[58] Ma J, Strub P, Xiao L, Lavori PW, Camargo Jr. CA, Wilson SR, et al. Behavioral weight loss and physical activity intervention in obese adults with asthma. A randomized trial. Ann Am Thorac Soc 2015;12(1):1–11.

[59] Dias-Junior SA, Reis M, de Carvalho-Pinto RM, Stelmach R, Halpern A, Cukier A. Effects of weight loss on asthma control in obese patients with severe asthma. Eur Respir J 2014;43(5):1368–77.

[60] Pakhale S, Baron J, Dent R, Vandemheen K, Aaron SD. Effects of weight loss on airway responsiveness in obese adults with asthma: does weight loss lead to reversibility of asthma? Chest 2015;147(6):1582–90.

[61] Hewitt S, Humerfelt S, Sovik TT, Aasheim ET, Risstad H, Kristinsson J, et al. Long-term improvements in pulmonary function 5 years after bariatric surgery. Obes Surg 2014;24(5):705–11.

[62] Chapman DG, Irvin CG, Kaminsky DA, Forgione PM, Bates JH, Dixon AE. Influence of distinct asthma phenotypes on lung function following weight loss in the obese. Respirology 2014;19(8):1170–7.

[63] Dixon AE, Subramanian M, DeSarno M, Black K, Lane L, Holguin F. A pilot randomized controlled trial of pioglitazone for the treatment of poorly controlled asthma in obesity. Respir Res 2015;16:143.

[64] Lang JE, Mougey EB, Allayee H, Blake KV, Lockey R, Gong Y, et al. Nutrigenetic response to omega-3 fatty acids in obese asthmatics (NOOA): rationale and methods. Contemp Clin Trials 2013;34(2):326–35.

[65] Lambert AA, Putcha N, Drummond MB, Boriek AM, Hanania NA, Kim V, et al. Obesity is associated with increased morbidity in moderate to severe COPD. Chest 2017;151(1):68–77.

[66] Lenartova P, Habanova M, Mrazova J, Chlebo P, Wyka J. Analysis of visceral fat in patients with chronic obstructive pulmonary disease (COPD). Rocz Panstw Zakl Hig 2016;67(2):189–96.

[67] Garg T, Rosas U, Rivas H, Azagury D, Morton JM. National prevalence, causes, and risk factors for bariatric surgery readmissions. Am J Surg 2016;212(1):76–80.

[68] Hanson C, LeVan T. Obesity and chronic obstructive pulmonary disease: recent knowledge and future directions. Curr Opin Pulm Med 2017;23(2):149–53.

[69] Garcia-Rio F, Soriano JB, Miravitlles M, Munoz L, Duran-Tauleria E, Sanchez G, et al. Impact of obesity on the clinical profile of a population-based sample with chronic obstructive pulmonary disease. PLoS One 2014;9(8):e105220.

[70] Koniski ML, Salhi H, Lahlou A, Rashid N, El Hasnaoui A. Distribution of body mass index among subjects with COPD in the Middle East and North Africa region: data from the BREATHE study. Int J Chron Obstruct Pulmon Dis 2015;10: 1685–94.

[71] Mitsiki E, Bania E, Varounis C, Gourgoulianis KI, Alexopoulos EC. Characteristics of prevalent and new COPD cases in Greece: the GOLDEN study. Int J Chron Obstruct Pulmon Dis 2015;10:1371–82.

[72] Liu Y, Pleasants RA, Croft JB, Lugogo N, Ohar J, Heidari K, et al. Body mass index, respiratory conditions, asthma, and chronic obstructive pulmonary disease. Respir Med 2015;109(7):851–9.

[73] van de Bool C, Rutten EP, Franssen FM, Wouters EF, Schols AM. Antagonistic implications of sarcopenia and abdominal obesity on physical performance in COPD. Eur Respir J 2015;46(2):336–45.

[74] Lutchmedial SM, Creed WG, Moore AJ, Walsh RR, Gentchos GE, Kaminsky DA. How common is airflow limitation in patients with emphysema on CT scan of the chest? Chest 2015;148(1):176–84.

[75] Koo HK, Park JH, Park HK, Jung H, Lee SS. Conflicting role of sarcopenia and obesity in male patients with chronic obstructive pulmonary disease: Korean National Health and Nutrition Examination Survey. PLoS One 2014;9(10):e110448.

[76] Breyer MK, Spruit MA, Hanson CK, Franssen FM, Vanfleteren LE, Groenen MT, et al. Prevalence of metabolic syndrome in COPD patients and its consequences. PLoS One 2014;9(6):e98013.

[77] Dimov D, Tacheva T, Koychev A, Ilieva V, Prakova G, Vlaykova T. Obesity in Bulgarian patients with chronic obstructive pulmonary disease. Chron Respir Dis 2013;10(4):215–22.

[78] Cazzola M, Calzetta L, Lauro D, Bettoncelli G, Cricelli C, Di Daniele N, et al. Asthma and COPD in an Italian adult population: role of BMI considering the smoking habit. Respir Med 2013;107(9):1417–22.

[79] Garcia-Olmos L, Alberquilla A, Ayala V, Garcia-Sagredo P, Morales L, Carmona M, et al. Comorbidity in patients with chronic obstructive pulmonary disease in family practice: a cross sectional study. BMC Fam Pract 2013;14:11.

[80] Rodriguez DA, Garcia-Aymerich J, Valera JL, Sauleda J, Togores B, Galdiz JB, et al. Determinants of exercise capacity in obese and non-obese COPD patients. Respir Med 2014;108(5):745–51.

[81] Park JH, Lee JK, Heo EY, Kim DK, Chung HS. The effect of obesity on patients with mild chronic obstructive pulmonary disease: results from KNHANES 2010 to 2012. Int J Chron Obstruct Pulmon Dis 2017;12:757–63.

[82] Colak Y, Marott JL, Vestbo J, Lange P. Overweight and obesity may lead to underdiagnosis of airflow limitation: findings from the Copenhagen City Heart Study. COPD 2015;12(1):5–13.

[83] Fernandez-Villar A, Lopez-Campos JL, Represas Represas C, Marin Barrera L, Leiro Fernandez V, Lopez Ramirez C, et al. Factors associated with inadequate diagnosis of COPD: on-sint cohort analysis. Int J Chron Obstruct Pulmon Dis 2015;10:961–7.

[84] Leidy NK, Kim K, Bacci ED, Yawn BP, Mannino DM, Thomashow BM, et al. Identifying cases of undiagnosed, clinically significant COPD in primary care: qualitative insight from patients in the target population. NPJ Prim Care Respir Med 2015;25:15024.

[85] van den Borst B, Gosker HR, Schols AM. Central fat and peripheral muscle: partners in crime in chronic obstructive pulmonary disease. Am J Respir Crit Care Med 2013;187(1):8–13.

[86] Martin M, Almeras N, Despres JP, Coxson HO, Washko GR, Vivodtzev I, et al. Ectopic fat accumulation in patients with COPD: an ECLIPSE substudy. Int J Chron Obstruct Pulmon Dis 2017;12:451–60.

[87] van de Bool C, Mattijssen-Verdonschot C, van Melick PP, Spruit MA, Franssen FM, Wouters EF, et al. Quality of dietary intake in relation to body composition in patients with chronic obstructive pulmonary disease eligible for pulmonary rehabilitation. Eur J Clin Nutr 2014;68(2):159–65.

[88] Moualla M, Qualls C, Arynchyn A, Thyagarajan B, Kalhan R, Smith LJ, et al. Rapid decline in lung function is temporally associated with greater metabolically active adiposity in a longitudinal study of healthy adults. Thorax 2017;72(12):1113–20.

[89] Behrens G, Matthews CE, Moore SC, Hollenbeck AR, Leitzmann MF. Body size and physical activity in relation to incidence of chronic obstructive pulmonary disease. CMAJ 2014;186(12):E457–69.

[90] Zhou Y, Wang D, Liu S, Lu J, Zheng J, Zhong N, et al. The association between BMI and COPD: the results of two population-based studies in Guangzhou, China. COPD 2013;10(5):567–72.

[91] Kim V, Criner GJ. The chronic bronchitis phenotype in chronic obstructive pulmonary disease: features and implications. Curr Opin Pulmon Med 2015;21(2):133–41.

[92] Konrad S, Hossain A, Senthilselvan A, Dosman JA, Pahwa P. Chronic bronchitis in aboriginal people—prevalence and associated factors. Chronic Dis Inj Can 2013;33(4):218–25.

[93] Pahwa P, Karunanayake C, Willson PJ, Hagel L, Rennie DC, Lawson JA, et al. Prevalence of chronic bronchitis in farm and nonfarm rural residents in Saskatchewan. J Occup Environ Med 2012;54(12):1481–90.

[94] Halldin CN, Doney BC, Hnizdo E. Changes in prevalence of chronic obstructive pulmonary disease and asthma in the US population and associated risk factors. Chron Respir Dis 2015;12(1):47–60.

[95] Wang M, Yi Y, Roebothan B, Colbourne J, Maddalena V, Wang PP, et al. Body mass index trajectories among middle-aged and elderly Canadians and Associated Health Outcomes. J Environ Public Health 2016;2016:7014857.

[96] Kearns K, Dee A, Fitzgerald AP, Doherty E, Perry IJ. Chronic disease burden associated with overweight and obesity in Ireland: the effects of a small BMI reduction at population level. BMC Public Health 2014;14:143.

[97] Perez T, Burgel PR, Paillasseur JL, Caillaud D, Deslee G, Chanez P, et al. Modified Medical Research Council scale vs Baseline Dyspnea Index to evaluate dyspnea in chronic obstructive pulmonary disease. Int J Chron Obstruct Pulmon Dis 2015;10:1663–72.

[98] Sievi NA, Senn O, Brack T, Brutsche MH, Frey M, Irani S, et al. Impact of comorbidities on physical activity in COPD. Respirology 2015;20(3):413–8.

[99] Hanson C, Rutten EP, Wouters EF, Rennard S. Influence of diet and obesity on COPD development and outcomes. Int J Chron Obstruct Pulmon Dis 2014;9:723–33.

[100] Sood A, Petersen H, Meek P, Tesfaigzi Y. Spirometry and health status worsen with weight gain in obese smokers but improve in normal-weight smokers. Am J Respir Crit Care Med 2014;189(3):274–81.

[101] O'Donnell DE, Ciavaglia CE, Neder JA. When obesity and chronic obstructive pulmonary disease collide. Physiological and clinical consequences. Ann Am Thorac Soc 2014;11(4):635–44.

[102] Li LS, Caughey GE, Johnston KN. The association between co-morbidities and physical performance in people with chronic obstructive pulmonary disease: a systematic review. Chron Respir Dis 2014;11(1):3–13.

[103] Mullerova H, Lu C, Li H, Tabberer M. Prevalence and burden of breathlessness in patients with chronic obstructive pulmonary disease managed in primary care. PLoS One 2014;9(1):e85540.

[104] Boiselle PM, Litmanovich DE, Michaud G, Roberts DH, Loring SH, Womble HM, et al. Dynamic expiratory tracheal collapse in morbidly obese COPD patients. COPD 2013;10(5):604–10.

[105] Jackson BE, Suzuki S, Coultas D, Singh KP, Bae S. Chronic obstructive pulmonary disease and health-related quality of life in the 2009 Texas Behavioral Risk Factor survey. Health Educ Behav 2013;40(4):469–79.

[106] McCormack MC, Belli AJ, Kaji DA, Matsui EC, Brigham EP, Peng RD, et al. Obesity as a susceptibility factor to indoor particulate matter health effects in COPD. Eur Respir J 2015;45(5):1248–57.

[107] Fattahi F, ten Hacken NH, Lofdahl CG, Hylkema MN, Timens W, Postma DS, et al. Atopy is a risk factor for respiratory symptoms in COPD patients: results from the EUROSCOP study. Respir Res 2013;14:10.

[108] Bhatt SP, Terry NL, Nath H, Zach JA, Tschirren J, Bolding MS, et al. Association between expiratory central airway collapse and respiratory outcomes among smokers. JAMA 2016;315(5):498–505.

[109] Kurnutala LN, Joshi M, Kamath H, Yarmush J. A surprising cause of wheezing in a morbidly obese patient: a case report. Int Med Case Rep J 2014;7:143–5.

[110] Represas-Represas C, Leiro-Fernandez V, Mallo-Alonso R, Botana-Rial MI, Tilve-Gomez A, Fernandez-Villar A. Excessive dynamic airway collapse in a small cohort of chronic obstructive pulmonary disease patients. Ann Thorac Med 2015;10(2):118–22.

[111] Zapatero A, Barba R, Ruiz J, Losa JE, Plaza S, Canora J, et al. Malnutrition and obesity: influence in mortality and readmissions in chronic obstructive pulmonary disease patients. J Hum Nutr Diet 2013;26(Suppl 1):16–22.

[112] Yamauchi Y, Hasegawa W, Yasunaga H, Sunohara M, Jo T, Takami K, et al. Paradoxical association between body mass index and in-hospital mortality in elderly patients with chronic obstructive pulmonary disease in Japan. Int J Chron Obstruct Pulmon Dis 2014;9:1337–46.

[113] Galesanu RG, Bernard S, Marquis K, Lacasse Y, Poirier P, Bourbeau J, et al. Obesity in chronic obstructive pulmonary disease: is fatter really better? Can Respir J 2014;21(5):297–301.

[114] Guo Y, Zhang T, Wang Z, Yu F, Xu Q, Guo W, et al. Body mass index and mortality in chronic obstructive pulmonary disease: a dose-response meta-analysis. Medicine 2016;95(28):e4225.

[115] Johannesdottir SA, Christiansen CF, Johansen MB, Olsen M, Xu X, Parker JM, et al. Hospitalization with acute exacerbation of chronic obstructive pulmonary disease and associated health resource utilization: a population-based Danish cohort study. J Med Econ 2013;16(7):897–906.

[116] Vanfleteren L, Spruit MA, Wouters EFM, Franssen FME. Management of chronic obstructive pulmonary disease beyond the lungs. Lancet Respir Med 2016;4(11):911–24.

[117] McDonald VM, Gibson PG, Scott HA, Baines PJ, Hensley MJ, Pretto JJ, et al. Should we treat obesity in COPD? The effects of diet and resistance exercise training. Respirology 2016;21(5):875–82.

[118] McNamara RJ, McKeough ZJ, McKenzie DK, Alison JA. Obesity in COPD: the effect of water-based exercise. Eur Respir J 2013;42(6):1737–9.

[119] Torres-Sanchez I, Valenza MC, Saez-Roca G, Cabrera-Martos I, Lopez-Torres I, Rodriguez-Torres J. Results of a multimodal program during hospitalization in obese COPD exacerbated patients. COPD 2016;13(1):19–25.

[120] Borel JC, Pepin JL, Pison C, Vesin A, Gonzalez-Bermejo J, Court-Fortune I, et al. Long-term adherence with non-invasive ventilation improves prognosis in obese COPD patients. Respirology 2014;19(6):857–65.

FURTHER READING

Mohanan S, Tapp H, McWilliams A, Dulin M. Obesity and asthma: pathophysiology and implications for diagnosis and management in primary care. Exp Biol Med (Maywood) 2014;239(11):1531–40.

GLOSSARY

COPD   Chronic obstructive pulmonary disease presents as emphysema and/or chronic bronchitis phenotypes

FEV$_1$/FVC ratio   Ratio of forced expiratory volume in 1 second to forced vital capacity, indicating airflow obstruction

Mechanistic insights from human studies of asthma

5

Deepa Rastogi*, Anne E. Dixon[†]

Albert Einstein College of Medicine, Bronx, NY, United States* Department of Medicine, Division of Pulmonary and Critical Care Medicine, University of Vermont, Burlington, VT, United States[†]

INTRODUCTION

Obesity is a global problem affecting both children and adults [1]. At present, 18.5% of children and 39.5% of all adults in the United States are obese [2]. Rates of obesity vary by gender, ethnicity, socioeconomic status, educational level, and geographic location [3,4]. For instance, Hispanic and African American children and adults have higher rates of obesity compared with their non-Hispanic White and Asian counterparts [3,4].

There is a higher prevalence of obesity among those with asthma than among the general population: nearly 60% of adults with severe asthma in the U.S. are obese [5]. This is likely because obesity is a risk factor for the development of asthma in children [6–12] and adults, particularly among adult women [13]; in the United States, the prevalence of asthma is 15% and 7% among obese women and men, respectively, but only 8% and 6% among lean women and men, respectively [14]. Obese children have two- to seven-fold higher risk of developing asthma than lean children. Racial and ethnic differences evident in the prevalence of obesity parallel those seen in asthma; asthma is more common among African Americans and Hispanics, compared with non-Hispanic Whites [15].

Obese people tend to suffer more severe disease than leaner individuals. Obese children have 1.6-fold and obese adults a three- to five-fold increased risk of being hospitalized for asthma compared with lean asthmatics [16,17]. Obese children and adults also have worse disease control [18–23]. Moreover, neither children nor adults respond as well to standard asthma medications as lean individuals, with decreased responsiveness to inhaled corticosteroids and to combination therapy with inhaled corticosteroids and long-acting beta agonists [24,25]. These clinical attributes, summarized in Table 5.1, result in increased healthcare expenditure and lower quality of life in obese asthmatics [22].

Obesity is a complex syndrome likely caused by interactions between environmental exposures and altered diet, including excess caloric intake that leads to changes in metabolism, immune function, and biomechanics. Asthma is also a

Mechanisms and Manifestations of Obesity in Lung Disease. https://doi.org/10.1016/B978-0-12-813553-2.00005-1

Table 5.1 Effects of Obesity on Clinical Presentation of Asthma

↑ Risk of developing asthma
↑ Asthma severity
↓ Asthma control
↓ Response to inhaled corticosteroids
↓ Response to inhaled corticosteroid-long acting β agonists

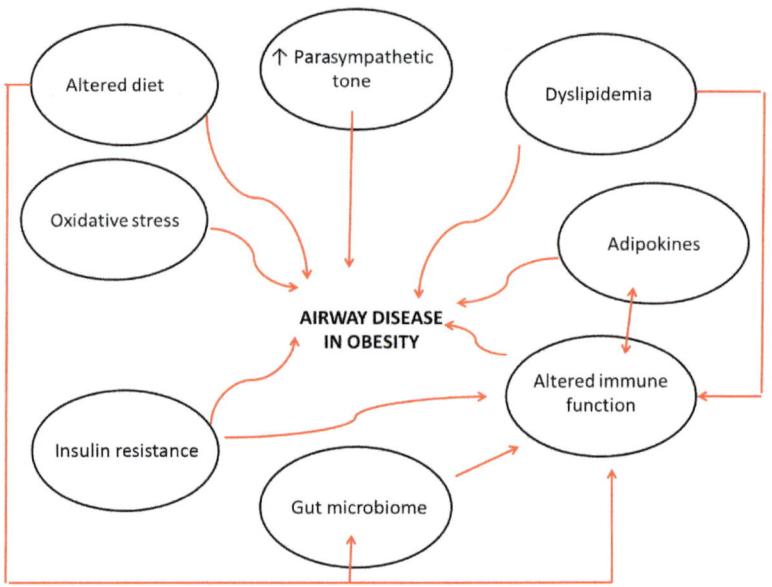

FIGURE 5.1

Multiple pathways that interact to produce the syndrome of asthma in obese patients.

complex syndrome in which environmental exposures, inflammation, remodeling, altered respiratory mechanics, and other factors affecting airway function interact to produce airway disease (Figure 5.1). Differences in the timing of onset of both obesity and asthma, as well as the varying effects of these putative causative factors, result in several phenotypes among obese asthmatics.

COMMON PATHWAYS PRODUCING ASTHMA AND OBESITY

Although obesity is certainly a risk factor for the development of incident asthma (obesity-associated asthma), in some individuals, asthma precedes and is worsened by the development of obesity [26]. For many years, there was speculation that the development of obesity in asthmatics could be related to impaired physical capacity

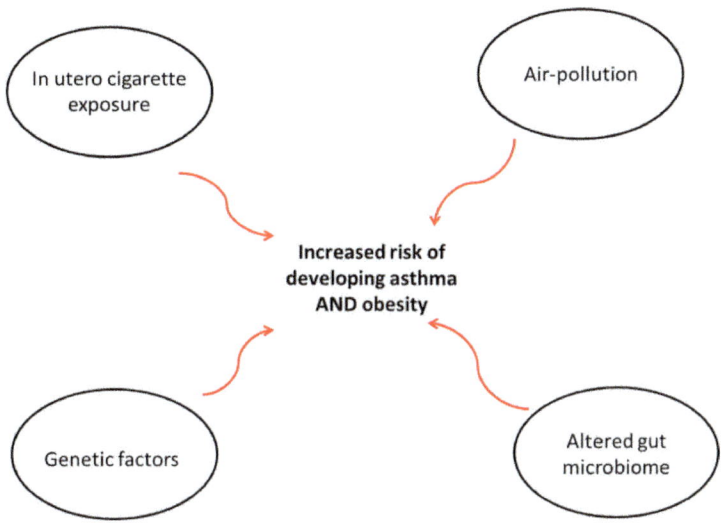

FIGURE 5.2

Common pathways likely lead to the development of asthma and obesity. These include common genetic factors (e.g., chitinase 3-like pathway), common risk factors (in utero exposure to tobacco smoke, exposure to ambient air pollution), and changes in the gut microbiome related to diet that increase the risk of developing both asthma and obesity.

and/or use of obesogenic medications such as corticosteroids, both of which might cause weight gain. However, there are data to suggest a number of common pathways could lead to *both* asthma and obesity (Figure 5.2).

Certain common genetic pathways could lead to the development of both asthma and obesity. Genetic polymorphisms in the chitinase 3-like 1 pathway are associated with an increased risk of allergic asthma; studies in mice show that chitinase 3-like 1 regulates both visceral fat accumulation *and* allergic airway inflammation, and that levels in serum and sputum are increased in obese people with asthma [27]. This suggests there may be common genetic factors that could lead to both asthma and obesity.

There may also be common environmental exposures that lead to the development of both disorders. Prenatal cigarette exposure is one such example. In utero and postnatal exposures to cigarette smoke increase the risk of developing both asthma and obesity. Maternal smoking leads to a 1.3- to 1.8-fold increased risk of childhood asthma [28], and a 1.6- to 1.8-fold increased risk of childhood obesity [29,30]. Ambient air pollution is another factor that has long been associated with the development of asthma [31,32] and recently also described as a risk factor for the development of obesity [33,34]. It is becoming apparent that there are both genetic and environmental factors that increase the risk of developing both asthma and obesity; diet and microbiome likely also fall into this category.

DIET, OBESITY, AND ASTHMA

Diet may play a role in obesity-related asthma via several mechanisms, including effects on circulating metabolic factors and the gut microbiome, as well as the effects of micronutrients on immune responses and oxidant signaling pathways. Studies have linked a Western lifestyle including high-fat snacks to increased risk of asthma [35]. The origins of diet-induced metabolic dysregulation and pulmonary morbidity may begin in utero, as children born to mothers with low intake of n-3-polyunsaturated fatty acids and α-linolenic acid are at higher risk of developing asthma, and those with high intake of saturated fatty acids, palmitic acid, and low arachidonic acid have decreased risk of developing asthma [36].

The effect of high fat on pulmonary physiology in those with established asthma can be observed within hours of exposure: neutrophilic airway inflammation and decreased responsiveness to bronchodilators were observed in adult asthmatics following a high-fat meal [37]. Elevated triglycerides and decreased HDL correlated with increased levels of FeNO in a study of healthy adults after consumption of a high-fat meal. However, pulmonary function did not change during that short time period [38]. These results suggest that increased airway inflammation and diminished bronchodilator response may be early responses to a high-fat diet.

High fat intake is also associated with decreased consumption of antioxidants, including micronutrients such as vitamins, which have a protective effect against asthma by counteracting oxidative damage to the airway [37]. Decreased antioxidants make the lung more susceptible to oxidative damage and inflammation [39,40]. Studies in humans suggest that a diet rich in antioxidant-high foods (but not rich in antioxidants as supplements) improves asthma outcomes [41], though the mechanisms for this are poorly understood.

MICROBIOME, OBESITY, AND ASTHMA

Recent literature has identified an important role for the gut microbiome in mediating obesity. Obese individuals have much lower gut bacterial diversity than normal-weight individuals [42], and similar associations have been observed in asthma as well [43]. Both diseases are associated with a lower ratio of Firmicutes to Bacteroidetes phyla. Alterations in the microbiome as early in life as in utero are associated with the development of obesity and asthma in the offspring [44]. As the infant's microbiome is determined initially by the maternal microbiome, high dietary fat intake alters maternal gut flora and consequently infant gut flora, setting the stage for childhood obesity [45] (see also Chapter 12). Furthermore, although fat intake affects gut microbiome more than carbohydrate intake [46], gut microbiome is directly related to metabolic health, including insulin sensitivity [47], suggesting diet may also drive metabolic effects on the airway through changes to the microbiome.

The full mechanisms by which the microbiome may influence pulmonary health in obese individuals are just beginning to be understood. Altered gut microbiome appears to influence immune responses by modifying the gut and lung mucosal immune system [48]. Murine studies have identified differences in NK T and Th17 cell accumulation in the gut between mice raised in a germ-free environment compared with a conventional environment [49], and alterations in gut immune cells are associated with altered cell accumulations in several other organ systems including the lung [50]. Decreased Firmicutes-to-Bacteroidetes ratio also leads to altered metabolism of dietary starches, leading to decreased production of short chain fatty acids (SCFA), such as acetate, propionate, and butyrate, as well as altered bile acid metabolism [51]. SCFA have several effects on the immune system: they impact T-cell differentiation, particularly T regulatory cells that ameliorate T-cell-dependent inflammatory diseases [52]; they enter the systemic circulation and influence the metabolome in other organs [53]; and lastly these metabolites are also associated with diurnal glucocorticoid secretion from intestinal epithelial cells and may influence development of metabolic abnormalities including insulin resistance and dyslipidemia [54,55]. Trompette *et al.* showed that altered microbiome and consequent decreased propionate related to a low-fiber diet increases the capacity of dendritic cells to promote Th2 responses, increasing airway allergic inflammation in a mouse model of asthma [56]. Thus changes in the gut microbiome associated with a Western diet might promote allergic airway disease, though this requires investigation in humans.

Future studies investigating the effects of diet on the gut microbiome, and how this related to the development of asthma in obesity, may provide important insights into the mechanisms linking obesity and asthma.

PULMONARY PHYSIOLOGY IN OBESITY AND ASTHMA

Mechanical fat load, particularly truncal adiposity, has been associated with the development or worsening of asthma among obese children and atopic asthma among obese adults [57,58]. Excess truncal adiposity renders a mechanical disadvantage to the diaphragm and chest wall, decreasing functional residual capacity (FRC), comprised of reduced residual volume (RV) and expiratory reserve volume (ERV) [59–63]. Lower FRC reduces bronchial smooth muscle stretch, especially at the end of tidal volume exhalation, which leads to a perception of increased respiratory effort with normal inspiration [64]. Decreased lung volume also reduces airway parenchymal tethering, which is thought to explain the increased airway reactivity induced by decreasing lung volume even in healthy individuals [65]. Thus truncal obesity predisposes to both dyspnea and airway hyperreactivity.

There are some inherent differences between adults and children in pulmonary function deficits associated with obesity-related asthma, which may relate to the phenotype of obese asthma (Table 5.2). Obese children tend to present with lower FEV_1/FVC ratio (compared with a higher, more normal-appearing ratio among adults with late onset obese asthma). However, these children have reduced RV with low lung volumes—similar to adults with obese asthma—rather than the elevated RV due to air trapping classically associated with childhood asthma [59]. The differences between

Table 5.2 Effects of Obesity on Lung Function in Different Types of Obese Asthma

Children	↑FEV_1, ↑↑ FVC, ↓FEV_1/FVC
Adults with early onset asthma	↓ FEV_1, ↓FEV_1/FVC
	Normal FEV_1/FVC
Adults with late onset asthma	↑ peripheral airway dysfunction
	↑ airway closure during bronchoconstriction

obese children and adults may be related to accelerated somatic growth coupled with earlier onset puberty in childhood obesity. Given the accelerated somatic growth, obese children have higher forced expiratory volume in 1 second (FEV_1) and forced vital capacity (FVC) but lower FEV_1/FVC ratio compared with normal-weight children [66,67]. We speculate that obesity in conjunction with somatic growth may influence lung development in obese children, likely leading to airway dysanapsis, the differential growth in airways compared with the lung [68]. These changes early in life may underlie persistent bronchial hyperresponsiveness and pulmonary function deficits into adulthood [69–73]. Thus a relationship likely exists between the developmental effects of obesity on lung structure and function, and asthma.

Obese adults with early onset asthma also have a greater decrement in FEV_1/FVC ratio than lean early onset asthmatics [17]; in fact, longitudinal studies show that FEV_1/FVC ratio falls as young adults with asthma develop obesity [26]. In those with later-onset nonallergic asthma, FEV_1 and FVC are both equally affected leading to preservation of FEV_1/FVC ratio often in the normal range [74]. The role of the mechanical effect of obesity is supported by the restoration of pulmonary function, with an increase in FRC and ERV, and a decrease in bronchial hyperreactivity, as well as decreased severity of exercise-induced bronchoconstriction with weight loss [75,76].

Factors other than mechanics contribute to the development of asthma in obesity, as not all obese people develop asthma. Compared with obese nonasthmatics, obese asthmatic adults with nonallergic late-onset asthma have increased peripheral lung elastance, which improves with weight loss. This suggests that asthma in these individuals is associated with peripheral airway dysfunction and increased airway closure [74]. Indeed, weight loss decreases airway reactivity particularly by decreasing sensitivity to airway closure in these individuals [77]. The causes for peripheral airway dysfunction in this phenotype of obese asthma are not known, nor are there corollary studies in children, hence the effect of obesity on peripheral airway dysfunction on the growing lung is not known. In adults, this peripheral airway dysfunction is associated with metabolic dysregulation and not BMI [74,78]. Truncal adiposity, associated with the mechanical effects of obesity, is also a risk factor for metabolic dysregulation [79]. We speculate that metabolic dysregulation may partly underlie the association of pulmonary function defects and asthma morbidity in the setting of truncal adiposity independent of any associated mechanical defects [57,80].

METABOLIC DYSREGULATION AND OBESITY-RELATED ASTHMA

INSULIN RESISTANCE

High rates of insulin resistance [79,81]—a diabetes precursor state associated with systemic hyperinsulinemia—are found among obese children and adults, particularly certain ethnic minorities who have greater truncal adiposity for the same body weight than Caucasians [82]. Insulin resistance likely contributes to obesity-related asthma in adults [83]; the presence of insulin resistance is predictive of asthma-like symptoms [84] and is associated with impaired lung function [85,86] and bronchial hyperresponsiveness [87] among adults. Children with asthma are more likely to have insulin resistance or acanthosis nigricans (a surrogate marker of insulin resistance) than their nonasthmatic counterparts [7,88]. Serum insulin levels are associated with a lower FEV_1/FVC ratio, increased bronchial hyperresponsiveness to mannitol challenge [89], and reduced lung volumes, independent of truncal and general adiposity in children [60]. Moreover, the higher prevalence of metabolic syndrome and insulin resistance among both pediatric and adult asthmatic patients [83,90–93] is associated with markers of nonatopic inflammation, such as increased levels of IL-6 and TNFα [94].

Insulin has antiinflammatory functions in addition to its role in glucose metabolism [95]. Inhibition of insulin signaling results in increased activation of Th1 cells and innate immune pathways involving macrophages, with elevated systemic levels of IL-1, IL-6, TNF, and IFNγ [94]. There is also evidence that insulin mediates the association of Th1 polarization with pulmonary function deficits among obese patients [96]. Acute lung injury is attenuated by supplemental insulin: levels of TNF, IL-1β, and IL-6 in the bronchoalveolar lavage fluid are lower, and there is decreased expression of Toll-like receptors and NF-κB with insulin-treated rats within a lipopolysaccharide model of acute lung injury [97]. Thus reduced insulin signaling due to insulin resistance may contribute to airway inflammation through a number of different mechanisms in obese asthma.

Another mechanism by which insulin resistance might mediate airway disease is through effects of insulin on airway smooth muscle (ASM) function. Increased smooth muscle contractility has been associated with insulin resistance [98–100]. Several mechanisms might be involved. Hyperinsulinemia results in increased free insulin-like growth factor (IGF-1) and reduced insulin-like growth factor binding proteins (IGFBP 1 and 3); these proteins are associated with increased ASM contractility and myofibroblast differentiation [101]. Additionally, increased laminin expression within bovine ASM cells, with activation of the phospho-inositide-3 kinase (PI3K)/Akt dependent pathway, has also been reported in hyperinsulinemic states [100]. Obesity might also affect ASM function through neurological pathways; in an obese rat model, parasympathetic activity is modulated by insulin [102]. These findings highlight the various mechanisms by which hyperinsulinemia secondary to insulin resistance might mediate airway hyperresponsiveness by increasing ASM contractility.

Together, these findings highlight the need for translational studies to determine the roles of these various pathways in obese individuals with asthma, and whether targeting insulin resistance might be a potential treatment for obesity-related asthma [103]. These studies are needed for both individuals who develop asthma as a consequence of obesity and those whose asthma worsens due to obesity.

DYSLIPIDEMIA

Dyslipidemia has been associated with wheezing in adults [104] and asthma diagnosis among children [7], although the literature supporting this association is less robust than that supporting the association between insulin resistance with asthma. Increased levels of HDL have a protective effect on pulmonary function among obese urban minority adolescents with asthma, independent of general and truncal adiposity [60], a finding mirrored in a Danish pediatric cohort [105]. Because HDL has been inversely associated with monocyte activation and Th1 polarization in obese children with asthma [96], we speculate that dyslipidemia may affect pulmonary disease through inflammatory pathways. Contrary to the protective effects of HDL, increased levels of LDL have been associated with a lower FEV_1/FVC ratio and increased aeroallergen sensitization in Danish children [105]. These findings show that dyslipidemia, irrespective of BMI, seems to be a risk factor for asthma [7].

OXIDATIVE STRESS

Another mechanism by which metabolic factors may affect asthma is by increasing oxidative stress. Obesity is associated with increased systemic and airway levels of oxidative stress; increased levels of 8-isoprostanes have been reported in exhaled breath condensates of obese asthmatics [106]. Oxidative stress could potentially contribute to oxidative signaling/damage in the airway and also contribute to the reduced levels of nitric oxide that have been described in obese asthmatic children [107] and obese asthmatic adults with late onset asthma. Production of nitric oxide in the airway requires the substrate arginine and is inhibited by asymmetric dimethyl arginine. Circulating levels of asymmetric dimethyl arginine are increased, whereas arginine is decreased in obese asthmatics [108], which may contribute to the decreased bioavailability of nitric oxide in the airway of these subjects. This likely has important physiological consequences: nitric oxide is an endogenous bronchodilator, and so it is conceivable that reduced nitric oxide could lead to reduced airway caliber. Indeed, reduced arginine/asymmetric dimethyl arginine ratios are associated with lower FEV_1 and FVC in obese asthmatics with late-onset asthma [109].

ADIPOKINES AND OBESITY-RELATED ASTHMA

Obesity is associated with changes in adipose tissue, discussed in detail later, which alter levels of circulating adipokines. Blood from obese pediatric and adult patients with asthma show increased leptin and reduced adiponectin levels [59,110–112], suggesting that obesity-induced changes in the adipokine environment may predispose obese patients to develop asthma [106,113–116]. Leptin itself may have direct effects on the airway. Visceral fat leptin expression closely correlates with airway reactivity in obese asthmatics [78], and leptin infusion in lean allergen-challenged mice increases airway reactivity but not airway inflammation [117]. The mechanisms linking leptin with airway reactivity are not well understood, but as leptin is involved in neonatal lung development and surfactant production [118], it may have direct effects on airway structure and surfactant production. Leptin may also affect ASM function by direct effects on parasympathetic signaling to the airway [119]. Lastly, leptin also has effects on immune cell function as discussed later.

Other mediators produced by adipose tissue could also lead to airway disease. Interleukin-6 (IL-6) is produced by adipose tissue and is significantly increased in obesity, particularly in those with metabolic dysfunction. Serum IL-6 is closely correlated with asthma severity [120]. The mechanisms linking asthma severity with serum IL-6 are not yet known, but IL-6 correlates with Th1/Th2 ratio in obese asthmatic children, suggesting that IL-6 could affect lymphocyte function directly. IL-6 could also have effects on numerous other cell types in the lung through a trans-signaling mechanism [121] and so is a potential mediator of airway disease in obesity.

IMMUNE RESPONSES AND OBESITY-RELATED ASTHMA
SYSTEMIC IMMUNE RESPONSES AND OBESITY-RELATED ASTHMA

Low-grade inflammation present in obesity has been associated with asthma prevalence and severity, as well as pulmonary function deficits [59,122]. Adipocyte hyperplasia and hypertrophy accompanied by delayed neovascularization secondary to rapid weight gain is the most potent known stimulus of adipose tissue inflammation [123,124]. The resulting relative hypoxia is associated with the release of leptin, which activates macrophages and T-helper (Th) cells [124,125]. Normally, the proinflammatory effect of leptin is offset by antiinflammatory adipokines such as adiponectin, omentin, and vaspin, as well as related antiinflammatory cytokines such as IL-10, interleukin-1 receptor antagonist (IL1-RA), transforming growth factor-β (TGF-β), and growth differentiation factor-15 (GDF-15) [126–128]. This balance of proinflammatory and antiinflammatory signaling is lost in obesity. Adipose-tissue inflammation in obesity also results in a shift of the macrophage pool from antiinflammatory M2 macrophages to proinflammatory M1 macrophages [124]. M1 macrophages and leptin together increase CD4+ T lymphocyte proliferation, driving differentiation into Th1 cells, which release interferon-gamma (IFN-γ), IL-6, and TNF. These cytokines suppress production and differentiation of Th2 cells as well as regulatory T cells. Thus leptin may augment nonallergic inflammation in the obese airway.

In support of this, serum leptin levels have been shown to correlate with Th1 inflammation, with higher IFN-γ, lower IL-4, and higher Th1/Th2 cell ratio among preadolescents and adolescents with obesity-related asthma, compared with their nonobese counterparts [59,96]. These nonatopic systemic inflammatory patterns directly correlate with lower airway obstruction among obese asthmatic children [59] and persist into adulthood [122].

In addition to Th1 skewing, obesity is also associated with effector altered lymphocyte function. Cytokine responses of peripheral blood CD4 cells appear to be dampened in obese asthmatics and increase with weight loss after bariatric surgery [129]. Similarly, in mice, mediastinal lymph node cells and splenocytes have decreased response to stimulation following a high-fat diet [130,131]. The reason for these suppressed cytokine responses are not known, but these findings suggest that CD4 T_H2 lymphocyte responses are less important in the pathogenesis of asthma in the setting of obesity than in lean asthmatics.

Abnormal adipokine milieu may also explain the impaired response to glucocorticoids observed among obese asthmatics. Corticosteroids function as antiinflammatory agents by inducing mitogen-activated protein kinase (MAPK) phosphatase, decreasing the activity of MAPK, thereby inhibiting gene expression of proinflammatory molecules [132]. In obesity, adipokines cause phosphorylation of the glucocorticoid receptor, making it insensitive to corticosteroids [133,134], diminishing the antiinflammatory effects of glucocorticoids [134]. Additionally, obese individuals have higher levels of TNF, TNF receptor 1, and TNF-converting enzymes in peripheral blood mononuclear cells. These increased levels are linked to glucocorticoid resistance among the obese asthmatic patients [134]. Given the inflammatory environment underlying glucocorticoid resistance, this patient population may respond better to antiinflammatory agents such as TNF decoy receptor (etanercept) [135,136].

Although the majority of studies have identified obesity to be associated with nonatopic asthma, there are studies that report increased atopic sensitization with eosinophilic airway inflammation among obese asthmatics, particularly in girls [137,138], and one study found higher levels of submucosal rather than luminal eosinophils in obese asthmatic adults [139]. How these findings relate to altered innate and adaptive immune pathways in obesity needs further investigation. We speculate that these discordant results may partly be explained by ethnic or racial differences and/or genetic and epigenetic differences in the pathways activated by obesity. In addition, there may be inherent differences in atopic inflammation among individuals who developed asthma secondary to obesity, compared with those with coexistent asthma and obesity, or those that develop obesity due to poorly controlled atopic asthma. These differences highlight the importance of distinguishing mechanisms underlying the different phenotypes of asthma in obesity [17,140].

AIRWAY INFLAMMATION IN OBESITY-RELATED ASTHMA

Altered airway inflammation might be related to different immune mechanisms associated with obese asthma, because the inflammation is driven systemically by obesity, rather than due to inhalational exposures that lead to activation of luminal immune responses. Thus the concept of "outside-in" inflammation may explain involvement of the airway due to circulating immune measures, including IL-6, rather than inflammation starting at the luminal surface [141].

Studies show decreased airway sputum eosinophils in obese asthmatics [114,142]. However, as noted earlier, airway wall biopsies of obese asthmatics suggest that eosinophils are increased in the wall compared with the airway lumen [139]; there is animal data to support this observation, with delayed trafficking of eosinophils to the airway reported in some mouse models of obese allergic asthma [143,144]. The reasons for this lower level of eosinophilia may be related to lower levels of surfactant protein A; obese asthmatics have lower levels of surfactant protein A in bronchoalveolar lavage than lean asthmatics, and surfactant protein A can inhibit allergic airway inflammation [145]. These studies all point to altered airway eosinophil function in obese asthma.

Increased and altered macrophage function in the airway may also contribute to obese asthma pathogenesis. Studies in bariatric surgery patients suggest an increase in airway macrophages in obese asthmatics, which appears to be associated with increased proinflammatory M1 macrophages in visceral adipose tissue [129,146]. The function of these macrophages also appear to be altered in the setting of obesity, with attenuated responses to lipopolysaccharide [78]. However, the role of macrophages in the pathogenesis of asthma in obesity requires further study.

Lastly, animal studies suggest that airway hyperreactivity in a high-fat diet model of obese asthma is related to activation of the NLRP3 inflammasome and recruitment of innate lymphoid cells producing IL-17 to the airway, though the significance of innate lymphoid cells in the airway in obese humans with asthma is not known [147].

CONCLUSION

Asthma and obesity are both complex syndromes. Multiple aspects of obesity may contribute to the development of airway disease and, together, orchestrate airway disease in the setting of obesity. These factors range from genetic and environmental factors (including diet and air pollution), growth and development, altered metabolism, changes in immune function, and neurologic perturbations. Given the increased incidence, severity, and refractoriness to standard treatments of asthma in obese patients, future studies to better understand the mechanistic relationship between airway disease and obesity will be essential to develop new therapeutic strategies to intervene in this patient population.

REFERENCES

[1] Collaborators GBDO. Health effects of overweight and obesity in 195 countries over 25 years. N Engl J Med 2017. https://doi.org/10.1056/NEJMoa1614362. PubMed PMID: 28604169; PMCID: PMC5479627.

[2] Hales CM, Carroll MD, Fryar CD, Ogden CL. Prevalence of obesity among adults and youth: United States, 2015–2016. NCHS Data Brief 2017(288):1–8. PubMed PMID: 29155689.

[3] Ogden CL, Carroll MD, Kit BK, Flegal KM. Prevalence of obesity and trends in body mass index among US children and adolescents, 1999–2010. JAMA 2012;307(5):483–90, 10.1001/jama.2012.40.

[4] Ogden CL, Carroll MD, Fryar CD, Flegal KM. Prevalence of obesity among adults and youth: United States 2011–2014. In: Services USDoHaH, Prevention CfDCa, editors. 2015. https://doi.org/10.1093/biosci/biv112.

[5] Schatz M, Hsu JW, Zeiger RS, Chen W, Dorenbaum A, Chipps BE, Haselkorn T. Phenotypes determined by cluster analysis in severe or difficult-to-treat asthma. J Allergy Clin Immunol 2014;133(6):1549–56. https://doi.org/10.1016/j.jaci.2013.10.006. PubMed PMID: 24315502.

[6] Matricardi PM, Gruber C, Wahn U, Lau S. The asthma-obesity link in childhood: open questions, complex evidence, a few answers only, Clin Exp Allergy 2007;37(4):476–84. [Epub 2007/04/14]. https://doi.org/10.1111/j.1365-2222.2007.02664.x. [pii] CEA2664. PubMed PMID: 17430342.

[7] Cottrell L, Neal WA, Ice C, Perez MK, Piedimonte G. Metabolic abnormalities in children with asthma. Am J Respir Crit Care Med 2011;183(4):441–8. https://doi.org/10.1164/rccm.201004-0603OC.

[8] Weiss ST. Obesity: insight into the origins of asthma. Nat Immunol 2005;6(6):537–9. [Epub 2005/05/24]. https://doi.org/10.1038/ni0605-537. [pii] ni0605-537; PubMed PMID: 15908930.

[9] Farah CS, Salome CM. Asthma and obesity: a known association but unknown mechanism. Respirology 2012;17:412–21. https://doi.org/10.1111/j.1440-1843.2011.02080.x.

[10] Flaherman V, Rutherford GW. A meta-analysis of the effect of high weight on asthma. Arch Dis Child 2006;91(4):334–9. https://doi.org/10.1136/adc.2005.080390. PubMed PMID: WOS:000236151900015.

[11] Chen YC, Dong GH, Lin KC, Lee YL. Gender difference of childhood overweight and obesity in predicting the risk of incident asthma: a systematic review and meta-analysis. Obes Rev 2013;14(3):222–31. https://doi.org/10.1111/j.1467-789X.2012.01055.x. PubMed PMID: 23145849.

[12] Rzehak P, Wijga AH, Keil T, Eller E, Bindslev-Jensen C, Smit HA, Weyler J, Dom S, Sunyer J, Mendez M, Torrent M, Vall O, Bauer CP, Berdel D, Schaaf B, Chen CM, Bergstrom A, Fantini MP, Mommers M, Wahn U, Lau S, Heinrich J, Cohorts G-WB. Body mass index trajectory classes and incident asthma in childhood: results from 8 European Birth Cohorts—a global allergy and asthma European network initiative. J Allergy Clin Immunol 2013;131(6):1528–36. https://doi.org/10.1016/j.jaci.2013.01.001. PubMed PMID: 23403049.

[13] Camargo CA, Jr., Weiss ST, Zhang S, Willett WC, Speizer FE. Prospective study of body mass index, weight change, and risk of adult-onset asthma in women. Arch Intern Med 1999;159(21):2582–8. PubMed PMID: 10573048, https://doi.org/10.1001/archinte.159.21.2582.

[14] Akinbami LJ, Fryar CD. Asthma prevalence by weight status among adults: United States, 2001–2014. NCHS data brief, no 239. Hyattsville, MD: National Center for Health Statistics 2016; 2016. NCHS Data Brief [Internet].

[15] Akinbami LJ, Moorman JE, Garbe PL, Sondik EJ. Status of childhood asthma in the United States, 1980–2007. Pediatrics 2009;123(Suppl 3):S131–45.

[16] Mosen DM, Schatz M, Magid DJ, Camargo CA Jr. The relationship between obesity and asthma severity and control in adults. J Allergy Clin Immunol 2008;122(3): 507-11e6 [Epub 2008/09/09]. https://doi.org/10.1016/j.jaci.2008.06.024. S0091–6749(08)01173–1 [pii]. PubMed PMID: 18774387.

[17] Holguin F, Bleecker ER, Busse WW, Calhoun WJ, Castro M, Erzurum SC, Fitzpatrick AM, Gaston B, Israel E, Jarjour NN, Moore WC, Peters SP, Yonas M, Teague WG, Wenzel SE.

Obesity and asthma: an association modified by age of asthma onset. J Allergy Clin Immunol 2011;127(6):1486–93e2. https://doi.org/10.1016/j.jaci.2011.03.036. PubMed PMID: 21624618; PMCID: PMC3128802.

[18] Belamarich PF, Luder E, Kattan M, Mitchell H, Islam S, Lynn H, Crain EF. Do obese inner-city children with asthma have more symptoms than nonobese children with asthma? Pediatrics 2000;106(6):1436–41. [Epub 2000/01/11]. PubMed PMID: 11099600, https://doi.org/10.1542/peds.106.6.1436.

[19] Kattan M, Kumar R, Bloomberg GR, Mitchell HE, Calatroni A, Gergen PJ, Kercsmar CM, Visness CM, Matsui EC, Steinbach SF, Szefler SJ, Sorkness CA, Morgan WJ, Teach SJ, Gan VN. Asthma control, adiposity, and adipokines among inner-city adolescents. J Allergy Clin Immunol 2010;125(3):584–92. https://doi.org/10.1016/j.jaci.2010.01.053.

[20] Quinto KB, Zuraw BL, Poon KT, Chen W, Schatz M, Christiansen SC. The association of obesity and asthma severity and control in children. J Aller Clin Immunol 2011;128:964–9. https://doi.org/10.1016/j.jaci.2011.06.031.

[21] Barros LL, Souza-Machado A, Correa LB, Santos JS, Cruz C, Leite M, Castro L, Coelho AC, Almeida P, Cruz AA. Obesity and poor asthma control in patients with severe asthma. J Asthma 2011;48(2):171–6. https://doi.org/10.3109/02770903.2011.554940. PubMed PMID: 21275851.

[22] Lavoie KL, Bacon SL, Labrecque M, Cartier A, Ditto B. Higher BMI is associated with worse asthma control and quality of life but not asthma severity. Respir Med 2006;100(4):648–57. https://doi.org/10.1016/j.rmed.2005.08.001. PubMed PMID: 16159709.

[23] Farah CS, Kermode JA, Downie SR, Brown NJ, Hardaker KM, Berend N, King GG, Salome CM. Obesity is a determinant of asthma control independent of inflammation and lung mechanics. Chest 2011;140(3):659–66. https://doi.org/10.1378/chest.11-0027. PubMed PMID: 21415135.

[24] Boulet LP, Franssen E. Influence of obesity on response to fluticasone with or without salmeterol in moderate asthma. Respir Med 2007;101(11):2240–7. [Epub 2007/08/10]. https://doi.org/10.1016/j.rmed.2007.06.031. PubMed PMID: 17686624.

[25] Peters-Golden M, Swern A, Bird SS, Hustad CM, Grant E, Edelman JM. Influence of body mass index on the response to asthma controller agents. Eur Respir J 2006;27(3):495–503. https://doi.org/10.1183/09031936.06.00077205. PubMed PMID: 16507848.

[26] Strunk RC, Colvin R, Bacharier LB, Fuhlbrigge A, Forno E, Arbelaez AM, Tantisira KG, Childhood Asthma Management Program Research G. Airway obstruction worsens in young adults with asthma who become obese. J Allergy Clin Immunol Pract 2015;3(5):765-71e2. https://doi.org/10.1016/j.jaip.2015.05.009. PubMed PMID: 26164807; PMCID: PMC4568157.

[27] Ahangari F, Sood A, Ma B, Takyar S, Schuyler M, Qualls C, Dela Cruz CS, Chupp GL, Lee CG, Elias JA. Chitinase 3-like-1 regulates both visceral fat accumulation and asthma-like Th2 inflammation. Am J Respir Crit Care Med 2015. https://doi.org/10.1164/rccm.201405-0796OC. PubMed PMID: 25629580.

[28] Zacharasiewicz A. Maternal smoking in pregnancy and its influence on childhood asthma. ERJ Open Res 2016;2(3). https://doi.org/10.1183/23120541.00042-2016. PubMed PMID: 27730206; PMCID: PMC5034599.

[29] Moller SE, Ajslev TA, Andersen CS, Dalgard C, Sorensen TI. Risk of childhood overweight after exposure to tobacco smoking in prenatal and early postnatal life. PLoS One 2014;9(10):e109184. https://doi.org/10.1371/journal.pone.0109184. PubMed PMID: 25310824; PMCID: PMC4195647.

[30] Koshy G, Delpisheh A, Brabin BJ. Dose response association of pregnancy cigarette smoke exposure, childhood stature, overweight and obesity. Eur J Public Health 2011;21(3):286–91. https://doi.org/10.1093/eurpub/ckq173. PubMed PMID: 21126981.

[31] Rosas-Salazar C, Hartert TV. Prenatal exposures and the development of childhood wheezing illnesses. Curr Opin Allergy Clin Immunol 2017;17(2):110–5. https://doi.org/10.1097/ACI.0000000000000342. PubMed PMID: 28079560.

[32] Khreis H, Kelly C, Tate J, Parslow R, Lucas K, Nieuwenhuijsen M. Exposure to traffic-related air pollution and risk of development of childhood asthma: a systematic review and meta-analysis. Environ Int 2017;100:1–31. https://doi.org/10.1016/j.envint.2016.11.012. PubMed PMID: 27881237.

[33] Li W, Dorans KS, Wilker EH, Rice MB, Schwartz J, Coull BA, Koutrakis P, Gold DR, Fox CS, Mittleman MA. Residential proximity to major roadways, fine particulate matter, and adiposity: the Framingham Heart Study. Obesity (Silver Spring) 2016;24(12):2593–9. https://doi.org/10.1002/oby.21630. PubMed PMID: 27804220; PMCID: PMC5125859.

[34] Barakat-Haddad C, Saeed U, Elliott S. A longitudinal cohort study examining determinants of overweight and obesity in adulthood. Can J Public Health 2017;108(1):e27-e35. 10.17269/cjph.108.5772. PubMed PMID: 28425896.

[35] Lee SC, Yang YH, Chuang SY, Liu SC, Yang HC, Pan WH. Risk of asthma associated with energy-dense but nutrient-poor dietary pattern in Taiwanese children. Asia Pac J Clin Nutr 2012;21(1):73–81.

[36] Lumia M, Luukkainen P, Tapanainen H, Kaila M, Erkkola M, Uusitalo L, Niinistö S, Kenward MG, Ilonen J, Simell O, Knip M, Veijola R, Virtanen SM. Dietary fatty acid composition during pregnancy and the risk of asthma in the offspring. Pediatr Allergy Immunol 2011;22(8):827–35. https://doi.org/10.1111/j.1399-3038.2011.01202.x.

[37] Wood LG, Gibson PG. Dietary factors lead to innate immune activation in asthma. Pharmacol Ther 2009;123(1):37–53. https://doi.org/10.1016/j.pharmthera.2009.03.015. PubMed PMID: WOS:000266966500004.

[38] Rosenkranz SK, Townsend DK, Steffens SE, Harms CA. Effects of a high-fat meal on pulmonary function in healthy subjects. Eur J Appl Physiol 2010;109(3):499–506. https://doi.org/10.1007/s00421-010-1390-1. PubMed PMID: 20165863.

[39] Devereux G, Seaton A. Diet as a risk factor for atopy and asthma. J Allergy Clin Immunol 2005;115(6):1109–17; quiz 18. https://doi.org/10.1016/j.jaci.2004.12.1139. PubMed PMID: 15940119.

[40] Rubin RN, Navon L, Cassano PA. Relationship of serum antioxidants to asthma prevalence in youth. Am J Respir Crit Care Med 2004;169(3):393–8. https://doi.org/10.1164/rccm.200301-055OC. PubMed PMID: 14630617.

[41] Wood LG, Garg ML, Smart JM, Scott HA, Barker D, Gibson PG. Manipulating antioxidant intake in asthma: a randomized controlled trial. Am J Clin Nutr 2012;96(3):534–43. https://doi.org/10.3945/ajcn.111.032623. PubMed PMID: 22854412.

[42] Turnbaugh PJ, Hamady M, Yatsunenko T, Cantarel BL, Duncan A, Ley RE, Sogin ML, Jones WJ, Roe BA, Affourtit JP, Egholm M, Henrissat B, Heath AC, Knight R, Gordon JI. A core gut microbiome in obese and lean twins. Nature 2009;457(7228):480–4. https://doi.org/10.1038/nature07540. PubMed PMID: 19043404; PMCID: PMC2677729.

[43] Huang YJ, Boushey HA. The microbiome in asthma. J Allergy Clin Immunol 2015;135(1):25–30. https://doi.org/10.1016/j.jaci.2014.11.011. PubMed PMID: 25567040; PMCID: 4287960.

[44] Stokholm J, Sevelsted A, Bonnelykke K, Bisgaard H. Maternal propensity for infections and risk of childhood asthma: a registry-based cohort study. Lancet Respir Med 2014;2(8):631–7. https://doi.org/10.1016/S2213-2600(14)70152-3. PubMed PMID: 25066330.

[45] Chu DM, Antony KM, Ma J, Prince AL, Showalter L, Moller M, Aagaard KM. The early infant gut microbiome varies in association with a maternal high-fat diet. Genome Med 2016;8(1):77. https://doi.org/10.1186/s13073-016-0330-z. PubMed PMID: 27503374; PMCID: PMC4977686.

[46] Mandal S, Godfrey KM, McDonald D, Treuren WV, Bjornholt JV, Midtvedt T, Moen B, Rudi K, Knight R, Brantsaeter AL, Peddada SD, Eggesbo M. Fat and vitamin intakes during pregnancy have stronger relations with a pro-inflammatory maternal microbiota than does carbohydrate intake. Microbiome 2016;4(1):55. https://doi.org/10.1186/s40168-016-0200-3. PubMed PMID: 27756413; PMCID: PMC5070355.

[47] Haro C, Montes-Borrego M, Rangel-Zuniga OA, Alcala-Diaz JF, Gomez-Delgado F, Perez-Martinez P, Delgado-Lista J, Quintana-Navarro GM, Tinahones FJ, Landa BB, Lopez-Miranda J, Camargo A, Perez-Jimenez F. Two healthy diets modulate gut microbial community improving insulin sensitivity in a human obese population. J Clin Endocrinol Metab 2016;101(1):233–42. https://doi.org/10.1210/jc.2015-3351. PubMed PMID: 26505825.

[48] Cho Y, Shore SA, Obesity, asthma, and the microbiome. Physiology (Bethesda) 2016;31(2):108–16. https://doi.org/10.1152/physiol.00045.2015. PubMed PMID: 26889016; PMCID: PMC4888975.

[49] Olszak T, An D, Zeissig S, Vera MP, Richter J, Franke A, Glickman JN, Siebert R, Baron RM, Kasper DL, Blumberg RS. Microbial exposure during early life has persistent effects on natural killer T cell function. Science 2012;336(6080):489–93. https://doi.org/10.1126/science.1219328. PubMed PMID: 22442383; PMCID: PMC3437652.

[50] Herbst T, Sichelstiel A, Schar C, Yadava K, Burki K, Cahenzli J, McCoy K, Marsland BJ, Harris NL. Dysregulation of allergic airway inflammation in the absence of microbial colonization. Am J Respir Crit Care Med 2011;184(2):198–205. https://doi.org/10.1164/rccm.201010-1574OC. PubMed PMID: 21471101.

[51] Shore SA, Cho Y., Obesity and asthma: microbiome-metabolome interactions. Am J Respir Cell Mol Biol 2016;54(5):609–17. https://doi.org/10.1165/rcmb.2016-0052PS. PubMed PMID: 26949916; PMCID: PMC4942201.

[52] Furusawa Y, Obata Y, Fukuda S, Endo TA, Nakato G, Takahashi D, Nakanishi Y, Uetake C, Kato K, Kato T, Takahashi M, Fukuda NN, Murakami S, Miyauchi E, Hino S, Atarashi K, Onawa S, Fujimura Y, Lockett T, Clarke JM, Topping DL, Tomita M, Hori S, Ohara O, Morita T, Koseki H, Kikuchi J, Honda K, Hase K, Ohno H. Commensal microbe-derived butyrate induces the differentiation of colonic regulatory T cells. Nature 2013;504(7480):446–50. https://doi.org/10.1038/nature12721. PubMed PMID: 24226770.

[53] Li JV, Ashrafian H, Bueter M, Kinross J, Sands C, le Roux CW, Bloom SR, Darzi A, Athanasiou T, Marchesi JR, Nicholson JK, Holmes E. Metabolic surgery profoundly influences gut microbial-host metabolic cross-talk. Gut 2011;60(9):1214–23. https://doi.org/10.1136/gut.2010.234708. PubMed PMID: 21572120; PMCID: PMC3677150.

[54] Hussain MM. Metabolism: gut microbiota modulates diurnal secretion of glucocorticoids. Nat Rev Endocrinol 2013;9(8):444–6. https://doi.org/10.1038/nrendo.2013.129. PubMed PMID: 23817288; PMCID: PMC4371859.

[55] Mukherji A, Kobiita A, Ye T, Chambon P. Homeostasis in intestinal epithelium is orchestrated by the circadian clock and microbiota cues transduced by TLRs. Cell 2013;153(4):812–27. https://doi.org/10.1016/j.cell.2013.04.020. PubMed PMID: 23663780.

[56] Trompette A, Gollwitzer ES, Yadava K, Sichelstiel AK, Sprenger N, Ngom-Bru C, Blanchard C, Junt T, Nicod LP, Harris NL, Marsland BJ. Gut microbiota metabolism of dietary fiber influences allergic airway disease and hematopoiesis. Nat Med 2014;20(2):159–66. https://doi.org/10.1038/nm.3444. PubMed PMID: 24390308.

[57] Musaad SM, Patterson T, Ericksen M, Lindsey M, Dietrich K, Succop P, Hershey GKK. Comparison of anthropometric measures of obesity in childhood allergic asthma: Central obesity is most relevant. J Allergy Clin Immunol 2009;123(6):1321–7. https://doi.org/10.1016/j.jaci.2009.03.023. PubMed PMID: WOS:000266799100024; PMCID: PMC2771544.

[58] Ma J, Xiao L. Association of general and central obesity and atopic and nonatopic asthma in US adults. J Asthma 2013;50(4):395–402. https://doi.org/10.3109/02770903.2013.770014. PubMed PMID: 23351029.

[59] Rastogi D, Canfield S, Andrade A, Hall CB, Isasi CR, Rubinstein A, Arens R. Obesity-associated asthma in children: a distinct entity. Chest 2012;141(4):895–905. [Epub Oct 6], https://doi.org/10.1378/chest.11-0930.

[60] Rastogi D, Bhalani K, Hall CB, Isasi CR. Association of pulmonary function with adiposity and metabolic abnormalities in urban minority adolescents. Ann Am Thorac Soc 2014;11(5):744–52.

[61] Davidson WJ, Mackenzie-Rife KA, Witmans MB, Montgomery MD, Ball GD, Egbogah S, Eves ND. Obesity negatively impacts lung function in children and adolescents. Pediatr Pulmonol 2014;49(10):1003–10. https://doi.org/10.1002/ppul.2291524167154.

[62] Li AM, Chan D, Wong E, Yin J, Nelson EA, Fok TF. The effects of obesity on pulmonary function. Arch Dis Child 2003;88(4):361–3. https://doi.org/10.1136/adc.88.4.361.

[63] Gibson N, Johnston K, Bear N, Stick S, Logie K, Hall G. Expiratory flow limitation and breathing strategies in overweight adolescents during submaximal exercise. Int J Obes (Lond) 2014;38(1):22–6. https://doi.org/10.1038/ijo.2013.13723897219.

[64] Salome CM, King GG, Berend N. Physiology of obesity and effects on lung function. J Appl Physiol 2010;108(1):206–11. https://doi.org/10.1152/japplphysiol.00694.2009.

[65] Skloot G, Permutt S, Togias A. Airway hyperresponsiveness in asthma: a problem of limited smooth muscle relaxation with inspiration. J Clin Invest 1995;96(5):2393–403. https://doi.org/10.1172/JCI118296. PubMed PMID: 7593627; PMCID: 185891.

[66] Spathopoulos D, Paraskakis E, Trypsianis G, Tsalkidis A, Arvanitidou V, Emporiadou M, Bouros D, Chatzimichael A. The effect of obesity on pulmonary lung function of school aged children in Greece. Pediatr Pulmonol 2009;44(3):273–80. [Epub 2009/02/12]. https://doi.org/10.1002/ppul.20995. PubMed PMID: 19208374.

[67] Tantisira KG, Litonjua AA, Weiss ST, Fuhlbrigge AL. Association of body mass with pulmonary function in the Childhood Asthma Management Program (CAMP). Thorax 2003;58(12):1036–41. https://doi.org/10.1136/thorax.58.12.1036.

[68] Forno E, Weiner DJ, Mullen J, Sawicki G, Kurland G, Han YY, Cloutier MM, Canino G, Weiss ST, Litonjua AA, Celedón JC. Obesity and airway Dysanapsis in children with and without asthma. Am J Resp Crit Care Med. 2017;195(3):314–23. https://doi.org/10.1164/rccm.201605-1039OC27552676.

[69] Yap JC, Watson RA, Gilbey S, Pride NB. Effects of posture on respiratory mechanics in obesity. J Appl Physiol 1995;79:1199–205. https://doi.org/10.1152/jappl.1995.79.4.1199.

[70] Sampson MG, Grassino AE. Load compensation in obese patients during quiet tidal breathing. J Appl Physiol 1983;55:1269–76. https://doi.org/10.1152/jappl.1983.55.4.1269.

[71] Celedon JC, Palmer LJ, Litonjua AA, Weiss ST, Wang B, Fang Z, Xu X. Body mass index and asthma in adults in families of subjects with asthma in Anqing, China. Am J Respir Crit Care Med 2001;164(10 Pt 1):1835–40. https://doi.org/10.1164/ajrccm.164.10.2105033. PubMed PMID: 11734432.

[72] Litonjua AA, Sparrow D, Celedon JC, DeMolles D, Weiss ST. Association of body mass index with the development of methacholine airway hyperresponsiveness in men: the Normative Aging Study. Thorax 2002;57(7):581–5. https://doi.org/10.1136/thorax.57.7.581. PubMed PMID: 12096199; PMCID: 1746377.

[73] Chinn S, Jarvis D, Burney P, European Community Respiratory Health S. Relation of bronchial responsiveness to body mass index in the ECRHS. European Community Respiratory Health Survey. Thorax 2002;57(12):1028–33. https://doi.org/10.1136/thorax.57.12.1028. PubMed PMID: 12454296; PMCID: 1758811.

[74] Al-Alwan A, Bates JH, Chapman DG, Kaminsky DA, DeSarno MJ, Irvin CG, Dixon AE. The nonallergic asthma of obesity. A matter of distal lung compliance. Am J Respir Crit Care Med 2014;189(12):1494–502. https://doi.org/10.1164/rccm.201401-0178OC. PubMed PMID: 24821412.

[75] van Leeuwen JC, Hoogstrate M, Duiverman EJ, Thio BJ. Effects of dietary induced weight loss on exercise-induced bronchoconstriction in overweight and obese children. Pediatr Pulmonol 2014. https://doi.org/10.1002/ppul.22932. PubMed PMID: 24166939.

[76] Jensen ME, Gibson PG, Collins CE, Hilton JM, Wood LG. Diet-induced weight loss in obese children with asthma: a randomized controlled trial. Clin Exp Allergy 2014;43(7):775–84.

[77] Chapman DG, Irvin CG, Kaminsky DA, Forgione PM, Bates JH, Dixon AE. Influence of distinct asthma phenotypes on lung function following weight loss in the obese. Respirology 2014;19(8):1170–7. https://doi.org/10.1111/resp.1236825138203.

[78] Sideleva O, Suratt BT, Black KE, Tharp WG, Pratley RE, Forgione P, Dienz O, Irvin CG, Dixon AE. Obesity and asthma: an inflammatory disease of adipose tissue not the airway. Am J Respir Crit Care Med 2012;186(7):598–605. https://doi.org/10.1164/rccm.201203-0573OC. PubMed PMID: 22837379; PMCID: 3480522.

[79] Grundy SM, Brewer Jr. HM, Cleeman JI, Smith SC, Lenfant C. Definition of metabolic syndrome: report of the National Heart, Lung, and Blood Institute/American Heart Association conference on scientific issues related to definition. Circulation 2004;109:433–8. https://doi.org/10.1161/01.CIR.0000111245.75752.C6.

[80] Chen Y, Rennie D, Cormier Y, Dosman JA. Waist circumference associated with pulmonary function in children. Pediatr Pulmonol 2009;44(3):216–21. https://doi.org/10.1002/ppul.20854.

[81] Kahn BB, Flier JS. Obesity and insulin resistance. J Clin Invest 2000;106(4):473–81. https://doi.org/10.1172/JCI10842.

[82] Messiah SE, Arheart KL, Lipshultz SE, Miller TL. Ethnic group differences in waist circumference percentiles among U.S. children and adolescents: estimates from the 1999-2008 national health and nutrition examination surveys. Metab Syndr Relat Disord 2011;9(4):297–303. https://doi.org/10.1089/met.2010.0127.

[83] Agrawal A, Prakash YS. Obesity, metabolic syndrome, and airway disease: a bioenergetic problem? Immunol Allergy Clin N Am 2014;34:785–96. https://doi.org/10.1016/j.iac.2014.07.004.

[84] Thuesen BH, Husemoen LLN, Hersoug LG, Pisinger C, Linneberg A. Insulin resistance as a predictor of incident asthma-like symptoms in adults. Clin Exp Allergy 2009;39:700–7. https://doi.org/10.1111/j.1365-2222.2008.03197.x.

[85] Lawlor DA, Ebrahim S, Smith GD. Associations of measures of lung function with insulin resistance and type 2 diabetes: findings from the British Women's Heart and Health Study. Diabetologia 2004;47(2):195–203. https://doi.org/10.1007/s00125-003-1310-6. PubMed PMID: 14704837.

[86] Engstrom G, Hedblad B, Nilsson P, Wollmer P, Berglund G, Janzon L. Lung function, insulin resistance and incidence of cardiovascular disease: a longitudinal cohort study. J Intern Med 2003;253(5):574–81. PubMed PMID: 12702035, https://doi.org/10.1046/j.1365-2796.2003.01138.x.

[87] Kim KM, Kim SS, Lee SH, Song WJ, Chang YS, Min KU, Cho SH. Association of insulin resistance with bronchial hyperreactivity. Asia Pac Allergy 2014;4(2):99–105. https://doi.org/10.5415/apallergy.2014.4.2.99. PubMed PMID: 24809015; PMCID: 4005343.

[88] Al-Shawwa BA, Al-Huniti NH, DeMattia L, Gershan W. Asthma and insulin resistance in morbidly obese children and adolescents. J Asthma 2007;44(6):469–73. https://doi.org/10.1080/02770900701423597. PubMed PMID: WOS:000248722900010.

[89] Karampatakis N, Karampatakis T, Galli-Tsinopoulou A, Kotanidou EP, Tsergouli K, Eboriadou-Petikopoulou M, Haidopoulou K. Impaired glucose metabolism and bronchial hyperresponsiveness in obese prepubertal asthmatic children. Pediatr Pulmonol 2017;52(2):160–6. https://doi.org/10.1002/ppul.23516. PubMed PMID: 27362543.

[90] Del-Rio-Navarro BE, Castro-Rodriguez JA, Garibay Nieto N, Berber A, Toussaint G, Sienra-Monge JJ, Romieu I. Higher metabolic syndrome in obese asthmatic compared to obese nonasthmatic adolescent males. J Asthma 2010;47(5):501–6. https://doi.org/10.3109/02770901003702808.

[91] Naveed B, Weiden MD, Kwon S, Gracely EJ, Comfort AL, Ferrier N, Kasturiarachchi KJ, Cohen HW, Aldrich TK, Rom WN, Kelly K, Prezant DJ, Nolan A. Metabolic syndrome biomarkers predict lung function impairment: a nested case-control study. Am J Resp Crit Care Med 2012;185(4):392–9. https://doi.org/10.1164/rccm.201109-1672OC.

[92] Brumpton BM, Camargo CA, Jr., Romundstad PR, Langhammer A, Chen Y, Mai XM. Metabolic syndrome and incidence of asthma in adults: the HUNT study. Eur Resp J. 2013;42(6):1495–502. [Epub Jul 11]. https://doi.org/10.1183/09031936.00046013. PubMed PMID: 23845717.

[93] Chen WL, Wang CC, Wu LW, Kao TW, Chan JY, Chen YJ, Yang YH, Chang YW, Peng TC. Relationship between lung function and metabolic syndrome. PloS One 2014;9(10):e108989, https://doi.org/10.1371/journal.pone.0111172.

[94] Lumeng CN, Saltiel AR. Inflammatory links between obesity and metabolic disease. J Clin Invest 2011;121(6):2111–7. https://doi.org/10.1172/JCI57132.

[95] Hyun E, Ramachandran R, Hollenberg MD, Vergnolle N. Mechanisms behind the anti-inflammatory actions of insulin. Crit Rev Immunol 2011;31(4):307–40. https://doi.org/10.1615/CritRevImmunol.v31.i4.30.

[96] Rastogi D, Fraser S, Oh J, Huber AM, Schulman Y, Bhagtani RH, Khan ZS, Tesfa L, Hall CB, Macian F. Inflammation, metabolic dysregulation and pulmonary function

among obese asthmatic urban adolescents. Am J Resp Crit Care Med 2015;191(2):149–60. https://doi.org/10.1164/rccm.201409-1587OC. PMCID: PMC4347436.

[97] Liu ML, Dong HY, Zhang B, Zheng WS, Zhao PT, Liu Y, Niu W, Xu DQ, Li ZC. Insulin reduces LPS-induced lethality and lung injury in rats. Pulm Pharmacol Ther 2012;25(6):472–7. https://doi.org/10.1016/j.pupt.2012.09.002. PubMed PMID: 22982561.

[98] Schaafsma D, Gosens R, Ris JM, Zaagsma J, Meurs H, Nelemans SA. Insulin induces airway smooth muscle contraction. Br J Pharmacol 2007;150(2):136–42. https://doi.org/10.1038/sj.bjp.0706985.

[99] Schaafsma D, McNeill KD, Stelmack GL, Gosens R, Baarsma HA, Dekkers BG, Frohwerk E, Penninks JM, Sharma P, Ens KM, Nelemans SA, Zaagsma J, Halayko AJ, Meurs H. Insulin increases the expression of contractile phenotypic markers in airway smooth muscle. Am J Physiol Cell Physiol 2007;293(1):C429–39. https://doi.org/10.1152/ajpcell.00502.2006.

[100] Dekkers BG, Schaafsma D, Tran T, Zaagsma J, Meurs H. Insulin-induced laminin expression promotes a hypercontractile airway smooth muscle phenotype. Am J Respir Cell Mol Biol 2009;41(4):494–504. https://doi.org/10.1165/rcmb.2008-0251OC.

[101] Lee H, Kim SR, Oh Y, Cho SH, Schleimer RP, Lee YC. Targeting insulin-like growth factor-I and insulin-like growth factor—binding Protein-3 signaling pathways. Am J Respir Cell Mol Biol 2014;50(4):667–77. https://doi.org/10.1165/rcmb.2013-0397TR.

[102] Nie Z, Jacoby DB, Fryer AD. Hyperinsulinemia potentiates airway responsiveness to parasympathetic nerve stimulation in obese rats. Am J Respir Cell Mol Biol 2014;51(2):251–61. https://doi.org/10.1165/rcmb.2013-0452OC.

[103] Bossé Y. Endocrine regulation of airway contractility is overlooked. J Endocrinol 2014;222(2):R61–73. https://doi.org/10.1530/JOE-14-0220.

[104] Fenger RV, Gonzalez-Quintela A, Linneberg A, Husemoen LL, Thuesen BH, Aadahl M, Vidal C, Skaaby T, Sainz JC, Calvo E. The relationship of serum triglycerides, serum HDL, and obesity to the risk of wheezing in 85,555 adults. Respir Med 2013;107(6):816–24. https://doi.org/10.1016/j.rmed.2013.02.001.

[105] Vinding RK, Stokholm J, Chawes BL, Bisgaard H. Blood lipid levels associate with childhood asthma, airway obstruction, bronchial hyperresponsiveness, and aeroallergen sensitization. J Allergy Clin Immmunol 2016;137(1):68–74.

[106] Komakula S, Khatri S, Mermis J, Savill S, Haque S, Rojas M, Brown L, Teague GW, Holguin F. Body mass index is associated with reduced exhaled nitric oxide and higher exhaled 8-isoprostanes in asthmatics. Respir Res 2007;8:32. https://doi.org/10.1186/1465-9921-8-32. PubMed PMID: 17437645; PMCID: 1855924.

[107] Han Y-Y, Forno E, Celedon JC. Adiposity, fractional exhaled nitric oxide, and asthma in US children. Am J Resp Crit Care Med 2014;190(1):32–9. https://doi.org/10.1164/rccm.201403-0565OC. PMCID: PMC4226031.

[108] Holguin F, Comhair SA, Hazen SL, Powers RW, Khatri SS, Bleecker ER, Busse WW, Calhoun WJ, Castro M, Fitzpatrick AM, Gaston B, Israel E, Jarjour NN, Moore WC, Peters SP, Teague WG, Chung KF, Erzurum SC, Wenzel SE. An association between L-arginine/asymmetric dimethyl arginine balance, obesity, and the age of asthma onset phenotype. Am J Respir Crit Care Med 2013;187(2):153–9. https://doi.org/10.1164/rccm.201207-1270OC.

[109] Holguin F, Comhair SA, Hazen SL, Powers RW, Khatri SS, Bleecker ER, Busse WW, Calhoun WJ, Castro M, Fitzpatrick AM, Gaston B, Israel E, Jarjour NN, Moore WC, Peters SP, Teague WG, Chung KF, Erzurum SC, Wenzel SE. An association between

L-arginine/asymmetric dimethyl arginine balance, obesity, and the age of asthma onset phenotype. Am J Respir Crit Care Med 2013;187(2):153–9. https://doi.org/10.1164/rccm.201207-1270OC. PubMed PMID: 23204252; PMCID: 3570651.

[110] Yuksel H, Sogut A, Yilmaz O, Onur E, Dinc G. Role of adipokines and hormones of obesity in childhood asthma. Allergy Asthma Immunol Res 2012;4(2):98–103. https://doi.org/10.4168/aair.2012.4.2.98. PubMed PMID: 22379605; PMCID: 3283800.

[111] Mai XM, Chen Y, Krewski D. Does leptin play a role in obesity-asthma relationship? Pediatr Allergy Immunol 2009;20(3):207–12. [Epub 2009/05/21]. PubMed PMID: 19455721, https://doi.org/10.1111/j.1399-3038.2008.00812.x.

[112] Sood A, Shore SA, Adiponectin, leptin, and resistin in asthma: basic mechanisms through population studies. J Allergy (Cairo) 2013;2013:785835. https://doi.org/10.1155/2013/785835. PubMed PMID: 24288549; PMCID: PMC3832971.

[113] Scott HA, Gibson PG, Garg ML, Pretto JJ, Morgan PJ, Callister R, Wood LG. Relationship between body composition, inflammation and lung function in overweight and obese asthma. Respir Res 2012;13:10. https://doi.org/10.1186/1465-9921-13-10. PubMed PMID: 22296721; PMCID: PMC3329414.

[114] Lessard A, Turcotte H, Cormier Y, Boulet LP. Obesity and asthma: a specific phenotype? Chest 2008;134(2):317–23. [Epub 2008/07/22]. Chest 07-2959 [pii]. https://doi.org/10.1378/chest.07-2959. PubMed PMID: 18641097.

[115] Sutherland TJ, Cowan JO, Young S, Goulding A, Grant AM, Williamson A, Brassett K, Herbison GP, Taylor DR. The association between obesity and asthma: interactions between systemic and airway inflammation. Am J Respir Crit Care Med 2008;178(5):469–75. [Epub 2008/06/21]. [pii] 200802-301OC, https://doi.org/10.1164/rccm.200802-301OC. PubMed PMID: 18565954.

[116] Todd DC, Armstrong S, D'Silva L, Allen CJ, Hargreave FE, Parameswaran K. Effect of obesity on airway inflammation: a cross-sectional analysis of body mass index and sputum cell counts. Clin Exp Allergy 2007;37(7):1049–54. PubMed PMID: WOS:000247398300011, https://doi.org/10.1111/j.1365-2222.2007.02748.x.

[117] Shore SA, Schwartzman IN, Mellema MS, Flynt L, Imrich A, Johnston RA. Effect of leptin on allergic airway responses in mice. J Allergy Clin Immunol 2005;115(1):103–9. https://doi.org/10.1016/j.jaci.2004.10.007. PubMed PMID: 15637554.

[118] Chen H, Zhang JP, Huang H, Wang ZH, Cheng R, Cai WB. Leptin promotes fetal lung maturity and upregulates SP-A expression in pulmonary alveoli type-II epithelial cells involving TTF-1 activation. PLoS One 2013;8(7):e69297. https://doi.org/10.1371/journal.pone.0069297. PubMed PMID: 23894445; PMCID: PMC3718688.

[119] Arteaga-Solis E, Zee T, Emala CW, Vinson C, Wess J, Karsenty G. Inhibition of leptin regulation of parasympathetic signaling as a cause of extreme body weight-associated asthma. Cell Metab 2013;17(1):35–48. https://doi.org/10.1016/j.cmet.2012.12.004.

[120] Peters MC, McGrath KW, Hawkins GS, Hastie AT, Levy BD, Israel E, Phillips BR, Mauger DT, Comhair SA, Erzurum SC, Johansson MW, Jarjour NN, Coverstone AM, Castro M, Holguin F, Wenzel SE, Woodruff PG, Bleecker E, Fahy JV. Plasma IL6 levels, metabolic dysfunction, and asthma severity: a cross-sectional analysis of two cohorts. Lancet Respir Med 2016; https://doi.org/10.1016/S2213-2600(16)30048-0.

[121] Dixon AE, Rincon M. Metabolic dysfunction: mediator of the link between obesity and asthma? Lancet Respir Med 2016;4(7):533–4. https://doi.org/10.1016/S2213-2600(16)30104-7. PubMed PMID: 27283229.

[122] Dixon AE, Johnson SE, Griffes LV, Raymond DM, Ramdeo R, Soloveichik A, Suratt BT, Cohen RI. Relationship of adipokines with immune response and lung function in obese

asthmatic and non-asthmatic women. J Asthma 2011;48(8):811–7. https://doi.org/10.31 09/02770903.2011.613507. PMCID: PMC4133269.

[123] Ye J, Gao X, Yin J, He Q. Hypoxia is a potential risk factor for chronic inflammation and adiponectin reduction in adipose tissue of ob/ob and dietary obese mice. Am J Physiol Endocrinol Metabol 2007;293(4):1118–28.

[124] Ferrante Jr. AW. The immune cells in adipose tissue. Diabetes Obes Metab 2013;15(Suppl 3):34–8. PMCID: PMC3777665.

[125] Martin-Romero C, Santos-Alvarez J, Goberna R, Sanchez-Margalet V. Human leptin enhances activation and proliferation of human circulating T lymphocytes. Cell Immunol 2000;199(1):15–24. https://doi.org/10.1006/cimm.1999.1594. PubMed PMID: 10675271.

[126] Leal Vde O, Mafra D. Adipokines in obesity. Clin Chim Acta Int J Clin Chem 2013;419:87–94. https://doi.org/10.1016/j.cca.2013.02.003. PubMed PMID: 23422739.

[127] Catli G, Anik A, Abaci A, Kume T, Bober E. Low omentin-1 levels are related with clinical and metabolic parameters in obese children. Exp Clin Endocrinol Diabetes 2013;121(10):595–600. https://doi.org/10.1055/s-0033-1355338. PubMed PMID: 24085389.

[128] Hong EG, Ko HJ, Cho YR, Kim HJ, Ma Z, Yu TY, Friedline RH, Kurt-Jones E, Finberg R, Fischer MA, Granger EL, Norbury CC, Hauschka SD, Philbrick WM, Lee CG, Elias JA, Kim JK. Interleukin-10 prevents diet-induced insulin resistance by attenuating macrophage and cytokine response in skeletal muscle. Diabetes 2009;58(11):2525–35. https://doi.org/10.2337/db08-1261. PubMed PMID: 19690064; PMCID: 2768157.

[129] Dixon AE, Pratley RE, Forgione PM, Kaminsky DA, Whittaker-Leclair LA, Griffes LA, Garudathri J, Raymond D, Poynter ME, Bunn JY, Irvin CG. Effects of obesity and bariatric surgery on airway hyperresponsiveness, asthma control, and inflammation, J Allergy Clin Immunol 2011;128(3):508-15e1-2. https://doi.org/10.1016/j.jaci.2011.06.009. PubMed PMID: 21782230; PMCID: 3164923.

[130] de Vries A, Hazlewood L, Fitch PM, Seckl JR, Foster P, Howie SE. High-fat feeding re-directs cytokine responses and decreases allergic airway eosinophilia. Clin Exp Allergy 2009;39(5):731–9. https://doi.org/10.1111/j.1365-2222.2008.03179.x. PubMed PMID: 19178536.

[131] Ather JL, Chung M, Hoyt LR, Randall MJ, Georgsdottir A, Daphtary NA, Aliyeva MI, Suratt BT, Bates JH, Irvin CG, Russell SR, Forgione PM, Dixon AE, Poynter ME. Weight loss decreases inherent and allergic methacholine hyperresponsiveness in mouse models of diet-induced obese asthma. Am J Respir Cell Mol Biol 2016;55(2):176–87. https://doi.org/10.1165/rcmb.2016-0070OC. PubMed PMID: 27064658; PMCID: PMC4979374.

[132] Clark AR. MAP kinase phosphatase 1: a novel mediator of biological effects of glucocorticoids? J Endocrinol 2003;178(1):5–12. https://doi.org/10.1677/joe.0.1780005, PubMed PMID: 12844330.

[133] Ismaili N, Garabedian MJ. Modulation of glucocorticoid receptor function via phosphorylation. Ann N Y Acad Sci 2004;1024:86–101. https://doi.org/10.1196/annals.1321.007. PubMed PMID: 15265775.

[134] Sutherland ER, Goleva E, Strand M, Beuther DA, Leung DY. Body mass and glucocorticoid response in asthma. Am J Respir Crit Care Med 2008;178(7):682–7. https://doi.org/10.1164/rccm.200801-076OC. PubMed PMID: 18635892; PMCID: 2556450.

[135] Holgate ST, Noonan M, Chanez P, Busse W, Dupont L, Pavord I, Hakulinen A, Paolozzi L, Wajdula J, Zang C, Nelson H, Raible D. Efficacy and safety of

etanercept in moderate-to-severe asthma: a randomised, controlled trial. Eur Respir J 2011;37(6):1352–9. https://doi.org/10.1183/09031936.0006351021109557.

[136] Berry MA, Hargadon B, Shelley M, Parker D, Shaw DE, Green RH, Bradding P, Brightling CE, Wardlaw AJ, Pavord ID. Evidence of a role of tumor necrosis factor alpha in refractory asthma. N Engl J Med 2006;354(7):697–708. https://doi.org/10.1056/NEJMoa050580. PubMed PMID: 16481637.

[137] Huang SL, Shiao G, Chou P. Association between body mass index and allergy in teenage girls in Taiwan. Clin Exp Allergy 1999;29(3):323–9. PubMed PMID: WOS:000079406600007, https://doi.org/10.1046/j.1365-2222.1999.00455.x.

[138] Schachter LM, Peat JK, Salome CM. Asthma and atopy in overweight children. Thorax 2003;58(12):1031–5. [Epub 2003/12/03]. https://doi.org/10.1136/thorax.58.12.1031. PubMed PMID: 14645967; PMCID: 1746556.

[139] Desai D, Newby C, Symon FA, Haldar P, Shah S, Gupta S, Bafadhel M, Singapuri A, Siddiqui S, Woods J, Herath A, Anderson IK, Bradding P, Green R, Kulkarni N, Pavord I, Marshall RP, Sousa AR, May RD, Wardlaw AJ, Brightling CE. Elevated sputum interleukin-5 and submucosal eosinophilia in obese individuals with severe asthma. Am J Respir Crit Care Med 2013;188(6):657–63. https://doi.org/10.1164/rccm.201208-1470OC. PubMed PMID: 23590263; PMCID: 3826183.

[140] Sutherland ER, Goleva E, King TS, Lehman E, Stevens AD, Jackson LP, Stream AR, Fahy JV, Leung DY, Asthma Clinical Research N. Cluster analysis of obesity and asthma phenotypes. PLoS One 2012;7(5):e36631. https://doi.org/10.1371/journal.pone.0036631. PubMed PMID: 22606276; PMCID: 3350517.

[141] Fahy JV. Type 2 inflammation in asthma, present in most, absent in many. Nat Rev Immunol 2012;15(1):57–65.

[142] van Veen IH, Ten Brinke A, Sterk PJ, Rabe KF, Bel EH. Airway inflammation in obese and nonobese patients with difficult-to-treat asthma. Allergy 2008;63(5):570–4. https://doi.org/10.1111/j.1398-9995.2007.01597.x. PubMed PMID: 18394131.

[143] Calixto MC, Lintomen L, Schenka A, Saad MJ, Zanesco A, Antunes E. Obesity enhances eosinophilic inflammation in a murine model of allergic asthma. Br J Pharmacol 2010;159(3):617–25. https://doi.org/10.1111/j.1476-5381.2009.00560.x. PubMed PMID: 20100278; PMCID: 2828025.

[144] Lintomen L, Calixto MC, Schenka A, Antunes E. Allergen-induced bone marrow eosinophilopoiesis and airways eosinophilic inflammation in leptin-deficient ob/ob mice. Obesity (Silver Spring) 2012;20(10):1959–65. https://doi.org/10.1038/oby.2012.93. PubMed PMID: 22513490.

[145] Lugogo N, Francisco D, Addison KJ, Manne A, Pederson W, Ingram JL, Green CL, Suratt BT, Lee JJ, Sunday ME, Kraft M, Ledford JG. Obese asthmatics have decreased surfactant protein-A levels: mechanisms and implications. J Allergy Clin Immunol 2017;141(3):918–926.e3. https://doi.org/10.1016/j.jaci.2017.05.02828624607.

[146] Periyalil HA, Wood LG, Wright TA, Karihaloo C, Starkey MR, Miu AS, Baines KJ, Hansbro PM, Gibson PG. . Obese asthmatics are characterized by altered adipose tissue macrophage activation. Clin Exp Allergy 2018;48(6):641–9. https://doi.org/10.1111/cea.1310929383778.

[147] Kim HY, Lee HJ, Chang YJ, Pichavant M, Shore SA, Fitzgerald KA, Iwakura Y, Israel E, Bolger K, Faul J, DeKruyff RH, Umetsu DT. Interleukin-17-producing innate lymphoid cells and the NLRP3 inflammasome facilitate obesity-associated airway hyperreactivity. Nat Med 2014;20(1):54–61. https://doi.org/10.1038/nm.3423. PubMed PMID: 24336249; PMCID: 3912313.

Obesity and asthma: What have we learned from animal models?

6

Richard A. Johnston[*,a], **Stephanie A. Shore**[†]

Division of Critical Care Medicine, Department of Pediatrics, McGovern Medical School at The University of Texas Health Science Center at Houston, Houston, TX, United States[*]
Molecular and Integrative Physiological Sciences Program, Department of Environmental Health, Harvard T.H. Chan School of Public Health, Boston, MA, United States[†]

ABBREVIATIONS

AHR	airway hyperresponsiveness
BAL	bronchoalveolar lavage
BMI	body mass index
Cpe	carboxypeptidase E
CXCL	chemokine (C-X-C motif) ligand
DIO	diet-induced obesity
FEF$_{25-75}$	forced expiratory flow between 25% and 75% of forced vital capacity
FEV$_1$	forced expiratory volume in 1 second
FVC	forced vital capacity
GRP	gastrin-releasing peptide
Grpr	gastrin-releasing peptide receptor
HDM	house dust mite
HDME	house dust mite extract
h	hours
Ig	immunoglobulin
IL	interleukin
ILC	innate lymphoid cells
ILC2	type-2 innate lymphoid cells
ILC3	type-3 innate lymphoid cells
mRNA	messenger ribonucleic acid
O$_3$	ozone
OVA	ovalbumin
PBS	phosphate-buffered saline
R$_L$	pulmonary resistance

[a] Richard A. Johnston is serving in his personal capacity. The views expressed are his own and do not necessarily represent the views of the Centers for Disease Control and Prevention or the United States Government.

Mechanisms and Manifestations of Obesity in Lung Disease. https://doi.org/10.1016/B978-0-12-813553-2.00006-3

RT-qPCR	reverse transcription-quantitative real-time polymerase chain reactions
ST2	interleukin 1 receptor-like 1
Th	T-helper type
TNF	tumor necrosis factor
TNFR	tumor necrosis factor receptor
wk	weeks

INTRODUCTION

Asthma is a pleiotropic disease with multiple phenotypes [1], and two of these phenotypes include obesity as an important feature of the disease. One of these "obese asthma" phenotypes predominantly includes females with late-onset nonatopic asthma. In fact, in this group of asthmatics, weight loss reduces the magnitude of airway hyperresponsiveness (AHR), which suggests that obesity is a causal factor for this asthma phenotype [2]. The other "obese asthma" phenotype predominantly consists of individuals with early-onset atopic asthma. In these asthmatics, weight loss improves asthma control and quality of life but does not reduce the magnitude of AHR [2]. Thus these data suggest that obesity is not a causal factor for early-onset atopic asthma even though it worsens the symptoms of this specific phenotype.

Asthma controller medications are less effective in obese asthmatics compared with nonobese asthmatics [3–5]. The generation of therapeutics that *are* effective in obese asthmatics will require a detailed understanding of the mechanistic basis for obese asthma, an understanding that can be facilitated by the use of animal models. As previously described, there are both nonatopic and atopic obese asthmatics. Consequently, we describe herein the results of experiments in which obese mice were exposed to either nonatopic or atopic asthma triggers. We also summarize what has been learned about the mechanistic basis for obesity-related differences in the pulmonary response to these asthma triggers.

OBESITY AND ANIMAL MODELS OF NONATOPIC ASTHMA
PULMONARY RESPONSES TO O₃ IN OBESE HUMANS

Ozone (O_3), a common air pollutant produced from automobile exhaust in the presence of sunlight, is a nonatopic asthma trigger [6]. Following days in which high ambient O_3 is present, emergency room visits and hospital admissions for asthma increase [7, 8]. Foster *et al.* [9] demonstrated that human subjects exposed to fixed concentrations of O_3 exhibit asthma symptoms, reductions in lung function, and increases in airway responsiveness, especially when O_3 exposure is coupled with exercise to increase minute ventilation, thus the inhaled dose of O_3.

Pulmonary responses to O_3 are greater in obese compared with lean human subjects. Dong *et al.* [10] evaluated the respiratory effects of exposure to air pollution in a cross-sectional study of >30,000 children living in seven cities in northern China, a region with relatively high levels of air pollution. Among the participants, approximately 12% were overweight and 14% were obese. The authors observed that the effects of air pollution on asthma symptoms, including cough, phlegm, and wheeze, were greater in obese and overweight children compared with normal weight children. Excessive body mass also augments changes in pulmonary function induced by O_3 exposure. Bennett *et al.* [11] retrospectively examined changes in spirometry induced by exposure to O_3 (0.42 ppm for 1.5 hours [h]) in 197 nonasthmatic young adult men and women. They observed greater O_3-induced decrements in forced expiratory volume in 1 second (FEV_1) and forced expiratory flow between 25% and 75% of forced vital capacity (FVC) (FEF_{25-75}) with increasing body mass index (BMI), particularly in women. Interestingly, although many of the subjects were overweight, few were actually obese, indicating that substantial changes in BMI are not necessary to observe the effect of adiposity on lung function. In a follow-up analysis in which both obese and normal weight subjects were prospectively examined, O_3-induced decrements in lung function were again greater in obese subjects compared with normal weight subjects [12]. Finally, Alexeeff *et al.* [13] examined 904 elderly men from the Veterans Administration Normative Aging Study whose lung function was measured repeatedly over a period of 10 years. The investigators observed that days of high ambient O_3 were followed by declines in FEV_1, and these declines were greater in obese than in nonobese subjects. Importantly, the effect of obesity on pulmonary responses to O_3 was exaggerated if the subjects also exhibited AHR, which suggests that the effect of obesity on responses to O_3 may be enhanced in asthmatics.

PULMONARY RESPONSES TO O_3 IN OBESE MICE

Pulmonary responses to O_3 are also greater in obese mice compared with lean wild-type mice. O_3-induced increases in basal pulmonary resistance (R_L) are greater in obese compared with lean mice, and the magnitude of O_3-induced AHR is also greater in obese versus lean mice [14–22]. These effects are observed in mice obese as a result of a genetic deficiency in leptin, a satiety hormone (*ob/ob* mice); a genetic deficiency in the long isoform of the leptin receptor (*db/db* mice); a genetic deficiency in carboxypeptidase E (Cpe), an enzyme involved in processing prohormones and proneuropeptides involved in satiety and energy expenditure (*Cpe^{fat}* mice); and the consumption of a high-fat diet. The magnitude of O_3-induced AHR is greater in obese compared with wild-type mice regardless of whether methacholine or serotonin is used as the bronchoconstricting agonist, an observation that demonstrates the nonspecific nature of AHR in obese mice [17].

The toxic effects of O_3 are initiated following epithelial cell injury, which is caused by free radicals generated after O_3 reacts with lipids in the lungs and airway lining fluid [23, 24]. Epithelial cell injury induces an inflammatory response that includes the generation of many acute-phase cytokines and chemokines. O_3 exposure increases bronchoalveolar lavage (BAL) fluid concentrations of these inflammatory mediators to a greater extent in obese mice compared with lean mice [14–22]. Many of these mediators promote neutrophil migration to the lungs, and O_3-induced increases in BAL neutrophils are typically greater in obese mice compared with lean mice [14–22, 25].

The magnitude and duration of obesity may both contribute to obesity-related changes in pulmonary responses to O_3. In *Cpe^{fat}* mice, even a 25% increase in body mass relative to age-matched, lean, wild-type controls is sufficient to increase BAL indices of pulmonary inflammation following acute O_3 exposure [16]. In contrast, in mice with diet-induced obesity (DIO) induced by the consumption of a high-fat diet for 16–19 weeks (wk), a 37% increase in body mass is insufficient to augment O_3-induced pulmonary inflammation. However, consuming a high-fat diet for a longer duration (26–35 wk) does augment the effects of O_3 on pulmonary responses [14], even though the additional weight gained during the more extended period of consuming a high-fat diet was very small (2%).

Interleukin (IL)-33, IL-17A, and tumor necrosis factor (TNF)-α are among the cytokines and chemokines released in the lungs by O_3 [18, 26]. In addition, BAL concentrations of many of these cytokines are greater in obese compared with lean mice following O_3 exposure [18, 19]. Because each of these cytokines also have the capacity to augment airway responsiveness and to promote neutrophil recruitment to the lungs [27–29], events that also occur following exposure to O_3 [30], a number of investigators have examined the roles of these cytokines in the augmented responses to O_3 observed in obese mice.

MECHANISTIC BASIS FOR INCREASED PULMONARY RESPONSES TO O_3 IN OBESE MICE

Role of IL-33

The IL-1 family cytokine, IL-33, and its receptor, IL-1 receptor-like 1 (ST2), are both genetically associated with asthma [31]. IL-33 is constitutively expressed in lung epithelial cells and is released following cell injury or cell death [32], including after O_3-induced lung injury [18, 33, 34]. In mice, intratracheal administration of IL-33 causes AHR, recruits neutrophils to the lungs, and induces expression of CXCL1, CXCL2, and IL-6 [28]. Each of these events also occurs after acute O_3 exposure [35], suggesting that IL-33 may contribute to the augmented responses to O_3 observed in obese mice. To examine this hypothesis, Mathews *et al.* [18] treated obese and lean mice with an anti-ST2 or isotype control antibody prior to O_3 exposure. In this study, BAL IL-33 was greater in obese compared with lean mice exposed to O_3. Furthermore, in mice treated with the isotype control antibody, O_3 (1) increased basal R_L in obese but not lean mice and (2) increased airway responsiveness and BAL neutrophils to a greater extent in obese compared with lean mice. Finally, in obese mice treated with an anti-ST2 antibody, O_3-induced increases in (1) basal R_L, (2) airway responsiveness (Figure 6.1), and (3) BAL neutrophils were all reduced. In contrast, treatment with an anti-ST2 antibody had no effect in lean mice. As such, these data indicate that IL-33 contributes to obesity-related increases in pulmonary responses to O_3.

How does IL-33 mediate these effects? Exogenously administered IL-33 causes the release of T-helper type (Th)-2 cytokines in the lungs [36]. Indeed, BAL IL-5 and IL-13 are greater in O_3-exposed obese compared with lean mice [18, 21], which are

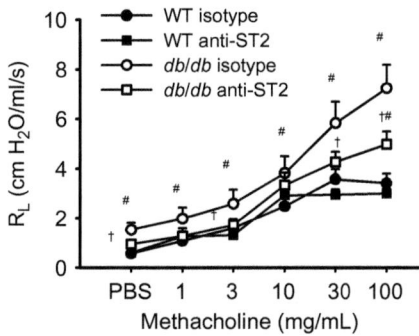

FIGURE 6.1

An anti-ST2 antibody reduces AHR in obese but not lean mice exposed to O_3. Shown are changes in pulmonary resistance (R_L) induced by inhaled aerosolized methacholine in lean wild-type (WT) and obese *db/db* female mice exposed to O_3 (2 ppm for 3 h). Mice were treated with either an isotype or anti-ST2 antibody prior to O_3 exposure. Results are mean $\pm$ SE of 4-8 mice per group. $^\#P<0.05$ versus lean mice with same antibody treatment; $^\dagger P<0.05$ versus isotype-treated mice of the same genotype.

This figure was reproduced from Mathews JA, Krishnamoorthy N, Kasahara DI, Cho Y, Wurmbrand AP,

Ribeiro L, et al. IL-33 drives augmented responses to ozone in obese mice. Environ Health Perspect

2017;125(2):246–53.

phenomena consistent with greater concentrations of BAL IL-33 in obese mice [18]. Furthermore, an anti-ST2 antibody attenuates O_3-induced increases in BAL Th-2 cytokines (e.g., IL-5 and IL-13) in obese mice [18]. It is likely that type-2 innate lymphoid cells (ILC2) contribute to the IL-33-dependent release of Th-2 cytokines following O_3 exposure. ILC2 express ST2, and the numbers of both IL-5$^+$ ILC2 and IL-13$^+$ ILC2 are greater in lungs of obese versus lean mice after O_3 exposure [18]. In addition, mice deficient in lymphoid cells have reduced Th-2 cytokine release in the nose and lungs after exposure to O_3, and adoptive transfer of ILC2 is sufficient to restore these responses [33, 34]. Mathews *et al.* [18] also demonstrated that $\gamma\delta$ T cells may also contribute to Th-2 cytokine release after O_3 exposure. First, via flow cytometry, these authors demonstrated that $\gamma\delta$ T cells isolated from lungs of O_3-exposed mice express ST2. Second, IL-13$^+$ $\gamma\delta$ T cells were increased in lungs of obese mice exposed to O_3. Third, the observation that O_3-induced increases in BAL IL-5 and BAL IL-13 are reduced in obese mice deficient in $\gamma\delta$ T cells also suggests a role for these cells in Th-2 cytokine production.

Th-2 cytokines contribute to the ability of exogenously administered IL-33 to promote airway narrowing [37]. Data from Williams *et al.* [21] suggest that the Th-2 cytokine, IL-13, also contributes to the IL-33-dependent effects of O_3 on basal pulmonary mechanics observed in obese mice. Not only is IL-13 induced by O_3 in an IL-33-dependent manner (see previous discussion), but administration of an anti-IL-13 antibody attenuates O_3-induced increases in basal pulmonary mechanics in obese mice [21] (Figure 6.2C). IL-33-dependent increases in IL-13 may also contribute to

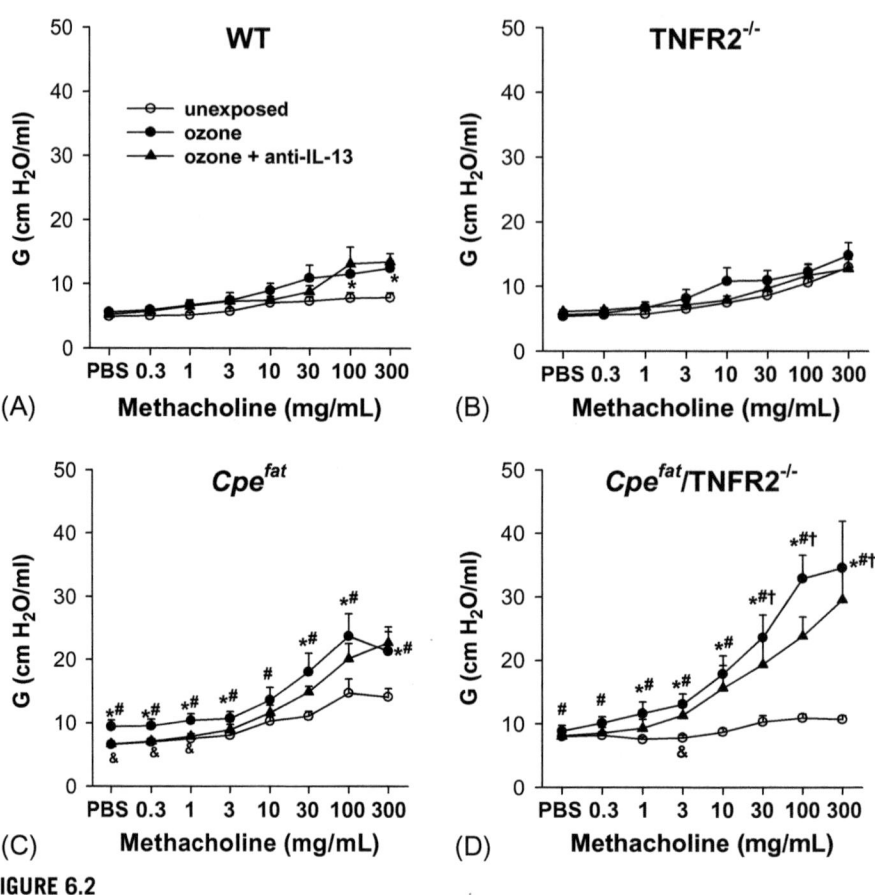

FIGURE 6.2

TNFR2 deficiency but not IL-13 augments obesity-related increases in O_3-induced AHR. Results are mean $\pm$ SE of data from 5 to 8 mice per group. Shown are data from (A) wild-type mice; (B) TNFR2-deficient mice; (C) obese Cpe^{fat} mice; and (D) obese Cpe^{fat} mice that are also deficient in TNFR2. In each case, mice were exposed to room air, O_3 (2 ppm for 3 h), or pretreated with an anti-IL-13 antibody then exposed to O_3. G: coefficient of lung tissue damping, a marker of changes in the mechanics of the lung periphery; *$P<0.05$ versus genotype-matched unexposed mice; #$P<0.05$ versus TNFR2 genotype-matched lean mice; †$P<0.05$ versus Cpe^{fat} TNFR2-sufficient mice; &$P<0.05$ versus Cpe^{fat} mice exposed to O_3 without treatment with an anti-IL-13 antibody.

This figure was reproduced from Williams AS, Mathews JA, Kasahara DI, Chen L, Wurmbrand AP, Si H, et al.
Augmented pulmonary responses to acute ozone exposure in obese mice: roles of TNFR2 and IL-13. Environ
Health Perspect 2013;121(5):551–7.

the observed role of IL-33 in neutrophil recruitment in obese mice following exposure to O_3 because an anti-IL-13 antibody reduced BAL neutrophils in obese mice exposed to O_3 [21]. In contrast, an anti-IL-13 antibody did not attenuate the magnitude of O_3-induced AHR in obese mice [21] (Figure 6.2C) even though an anti-ST2 antibody did [18]. These data demonstrate that other IL-33-inducible factors, rather than IL-13, must be responsible for the effects of IL-33 on O_3-induced AHR. CXCL1 and IL-6 may be among these factors. Intratracheal administration of IL-33 increases BAL concentrations of both of these cytokines [28]. Furthermore, when compared with lean mice, O_3-induced increases in BAL CXCL1 and IL-6 are greater in obese mice but attenuated after administration of an anti-ST2 antibody [18]. Importantly, inhibition of CXCL1 signaling reduces AHR induced by exogenous administration of IL-33 [28]. Thus CXCL1 could also contribute to O_3-induced AHR induced by endogenously released IL-33 [18]. Antibody blocking and/or genetic deletion experiments demonstrate that CXCL1 and IL-6 also contribute to O_3-induced increases in BAL neutrophils, including in obese mice [25, 30, 35]. Hence, IL-33-induced release of CXCL1 and/or IL-6 may also explain the role of IL-33 in O_3-induced neutrophil recruitment. Whereas ILC2 and $\gamma\delta$ T cells likely contribute to the IL-33-dependent release of Th-2 cytokines in O_3-exposed obese mice (see previous discussion), airway epithelial cells and macrophages are likely to be the cells responsible for IL-33-dependent increases in CXCL1 and IL-6 following O_3 exposure. Both cell types produce CXCL1 and IL-6 after O_3 exposure and express ST2 [32, 38–41].

In summary, IL-33 contributes to augmented responses to O_3 observed in obese mice at least, in part, via its ability to induce Th-2 cytokine production in ILC2 and $\gamma\delta$ T cells, although other cellular targets of IL-33, such as airway epithelial cells and macrophages, may also be involved. Therapeutics directed against IL-33 are in development and may prove to be effective in obese asthmatics, at least against nonatopic asthma triggers.

Role of IL-17A

As discussed previously, blocking IL-33 signaling with an anti-ST2 antibody reduced obesity-related increases in pulmonary responses to O_3 [18]. However, although it reduced responses to O_3, anti-ST2 antibody treatment did not completely abolish obesity-related differences in responses to O_3. Thus other factors must also be involved. IL-17A may be one of these factors. First, IL-17A causes AHR and neutrophil recruitment, which also occur following O_3 exposure when IL-33 and IL-13 are also elevated [27, 28]. Second, sputum IL-17A is greater in obese versus lean asthmatics [42]. Third, IL-17A expressing cells in the lung are also increased in number after O_3 exposure [26, 43], especially in obese mice [19]. Fourth, O_3-induced increases in BAL IL-23, a cytokine that contributes to IL-17A expression [44], and in BAL CCL20, a chemotactic factor for CCR6$^+$ IL-17A$^+$ cells, are also greater in obese than lean mice [19, 38]. Thus obesity creates an environment conducive to the pulmonary recruitment and activation of IL-17A$^+$ cells following acute O_3 exposure.

To test the hypothesis that IL-17A contributes to obesity-related differences in the response to O_3, Mathews *et al.* [19] treated *db/db* and wild-type mice with an anti-IL-17A or isotype control antibody prior to O_3 exposure. As previously described, O_3-induced increases in BAL neutrophils were greater in *db/db* than in wild-type mice. The anti-IL-17A antibody reduced BAL neutrophils in both *db/db* and wild-type mice. IL-17A recruits neutrophils not directly but via its ability to produce other neutrophil chemotactic and survival factors [45]. As previously discussed, many such factors are elevated in lungs of O_3-exposed lean and obese mice [14, 15, 17, 18, 21, 22, 46], but only CXCL1 is also reduced by the anti-IL-17A antibody in O_3-exposed *db/db* mice [19], which suggests that CXCL1 may mediate the effects of IL-17A on neutrophil recruitment.

In contrast to the effects of the anti-ST2 antibody previously described [18], obesity-related increases in the effect of O_3 on basal R_L were not attenuated in anti-IL-17A compared with isotype control antibody-treated mice [19]. However, obesity-related increases in the magnitude of O_3-induced AHR were attenuated, though not abolished by an anti-IL-17A antibody [19]. To examine the mechanism underlying the effect of IL-17A on O_3-induced AHR in obese mice, Mathews *et al.* [19] performed a microarray analysis. Among the genes affected both by O_3 and by an anti-IL-17A antibody in the lungs of *db/db* mice, Mathews *et al.* [19] identified *Grpr*, the gene for the receptor of gastrin-releasing peptide (GRP), a bronchoconstricting peptide found in pulmonary neuroendocrine cells [47]. RT-qPCR confirmed that O_3-induced increases in the pulmonary abundance of *Grpr* messenger ribonucleic acid (mRNA) were greater in obese compared with lean mice, and that *Grpr* expression was reduced by an anti-IL-17A antibody in O_3-exposed obese but not lean mice. Importantly, O_3-induced increases in BAL GRP were greater in obese versus lean mice, and similar to treatment with an anti-IL-17A antibody, treatment with an anti-GRP antibody reduced obesity-related increases in O_3-induced AHR [19].

Taken together, these data demonstrate that IL-17A contributes to obesity-related increases in the pulmonary response to O_3, and that GRP may be involved in mediating the effects of IL-17A. Finally, in a recent asthma clinical trial, IL-17A-neutralizing antibodies were found to have little beneficial effect [48]. However, subjects in that trial were recruited without regard to BMI or allergic status. Nevertheless, it is conceivable that targeting IL-17A would be beneficial in obesity-related asthma, particularly against asthma triggers such as O_3 that involve oxidative stress.

Role of TNF-α

Serum TNF-α is elevated in both human and murine obesity [49, 50], likely as a result of adipose tissue inflammation. Moreover, in mouse models, TNF-α plays a role in the development of some obesity-related conditions, including insulin resistance [51–55]. The risk of nonatopic asthma is greater among obese individuals with TNF-α polymorphisms that promote TNF-α expression [56]. Furthermore, exogenous administration of TNF-α induces AHR [29], and experiments in mice genetically deficient in either or both of the TNF-α receptors, tumor necrosis factor receptor (TNFR) 1 and 2, indicate that, in lean mice, TNF-α contributes to O_3-induced AHR [57, 58]. These data are consistent with a role for TNF-α in the augmented responses to O_3 characteristic of obese mice. However, results of experiments using obese mice genetically deficient in TNF-α or TNFR2 indicate that this role is quite complex.

Genetic deficiency in either TNF-α or TNFR2 attenuates O_3-induced neutrophil recruitment to the lungs in obese mice but has no effect in lean mice [21, 22], thus abolishing obesity-related increases in O_3-induced neutrophil recruitment to the lungs between lean and obese mice. However, TNF-α and TNFR2 deficiency each cause an *increase* in the magnitude of O_3-induced AHR (compare Figure 6.2C and D) in obese mice, even though TNFR2 deficiency reduces O_3-induced AHR in lean mice (compare Figure 6.2A and B) [21, 22]. These results indicate that TNF-α limits the severity of O_3-induced AHR but promotes neutrophil recruitment in obese mice. In addition, these results underscore the different roles for TNF-α in lean and obese mice. Finally, these results also indicate that O_3-induced AHR and inflammation are mechanistically dissociated, at least in obese mice: neutrophils can decrease even as airway responsiveness increases.

How might the protective effect of TNF-α against AHR in O_3-exposed obese mice be mediated? As previously described, both IL-33 and IL-17A contribute to O_3-induced AHR in obese mice. Hence, it is possible that TNF-α acting via TNFR2 could be exerting its protective effects against O_3-induced AHR in obese mice by affecting IL-33 or IL-17A release. Obesity-related increases in serum IL-17A were reduced in TNF-α deficient versus TNF-α sufficient obese mice, and a similar trend was observed for O_3-induced pulmonary expression of IL-17A [22]. These events would be expected to reduce O_3-induced AHR in the TNF-α deficient obese mice, not augment it, as observed [22]. In contrast, the observations that BAL IL-5 and IL-13 are augmented by TNF-α or TNFR2 deficiency in O_3-exposed obese mice [21, 22], coupled with the observation that O_3-induced increases in BAL concentrations of Th-2 cytokines are IL-33-dependent, supports the hypothesis that TNF-α may be protecting against O_3-induced AHR in obese mice by suppressing IL-33 release or action [18]. In obese mice, O_3-induced increases in BAL protein carbonyls, a marker of oxidative stress, are elevated in obese mice genetically deficient in TNF-α or TNFR2 [21, 22]. Greater oxidative stress would lead to more epithelial injury, more IL-33 release, and consequently, more Th-2 cytokine release, as observed [21, 22]. Therefore TNF-α may be acting to protect against O_3-induced AHR in obese mice by inducing expression of antioxidant enzyme mRNAs. The antioxidants *Hmox1*, *Mt1*, and *Mt2* are each induced by O_3 exposure, *but* genetic deficiencies in the TNF-α pathway do not alter the induction of these antioxidants by O_3 in obese mice [21, 22]. However, TNF-α may contribute to the induction of other antioxidant enzymes following O_3 exposure. For example, glutamate–cysteine ligase (*Gclc*), an enzyme that catalyzes the rate-limiting step in the synthesis of glutathione, a potent antioxidant, is induced to a greater extent in lungs of obese versus lean mice [59]. TNF-α is known to induce the expression of *Gclc* [60]. If TNF-α contributes to the induction of *Gclc* following O_3 exposure, then genetic deficiency in TNF-α signaling would be expected to result in greater O_3-induced oxidative stress and greater O_3-induced AHR, as observed [21, 22].

In summary, TNF-α limits the severity of O_3-induced AHR in obese mice, perhaps by suppressing oxidative stress. Etanercept, a recombinant human dimeric fusion protein that prevents TNF-α from binding to its cell surface receptors (TNFR1 and 2), is routinely used in the treatment of inflammatory disorders [61], but our data suggest that TNF-α inhibitors might actually worsen symptoms in patients with obese asthma that has a nonatopic component.

OBESITY AND ANIMAL MODELS OF ATOPIC ASTHMA

In atopic obese asthmatics, Dixon *et al.* [2] demonstrated that weight loss improves asthma control and quality of life yet does not reduce the magnitude of AHR induced by methacholine. Based on these data, Dixon *et al.* [2] concluded that obesity worsens the symptoms of atopic asthma but is not a causal factor for this particular asthma phenotype. Nevertheless, because atopic asthma is the most common asthma phenotype [62], it is important to understand the biological mechanisms by which obesity worsens the symptoms of atopic asthma. Consequently, in the remainder of this chapter, we shall summarize results from mouse studies that describe the effect of obesity on antigen-induced lung inflammation and AHR. We shall also discuss potential mechanisms whereby obesity may worsen the symptoms of atopic asthma. Finally, in our summary we only include those investigations in which mice designated as "obese" weighed twenty percent more than lean controls.

OBESITY AND ANTIGEN-INDUCED LUNG INFLAMMATION

Chronic lung inflammation, a notable feature of atopic asthma, is characterized, in part, by the presence of basophils, eosinophils, mast cells, Th-2 lymphocytes and cytokines, and ILC2 [63, 64]. Th-2 lymphocytes and ILC2 are especially important in sustaining lung inflammation in atopic asthmatics as these cells release a plethora of proinflammatory Th-2 cytokines. For example, Th-2 lymphocytes release IL-4, IL-5, IL-9, and IL-13, whereas ILC2 release IL-5, IL-9, and IL-13 [63, 65]. Expression of these cytokines increases in the lungs of asthmatics and incite many of the characteristic features of atopic asthma, including migration of eosinophils to the lungs, goblet cell hyperplasia, and AHR [66–71]. As described earlier in this chapter, acute exposure to O_3 exacerbates lung inflammation in obese mice regardless of the modality of obesity induction. However, the severity of lung inflammation in obese mice following antigen sensitization and challenge is quite variable and appears to be dependent on a number of factors, including the modality of obesity induction, the antigen sensitization and challenge protocol, mouse strain, etc.

To date, antigen-induced lung inflammation has been assessed in three strains of genetically obese mice: *db/db*, *ob/ob*, and *Cpe^{fat}*. In these specific investigations, which are summarized in Table 6.1, *db/db*, *ob/ob*, and *Cpe^{fat}* mice weighed, respectively, 61%, 87%, and 73% more than age- and gender-matched wild-type mice. In mice, Johnston *et al.* [72] were the first to report the effect of genetically induced obesity on lung inflammation following antigen (ovalbumin; OVA) sensitization and challenge, which leads to a pulmonary phenotype that mimics many of the characteristic features of atopic asthma in humans [75]. With specific regard to BAL leukocytes, Johnston *et al.* [72] reported that there were fewer BAL eosinophils and lymphocytes in *db/db* and *ob/ob* compared with wild-type mice following OVA sensitization and challenge. Figure 6.3 illustrates the number of BAL leukocytes recovered from wild-type and *ob/ob* mice sensitized to OVA and challenged with either phosphate-buffered saline (PBS) or OVA. Johnston *et al.* [72]

Table 6.1 Effect of Genetically Induced Obesity in Mice on Indices of Atopic Asthma[a]

Study	Genotype	Percent Difference in Body Mass[b]	BAL Eosinophils	Lung Eotaxin	Lung IL-5	Lung Tissue Inflammatory Cell Infiltrates	Basal Lung Mechanics and/or Airway Responsiveness	Serum Total or Antigen-Specific IgE
Johnston et al. [72]	db/db	61	Less than wild-type	–	–	–	Greater than wild-type	–
Johnston et al. [72]	ob/ob	87	Less than wild-type	Greater than wild-type	No difference	Less than wild-type	Greater than wild-type	Greater than wild-type
Lintomen et al. [73]	ob/ob	–[c]	Less than wild-type	No difference	No difference	Greater than wild-type	–	–
Dahm et al. [74]	Cpe^fat	73	Greater than wild-type	Greater than wild-type	Undetectable[d]	No difference	Greater than wild-type	No difference

[a] Qualitative comparisons were made between antigen-sensitized and challenged genetically obese and lean wild-type mice.
[b] Percent difference in body mass between genetically obese and lean wild-type mice.
[c] Indice was not measured.
[d] The investigators did measure IL-5 in bronchoalveolar lavage (BAL) fluid. However, IL-5 was undetectable in BAL fluid when analyzing samples with an enzyme-linked immunosorbent assay.

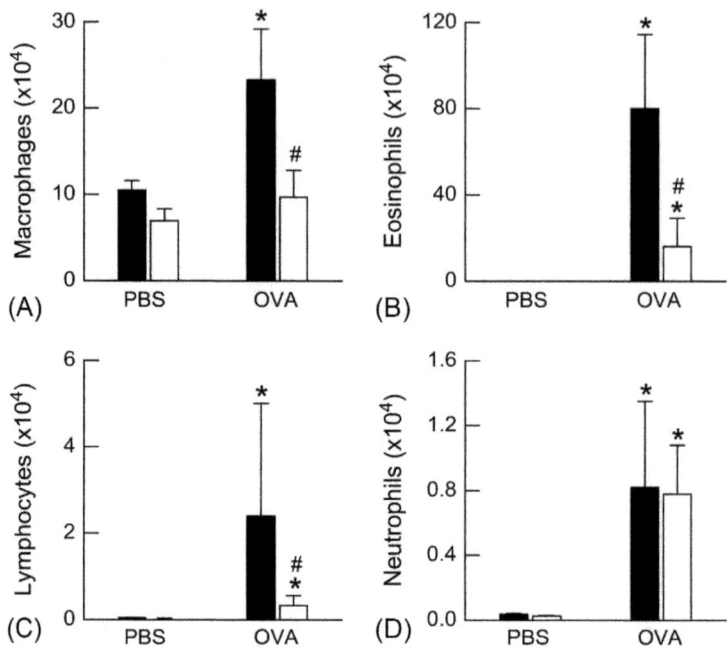

FIGURE 6.3

Total number of (A) macrophages, (B) eosinophils, (C) lymphocytes, and (D) neutrophils in bronchoalveolar lavage (BAL) fluid of ovalbumin (OVA)-sensitized wild-type (C57BL/6) and *ob/ob* mice challenged with aerosols of either phosphate-buffered saline (PBS) or OVA. BAL was performed 24 h after cessation of the final aerosol challenge. $n = 10$–14 mice for each group. *Solid bars*, wild-type; *open bars*, *ob/ob*. *$P < 0.05$ compared with genotype-matched, PBS-challenged controls; #$P < 0.05$ compared with wild-type mice with an identical exposure.

Reprinted with permission of the American Thoracic Society. Copyright © 2018 American Thoracic Society. Johnston RA, Zhu M, Rivera-Sanchez YM, Lu FL, Theman TA, Flynt L, Shore SA. Allergic airway responses in obese mice. Am J Respir Crit Care Med 2007;176:650–8. The American Journal of Respiratory and Critical Care Medicine is an official journal of the American Thoracic Society.

also reported that lung infiltrates of inflammatory cells were lower in *ob/ob* compared with wild-type mice. Consistent with the observations of Johnston *et al.* [72], Lintomen *et al.* [73] reported that the number of BAL eosinophils were lower in *ob/ob* compared with wild-type mice following OVA sensitization and challenge. However, in contrast to Johnston *et al.* [72], Lintomen *et al.* [73] reported that lung infiltrates of inflammatory cells were greater in *ob/ob* compared with wild-type mice. Because Johnston *et al.* [72] and Lintomen *et al.* [73] also reported that eotaxin, an eosinophil chemotactic cytokine, and IL-5, an eosinophil maturation cytokine, were either unaffected by obesity or actually greater in *ob/ob* mice following OVA sensitization and challenge [76, 77], it is doubtful that fewer BAL eosinophils in *ob/ob* mice could be attributed to genotype-related differences in either of these cytokines. Furthermore, Lintomen *et al.* [73] demonstrated that the number of bone marrow-derived and circulating eosinophils were greater in *ob/ob* compared with wild-type mice. Therefore it is also improbable that fewer BAL eosinophils in *ob/ob* compared with wild-type mice result from differences in either eosinophil maturation or survival. In contrast to the results obtained from OVA-sensitized and -challenged *db/db* and *ob/ob* mice [72, 73], Dahm *et al.* [74] reported that BAL eosinophils were greater in OVA-sensitized and -challenged Cpe^{fat} compared with wild-type mice (Figure 6.4C). However, differences in BAL eosinophils between Cpe^{fat} and wild-type mice occurred contemporaneously with no genotype-related differences in lung infiltrates of inflammatory cells. Furthermore, dissimilar to OVA-sensitized and -challenged *ob/ob* mice, OVA-sensitized and -challenged Cpe^{fat} mice had more BAL eotaxin, IL-4, and IL-13 than OVA-sensitized and -challenged wild-type mice [74]. Because eotaxin, IL-4, and IL-13 contribute to eosinophil migration [76, 78, 79], it is reasonable to assume that more BAL eosinophils in Cpe^{fat} compared with wild-type mice are a result of increased levels of BAL eosinophil chemotactic cytokines. For a summary of the differences in antigen-induced lung inflammation between genetically obese mice and their respective wild-type controls, please refer to Table 6.1.

Because of qualitative differences in BAL eosinophils among *db/db*, *ob/ob*, and Cpe^{fat} mice following OVA sensitization and challenge, it is necessary to consider potential explanations for this phenomenon. This is especially important to understand if investigators use these mice to comprehend the mechanisms by which obesity worsens the symptoms of atopic asthma. Therefore in the following text, we discuss four potential explanations for qualitative differences in BAL eosinophils among genetically obese mice. First, previous investigators demonstrated that the OVA-sensitization and -challenge protocol influences the development of OVA-induced lung inflammation [78, 80]. However, it is doubtful that the OVA-sensitization and -challenge protocol accounts for qualitative differences in BAL eosinophils among *db/db*, *ob/ob*, and Cpe^{fat} mice since similar sensitization and challenge protocols were used in each of these studies [72, 74]. Second, because *db/db* and *ob/ob* mice, respectively, are genetically deficient in the long isoform of the leptin receptor or leptin, leptin signaling is impaired in these animals [81], yet in Cpe^{fat} mice, leptin signaling is intact [81]. Kato *et al.* [82] and Wong *et al.* [83] previously reported that leptin enhances eosinophil chemotaxis in humans. Thus anomalies in leptin signaling in *db/db* and *ob/ob* mice could lead to impaired eosinophil chemotaxis, which could certainly explain the reduced number of BAL eosinophils in these animals. However, as discussed later in this chapter, Shore *et al.* [84] reported that there were no differences in BAL eosinophils between OVA-sensitized and -challenged wild-type mice infused with either leptin or a vehicle control. Third, the degree of OVA-induced lung inflammation varies dramatically among inbred mouse strains [85], yet this too is a questionable explanation as to the differences in BAL eosinophils among these genetically obese mice because *db/db*, *ob/ob*, and Cpe^{fat} mice are each in a C57BL/6J genetic background [72–74]. Finally, based on the first three considerations, it is more probable that other mechanisms contribute

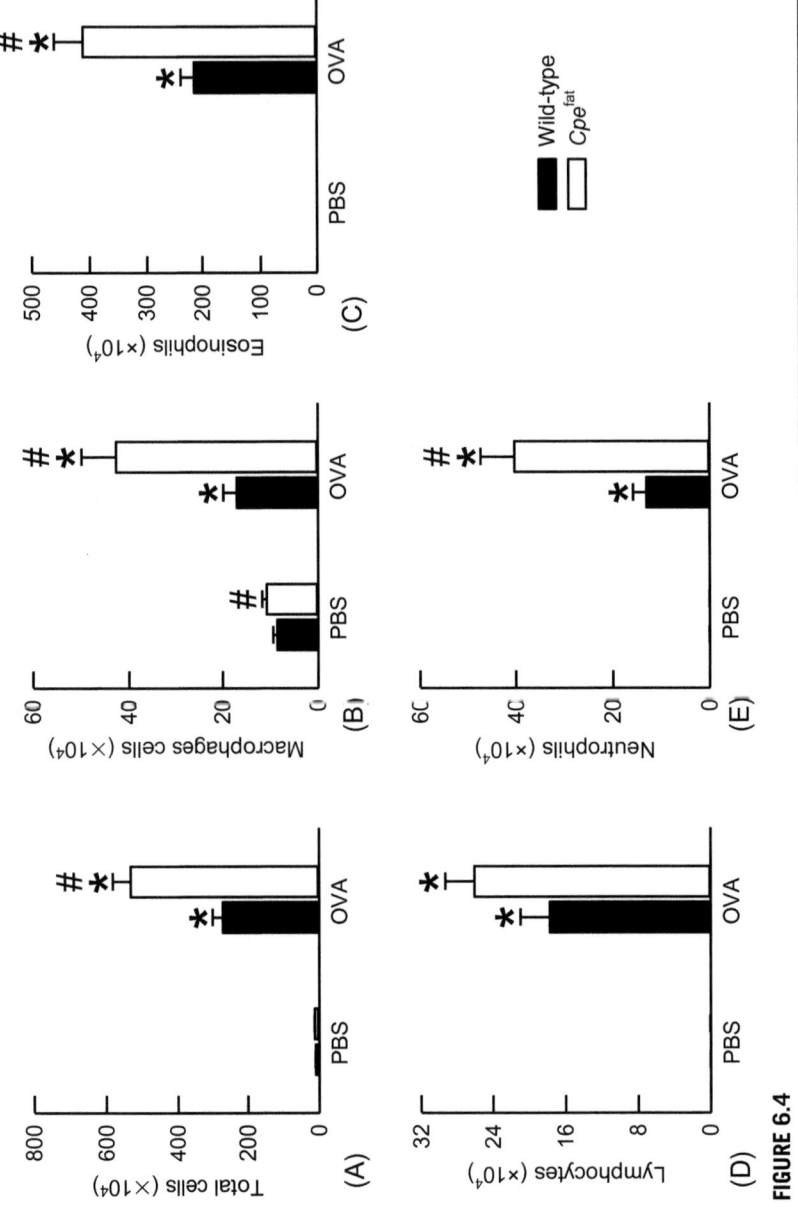

FIGURE 6.4

The number of total cells (A), macrophages (B), eosinophils (C), lymphocytes (D), and neutrophils (E) in bronchoalveolar lavage (BAL) fluid obtained from ovalbumin (OVA)-sensitized wild-type (C57BL/6) and Cpe^{fat} mice challenged with aerosols of either phosphate-buffered saline (PBS) or OVA. BAL fluid was obtained from mice 24 h following cessation of the final aerosol challenge. Each value is expressed as the mean ± SE. $n=9$–13 mice in each group. * $P<0.05$ compared with genotype-matched mice challenged with PBS. # $P<0.05$ compared with wild-type mice with an identical challenge.

This figure was taken from Dahm PH, Richards JB, Karmouty-Quintana H, Cromar KR, Sur S, Price RE, et al. Effect of antigen sensitization and challenge on oscillatory mechanics of the lung and pulmonary inflammation in obese carboxypeptidase E-deficient mice. Am J Physiol Integr Comp Physiol 2014;307(6):R621–33.

to qualitative differences in BAL eosinophils between genetically obese mice. One such mechanism may be the differential expression of endogenous antiinflammatory hormones among *db/db*, *ob/ob*, and *Cpe^{fat}* mice previously discussed by Dahm *et al.* [74]. In brief, because of a genetic deficiency in Cpe, *Cpe^{fat}* mice are unable to process prohormones [e.g., proopiomelanocortin and procholecystokinin] into biologically active antiinflammatory peptides (α-melanocyte-stimulating hormone and cholecystokinin, respectively) [81, 86, 87]. Because *db/db* and *ob/ob* mice express Cpe [81], these mice retain the ability to generate the aforementioned antiinflammatory hormones. Therefore qualitative differences in BAL eosinophils among these genetically obese mice may arise due to either the presence or absence of biologically active antiinflammatory hormones generated by the enzymatic activity of Cpe.

The effect of DIO on antigen-induced lung inflammation has also been assessed in mice, and Table 6.2 provides information pertinent to these investigations. There are 11 studies in Table 6.2 that differ according to (1) mouse strain (A/J, AKR/OlaHsd, BALB/c, and C57BL/6), (2) fat content of the high-fat diet (36%–60%), (3) percent difference in body mass between mice fed a high-fat compared to mice fed a standard diet (20%–75%), (4) the the number of wk consuming a high-fat diet (9–18 wk), and (5) the degree of antigen-induced lung inflammation. These studies also differed with respect to the antigen (cockroach allergen, house dust mite extract [HDME], or OVA) and/or the antigen sensitization and challenge protocol (data not shown). In eight of the studies in Table 6.2, investigators used C57BL/6 mice. Nevertheless, there were substantial qualitative differences in antigen-induced lung inflammation between C57BL/6 mice fed a high-fat diet and C57BL/6 mice fed a standard diet. For example, in C57BL/6 mice with DIO, BAL eosinophils were lower, higher, or not different compared with C57BL/6 mice fed a standard diet. In addition, DIO had variable effects on lung infiltrates of inflammatory cells in antigen-sensitized and -challenged C57BL/6 mice. A/J, AKR/OlaHsd, and BALB/c mice also had substantial qualitative differences in antigen-induced lung inflammation among each other and with C57BL/6 mice. Because antigen-induced lung inflammation is strain-dependent [85], heterogeneity in inflammatory responses among the aforementioned strains could arise from genetic differences. However, it is important to consider potential mechanisms for qualitative differences in antigen-induced lung inflammation that occur within the C57BL/6 strain as these mice are often used to model asthma.

Qualitative differences in antigen-induced lung inflammation among C57BL/6 mice with dietary obesity could arise for a number of reasons. First, because diets with varying amounts of fat differentially influence the development of inflammation [99], heterogeneity in antigen-induced lung inflammation among C57BL/6 mice with DIO may be consequent to the percentage of fat in the high-fat diet. Indeed, as demonstrated in Table 6.2, C57BL/6 mice consumed high-fats diets with different amounts of fat, and among these mice, there were variable degrees of lung inflammation. Second, Johnston *et al.* [14] demonstrated that C57BL/6 mice must consume a high-fat diet for at least 26 weeks before O_3-induced lung inflammation is greater than mice fed a standard diet, and the studies in Table 6.2

Table 6.2 Effect of Diet-Induced Obesity in Mice on Indices of Atopic Asthma[a]

Study	Strain	High-Fat Diet (% Fat)	Percent Difference in Body Mass[b]	Length of Time on High Fat Diet (wk)	BAL Eosinophils	Lung Tissue Inflammatory Cell Infiltrates	Airway Responsiveness	Total or Antigen Specific IgE
Mito et al. [88]	C57BL/6J	50	28	16	–[d]	Greater than standard diet	–	Less than standard diet
Calixto et al. [89]	C57BL/6J	55	32	10	Less than standard diet	Greater than standard diet	–	–
Ge et al. [90]	C57BL/6J	60	32[c]	15	Less than standard diet	Less than standard diet	–	–
Calixto et al. [91]	C57BL/6J	55	36[c]	10	Less than standard diet	Greater than standard diet	–	–
Kim et al. [92]	C57BL/6	60	75	16	No difference	–	Greater than standard diet	Greater than standard diet
Ather et al. [93]	C57BL/6J	60	35	16	No difference	–	Greater than standard diet	–
Everaere et al. [94]	C57BL/6J	60	40	11	Greater than standard diet	Greater than standard diet	Greater than standard diet	Greater than standard diet
Han et al. [95]	C57BL/6J	60	>20	15	Less than standard diet	Greater than standard diet	–	–
Saraiva et al. [96]	A/J	36	48	18	Greater than standard diet	Greater than standard diet	Greater than standard diet	–
Dietze et al. [97]	AKR/OlaHsd	40	35	9	–	Greater than standard diet	No difference	Greater than standard diet
Silva et al. [98]	BALB/c	60	21	10	Less than standard diet	Variable[e]	–	Less than standard diet

[a] Qualitative comparisons were made between antigen-sensitized and challenged wild-type mice fed either a high fat or standard diet.

[b] Percent difference in body mass between wild-type mice fed a high-fat compared to a standard diet.

[c] Percent difference in body mass between wild-type mice fed a high-fat compared to a standard diet had to be estimated from bar graphs since no body masses were provided in the text.

[d] Indice was not measured.

[e] The degree of lung tissue inflammatory cell infiltration varied according to the cell type (eosinophil, mast cell, neutrophil).

that used C57BL/6 mice report that these animals only consumed high-fat diets for 10–16 wk. Therefore, the effect of dietary obesity on antigen-induced lung inflammation in C57BL/6 mice might have been more consistent if the mice consumed a high fat diet for a longer period of time. Third, Wang *et al.* [100] demonstrated that different sources of fat influence the degree of adipose tissue inflammation. Thus the source of fat within high fat diets could account for qualitative differences in antigen-induced lung inflammation in C57BL/6 mice with dietary obesity. Indeed, Mito *et al.* [88] demonstrated differences in airway inflammation among OVA-sensitized and -challenged C57BL/6 mice fed diets in which fat was derived from either lard or soybean oil. Fourth, as mentioned earlier, the antigen sensitization and challenge protocol influences the development of antigen-induced lung inflammation [78, 80], and among the studies found in Table 6.2, antigen sensitization and challenge protocols varied dramatically (data not shown).

In summary, the effect of either genetic or dietary obesity on antigen-induced lung inflammation in mice is quite variable. However, the effect of obesity on biomarkers of antigen-induced lung inflammation in human asthmatic subjects is also very heterogeneous. For example, van Veen *et al.* [101] and Lessard *et al.* [102] reported an inverse relationship between sputum eosinophils and BMI, whereas Desai *et al.* [103], Sutherland *et al.* [104], and Todd *et al.* [105] reported no relationship between BMI and sputum eosinophils. Although Desai *et al.* [103] reported no effect of BMI on sputum eosinophils, these investigators did demonstrate that obese asthmatics had more submucosal eosinophils in the lungs. Thus even among obese human asthmatics, there are qualitative differences in lung eosinophils, which suggest the existence of subpopulations of atopic obese asthmatics with variable degrees of lung inflammation. Therefore, when attempting to model these subpopulations of human asthmatics in obese mice, it is imperative to select the appropriate mouse model.

OBESITY AND ANTIGEN-INDUCED AHR

AHR to nonspecific bronchoconstrictors, including histamine and methacholine, is a characteristic feature of asthma [106]. However, to the best of our knowledge, no investigators have compared airway responsiveness in obese and nonobese human asthmatics who are exclusively atopic. Nevertheless, several investigators have examined the effect of obesity on airway responsiveness or basal lung mechanics in mice following sensitization to and challenge with antigen (HDME or OVA). Johnston *et al.* [72] were the first to report that OVA-sensitized and -challenged *db/db* and *ob/ob* mice display greater airway responsiveness to methacholine compared with OVA-sensitized and PBS-challenged genotype-matched mice and OVA-sensitized and -challenged wild-type mice (Figure 6.5). Subsequent to the report of Johnston *et al.* [72], Kim *et al.* [92], Ather *et al.* [93], Saraiva, *et al.* [96], and Everaere *et al.* [94] made similar observations with regard to airway responsiveness in antigen-sensitized and -challenged mice with DIO. Furthermore,

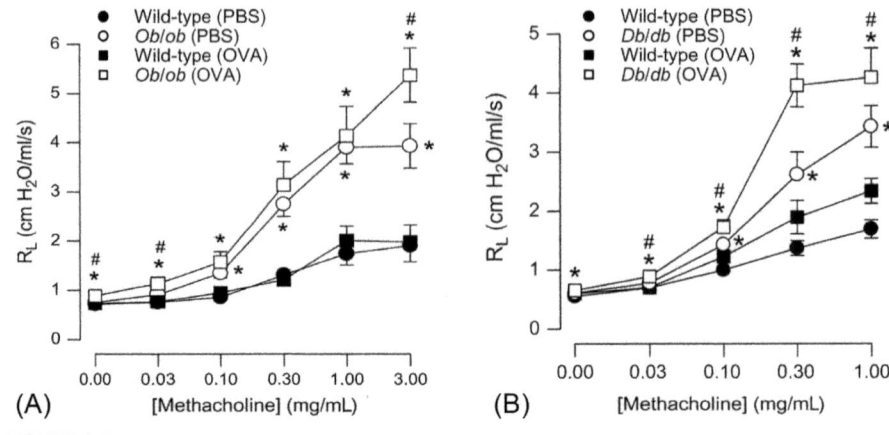

FIGURE 6.5

Changes in pulmonary resistance (R_L) induced by intravenous methacholine in ovalbumin (OVA)-sensitized wild-type (C57BL/6) and (A) *ob/ob* and (B) *db/db* mice challenged with aerosols of either phosphate-buffered saline (PBS) or OVA. Responses were measured 24 h after cessation of the final aerosol challenge. $n=6$–10 mice for each group. *$P<0.05$ compared with wild-type mice with an identical exposure; #$P<0.05$ compared with genotype-matched, PBS-challenged controls.

These figures were adapted and reprinted with permission of the American Thoracic Society. Copyright © 2018 American Thoracic Society. Johnston RA, Zhu M, Rivera-Sanchez YM, Lu FL, Theman TA, Flynt L, Shore SA. Allergic airway responses in obese mice. Am J Respir Crit Care Med 2007;176:650–8. The American Journal of Respiratory and Critical Care Medicine is an official journal of the American Thoracic Society.

Kim *et al.* [92] and Ather *et al.* [93] reported that weight loss in mice with DIO decreased airway responsiveness, which is consistent with data from human subjects [107]. In contrast to the aforementioned studies, Dietze *et al.* [97] observed no difference in airway responsiveness between OVA-sensitized and -challenged mice fed a high-fat compared with a standard diet. Finally, Dahm *et al.* [74] measured basal lung mechanics in *Cpe^fat* and wild-type mice following OVA sensitization and challenge. These investigators reported that increases in basal airway and parenchymal oscillation mechanics following OVA sensitization and challenge were greater in *Cpe^fat* compared with wild-type mice. As discussed in the preceding text, the effect of obesity on antigen-induced lung inflammation is variable, yet obesity consistently enhances antigen-induced increases in airway obstruction in mice. For a summary of the effects of genetic or dietary obesity on antigen-induced airway obstruction, please refer to Tables 6.1 and 6.2. Finally, when attempting to model subpopulations of obese human asthmatics with an atopic component, investigators should choose a mouse model based on the inflammatory phenotype of interest since increases in antigen-induced airway obstruction are enhanced in obese mice regardless of the modality of obesity induction or degree of lung inflammation.

POTENTIAL MECHANISMS FOR GREATER ASTHMA SEVERITY IN ATOPIC OBESE ASTHMATICS: INTERPRETATIONS FROM ANIMAL STUDIES

As discussed earlier in this chapter, investigators have examined the role of IL-33, IL-17A, and TNF-α as potential mediators to enhance lung inflammation and increases in airway responsiveness in obese mice following acute exposure to O_3, a model of irritant-induced nonatopic asthma. Thus far, and to the best of our knowledge, investigators have only examined the effects of TNF-α and ILC as molecular mechanisms by which increases in airway responsiveness are enhanced in obese mice following sensitization to and challenge with antigen. Nevertheless, other investigators have reported the effects of immunoglobulin E (IgE) or adipokines (adiponectin and leptin) on antigen-induced lung inflammation and AHR in wild-type mice. Therefore in the remaining portion of this chapter, we shall summarize the results of these studies that may ultimately lead to therapies to improve asthma symptoms and severity in atopic obese asthmatics.

Role of TNF-α

Based on data from both human and animal studies, there is strong evidence to suggest TNF-α may contribute to enhanced lung inflammation and increases in airway responsiveness observed in obese mice following antigen sensitization and challenge. First, exogenous administration of TNF-α to human subjects causes lung inflammation and AHR [29], which are characteristic features of asthma. Second, chronic systemic inflammation in obesity, which is characterized, in part, by increased levels of circulating TNF-α, occurs temporally with the development of innate AHR in obese mice [16, 50]. Third, BAL or lung tissue TNF-α is increased to a greater extent in obese compared with lean mice following OVA sensitization and challenge [73, 91, 92, 108], and furthermore, genetic deficiency of either TNF-α or TNFR1 prevents the development of OVA-induced inflammation and AHR in wild-type mice [109, 110]. Fourth, administration of etanercept, a TNF-α inhibitor, reduced airway responsiveness to methacholine in humans with refractory asthma, which is common in obesity and which has an atopic component [111–115]. With that said, two separate groups of investigators examined the impact of TNF-α inhibition on OVA-induced inflammation and/or AHR in mice with DIO.

Calixto *et al.* [91] administered a monoclonal TNF-α neutralizing antibody to OVA-sensitized lean and obese mice 1 hour prior to two separate intranasal OVA challenges. In OVA-sensitized and -challenged lean and obese mice treated with an isotype control IgG antibody, the total number of eosinophils were greater in the lung tissue of obese compared with lean mice. Administration of the TNF-α antibody significantly decreased the total number of eosinophils in the lung tissue of lean and obese mice [91]. Kim *et al.* [92] reported that a polyclonal TNF-α neutralizing antibody significantly reduced airway responsiveness to methacholine in OVA-sensitized and -challenged mice with DIO. As described earlier, the role of TNF-α and TNFR2 in the development of O_3-induced lung inflammation and AHR in obese mice is quite complex. However, inhibition of TNF-α signaling reduces lung inflammation and

AHR in antigen-sensitized and -challenged obese mice. Given that acute exposure to O_3 induces an innate immune response whereas OVA sensitization and challenge induces an adaptive immune response, the role of TNF-α and/or its receptors in innate versus adaptive immunity in obese mice may be different. Consequently, TNF-α inhibitors, such as etanercept, may be useful in treating obese asthmatics with a substantial atopic component.

Role of ILC

Kim *et al.* [116] previously reported that IL-17A$^+$ ILC3 contribute to the development of innate AHR in mice with DIO, and as discussed earlier in this chapter, it is probable that release of Th-2 cytokines from ILC2 in response to IL-33 enhances O_3-induced lung inflammation in obese mice [18]. Consistent with these data, Everaere *et al.* [94] reported that depletion of ILC with an anti-CD90.2 monoclonal antibody reduced lung inflammation and increases in airway responsiveness in OVA-sensitized and -challenged C57BL/6J mice with DIO. In addition, because the anti-CD90.2 monoclonal antibody can deplete T cells, Everaere *et al.* [94] demonstrated through a series of reconstitution experiments that ILC and not T cells exacerbate antigen-induced lung inflammation and increases in airway responsiveness in mice with DIO. Given that ILC appear to contribute to the exacerbation of lung inflammation and increases in airway responsiveness in obese mouse models of atopic and nonatopic asthma, these cells appear to be a promising target for therapies designed to specifically treat obese asthmatics regardless of their phenotype.

Role of IgE

Atopy is a significant risk factor for asthma [117–119], and in humans, there is a positive correlation between serum levels of total or antigen-specific IgE and asthma severity [120–123]. Thus it is reasonable to speculate that obesity may increase the severity of antigen-induced lung inflammation and increases in airway responsiveness via IgE. However, there is conflicting data as to whether obesity is positively associated with atopy or serum IgE [124–132]. Furthermore, the contribution of IgE to antigen-induced AHR and/or lung inflammation in wild-type mice is unclear [133–136]. In seven of the studies found in either Table 6.1 or 6.2, the effect of genetic or dietary obesity on antigen-induced increases in serum total or antigen-specific IgE was assessed. Consistent with the effect of obesity on antigen-induced lung inflammation, obesity had variable effects on antigen-induced increases in either serum total or antigen-specific IgE, which did not necessarily correlate with changes in indices of antigen-induced lung inflammation or airway obstruction. Thus the ability of obesity to worsen asthma symptoms in atopic asthmatics via an IgE-dependent mechanism remains unclear, and further research into the role of IgE in atopic obese asthmatics needs to be initiated.

Role of adiponectin

Circulating levels of adiponectin—an adipocyte-derived, insulin-sensitizing, antiinflammatory hormone—decrease in obesity [17, 137–141]. Because adiponectin decreases in obesity and primarily has antiinflammatory effects, this prompted investigators to examine the effects of adiponectin or T-cadherin, an extracellular adiponectin binding protein, on antigen-induced lung inflammation and AHR in wild-type mice [142–146]. Although none of the subsequently described experiments included obese mice, these studies do provide important information with regard to the use of adiponectin or adiponectin analogues as potential therapeutics to treat atopic lean and obese asthmatics.

Shore *et al.* [144] were the first to report the effects of adiponectin on lung inflammation and AHR following antigen sensitization and challenge. Specifically, administration of exogenous adiponectin to OVA-sensitized and -challenged BALB/c mice reduced antigen-induced lung inflammation and prevented OVA-induced AHR. Consistent with these results, Verbout *et al.* [146] reported that OVA-induced lung inflammation was reduced in transgenic mice that overexpress adiponectin. Nevertheless, genetic deficiency of adiponectin had no effect on lung inflammation following sensitization to and acute challenge with OVA [143, 145]. Following chronic challenge with OVA, however, Medoff *et al.* [145] demonstrated that OVA-induced lung inflammation and decreases in respiratory system compliance were exacerbated in mice genetically deficient in adiponectin. Finally, Williams *et al.* [143] reported that elevations in serum adiponectin via genetic deficiency of T-cadherin reduced OVA-induced lung inflammation and prevented the development of OVA-induced AHR. The protective effects of T-cadherin deficiency were abolished in mice genetically deficient in both T-cadherin and adiponectin [143], which suggests that the beneficial effects of T-cadherin deficiency required adiponectin. Human data demonstrate that high serum or sputum adiponectin are associated with a lower prevalence of asthma or fewer asthma symptoms and exacerbations [147, 148]. Furthermore, serum adiponectin is positively associated with FEF_{25-75} and FEV_1/FVC yet negatively associated with maximum decreases in FEV_1 following exercise [147, 149, 150]. Taken together, decreases in adiponectin in both humans and mice appear to be associated with worse outcomes related to asthma. Therefore increasing adiponectin in serum or lungs may be a beneficial treatment for lean and obese asthmatics with an atopic component.

Role of leptin

Serum levels of leptin, a satiety hormone and proinflammatory cytokine, are elevated in human and murine obesity [74, 151, 152]. These observations led Shore *et al.* [84] to investigate the potential contribution of leptin to the pathogenesis of asthma secondary to obesity. To that end, Shore *et al.* [84] administered exogenous leptin to OVA-sensitized and -challenged wild-type mice and demonstrated that leptin enhanced OVA-induced increases in airway responsiveness and serum IgE but had no effect on BAL eosinophils or Th-2 cytokines. Although leptin enhanced OVA-induced increases in airway responsiveness in wild-type mice, it is doubtful that leptin exerts the same effect in obese mice since OVA-induced increases in airway responsiveness are enhanced in obese mice with impaired leptin signaling (*db/db* and *ob/ob* mice) [72]. In humans, serum or plasma leptin inversely correlates with FEV_1 and/or FEF_{25-75} and positively associates with asthma or asthma symptoms and severity [149, 150, 153–157]. Furthermore, in asthmatics, Sideleva *et al.* [158] demonstrated that, as visceral fat leptin expression increases, sensitivity to methacholine also increases. However, this relationship was independent of BMI. Taken together, it is improbable that leptin contributes to the exacerbation of antigen-induced increases in airway responsiveness in obese mice yet may play a role in the development of AHR in human asthmatics.

CONCLUSIONS

Animal models of atopic and nonatopic asthma recapitulate many of the characteristic features of these asthma phenotypes in humans. Furthermore, increases in airway responsiveness induced by either atopic or nonatopic stimuli are exacerbated in obese mice, which is a phenotypic characteristic of obese human asthmatics. Consequently, obese mice are useful tools to enhance our understanding of the relationship between obesity and asthma. Because acceptable asthma control is hard to attain in obese asthmatics, preclinical animal studies will help identify new therapeutics to treat obese asthmatics. As discussed earlier in this chapter, animal studies demonstrate that effective therapies for obese asthmatics may differ according to atopic status. For example, therapeutics directed against IL-17A and IL-33 may ultimately prove effective in nonatopic obese asthmatics, whereas adiponectin or anti-TNF analogues may be useful to treat atopic obese asthmatics. Nevertheless, animal studies demonstrate that ILC appear to be important in the pathogenesis of atopic and nonatopic obese asthma. Therefore therapies directed against ILC may be successful in treating obese asthmatics regardless of the presence or absence of atopy.

ACKNOWLEDGMENTS

A portion of the research presented in this chapter was supported by the National Institute of Environmental Health Sciences of the National Institutes of Health under award numbers R03ES022378 (R.A. Johnston), R21ES024032 (S.A. Shore), R01ES013307 (S.A. Shore), and P30ES000002 (D.W. Dockery), and by the National Institute of Allergy and Infectious Diseases of the National Institutes of Health under award number R03AI107432 (R.A. Johnston).

REFERENCES

[1] Wenzel SE. Emergence of biomolecular pathways to define novel asthma phenotypes. Type-2 immunity and beyond. Am J Respir Cell Mol Biol 2016;55(1):1–4.
[2] Dixon AE, Pratley RE, Forgione PM, Kaminsky DA, Whittaker-Leclair LA, Griffes LA, et al. Effects of obesity and bariatric surgery on airway hyperresponsiveness, asthma control, and inflammation. J Allergy Clin Immunol 2011;128(3):508–15. e1–2.

[3] Dixon AE, Shade DM, Cohen RI, Skloot GS, Holbrook JT, Smith LJ, et al. Effect of obesity on clinical presentation and response to treatment in asthma. J Asthma 2006;43(7):553–8.

[4] Pradeepan S, Garrison G, Dixon AE. Obesity in asthma: approaches to treatment. Curr Allergy Asthma Rep 2013;13(5):434–42.

[5] Sutherland ER, Goleva E, Strand M, Beuther DA, Leung DY. Body mass and glucocorticoid response in asthma. Am J Respir Crit Care Med 2008;178(7):682–7.

[6] Stupfel M. Recent advances in investigations of toxicity of automotive exhaust. Environ Health Perspect 1976;17:253–85.

[7] Gent JF, Triche EW, Holford TR, Belanger K, Bracken MB, Beckett WS, et al. Association of low-level ozone and fine particles with respiratory symptoms in children with asthma. JAMA 2003;290(14):1859–67.

[8] Ji M, Cohan DS, Bell ML. Meta-analysis of the association between short-term exposure to ambient ozone and respiratory hospital admissions. Environ Res Lett 2011;6(2).

[9] Foster WM, Brown RH, Macri K, Mitchell CS. Bronchial reactivity of healthy subjects: 18-20 h postexposure to ozone. J Appl Physiol (1985) 2000;89(5):1804–10.

[10] Dong GH, Qian Z, Liu MM, Wang D, Ren WH, Fu Q, et al. Obesity enhanced respiratory health effects of ambient air pollution in Chinese children: the seven northeastern cities study. Int J Obes (Lond) 2013;37(1):94–100.

[11] Bennett WD, Hazucha MJ, Folinsbee LJ, Bromberg PA, Kissling GE, London SJ. Acute pulmonary function response to ozone in young adults as a function of body mass index. Inhal Toxicol 2007;19(14):1147–54.

[12] Bennett WD, Ivins S, Alexis NE, Wu J, Bromberg PA, Brar SS, et al. Effect of obesity on acute ozone-induced changes in airway function, reactivity, and inflammation in adult females. PLoS One 2016;11(8):e0160030.

[13] Alexeeff SE, Litonjua AA, Suh H, Sparrow D, Vokonas PS, Schwartz J. Ozone exposure and lung function: effect modified by obesity and airways hyperresponsiveness in the VA normative aging study. Chest 2007;132(6):1890–7.

[14] Johnston RA, Theman TA, Lu FL, Terry RD, Williams ES, Shore SA. Diet-induced obesity causes innate airway hyperresponsiveness to methacholine and enhances ozone-induced pulmonary inflammation. J Appl Physiol (1985) 2008;104(6):1727–35.

[15] Johnston RA, Theman TA, Shore SA. Augmented responses to ozone in obese carboxypeptidase E-deficient mice. Am J Physiol Regul Integr Comp Physiol 2006;290(1):R126–33.

[16] Johnston RA, Zhu M, Hernandez CB, Williams ES, Shore SA. Onset of obesity in carboxypeptidase E-deficient mice and effect on airway responsiveness and pulmonary responses to ozone. J Appl Physiol (1985) 2010;108(6):1812–9.

[17] Lu FL, Johnston RA, Flynt L, Theman TA, Terry RD, Schwartzman IN, et al. Increased pulmonary responses to acute ozone exposure in obese db/db mice. Am J Physiol Lung Cell Mol Physiol 2006;290(5):L856–65.

[18] Mathews JA, Krishnamoorthy N, Kasahara DI, Cho Y, Wurmbrand AP, Ribeiro L, et al. IL-33 drives augmented responses to ozone in obese mice. Environ Health Perspect 2017;125(2):246–53.

[19] Mathews JA, Krishnamoorthy N, Kasahara DI, Hutchinson J, Cho Y, Brand JD, et al. Augmented responses to ozone in obese mice require IL-17A and gastrin-releasing peptide. Am J Respir Cell Mol Biol 2018;58(3):341–51.

[20] Shore SA, Rivera-Sanchez YM, Schwartzman IN, Johnston RA. Responses to ozone are increased in obese mice. J Appl Physiol (1985) 2003;95(3):938–45.

[21] Williams AS, Mathews JA, Kasahara DI, Chen L, Wurmbrand AP, Si H, et al. Augmented pulmonary responses to acute ozone exposure in obese mice: roles of TNFR2 and IL-13. Environ Health Perspect 2013;121(5):551–7.

[22] Williams AS, Mathews JA, Kasahara DI, Wurmbrand AP, Chen L, Shore SA. Innate and ozone-induced airway hyperresponsiveness in obese mice: role of TNF-alpha. Am J Physiol Lung Cell Mol Physiol 2015;308(11):L1168–77.

[23] Bromberg PA. Mechanisms of the acute effects of inhaled ozone in humans. Biochim Biophys Acta 2016;1860(12):2771–81.

[24] Pino MV, Levin JR, Stovall MY, Hyde DM. Pulmonary inflammation and epithelial injury in response to acute ozone exposure in the rat. Toxicol Appl Pharmacol 1992;112(1):64–72.

[25] Lang JE, Williams ES, Mizgerd JP, Shore SA. Effect of obesity on pulmonary inflammation induced by acute ozone exposure: role of interleukin-6. Am J Physiol Lung Cell Mol Physiol 2008;294(5):L1013–20.

[26] Kasahara DI, Mathews JA, Park CY, Cho Y, Hunt G, Wurmbrand AP, et al. ROCK insufficiency attenuates ozone-induced airway hyperresponsiveness in mice. Am J Physiol Lung Cell Mol Physiol 2015;309(7):L736–46.

[27] Lajoie S, Lewkowich IP, Suzuki Y, Clark JR, Sproles AA, Dienger K, et al. Complement-mediated regulation of the IL-17A axis is a central genetic determinant of the severity of experimental allergic asthma. Nat Immunol 2010;11(10):928–35.

[28] Mizutani N, Nabe T, Yoshino S. IL-17A promotes the exacerbation of IL-33-induced airway hyperresponsiveness by enhancing neutrophilic inflammation via CXCR2 signaling in mice. J Immunol 2014;192(4):1372–84.

[29] Thomas PS, Yates DH, Barnes PJ. Tumor necrosis factor-alpha increases airway responsiveness and sputum neutrophilia in normal human subjects. Am J Respir Crit Care Med 1995;152(1):76–80.

[30] Johnston RA, Mizgerd JP, Shore SA. CXCR2 is essential for maximal neutrophil recruitment and methacholine responsiveness after ozone exposure. Am J Physiol Lung Cell Mol Physiol 2005;288(1):L61–7.

[31] Moffatt MF, Gut IG, Demenais F, Strachan DP, Bouzigon E, Heath S, et al. A large-scale, consortium-based genomewide association study of asthma. N Engl J Med 2010;363(13):1211–21.

[32] Cayrol C, Girard JP. IL-33: an alarmin cytokine with crucial roles in innate immunity, inflammation and allergy. Curr Opin Immunol 2014;31:31–7.

[33] Kumagai K, Lewandowski R, Jackson-Humbles DN, Li N, Van Dyken SJ, Wagner JG, et al. Ozone-induced nasal type 2 immunity in mice is dependent on innate lymphoid cells. Am J Respir Cell Mol Biol 2016;54(6):782–91.

[34] Yang Q, Ge MQ, Kokalari B, Redai IG, Wang X, Kemeny DM, et al. Group 2 innate lymphoid cells mediate ozone-induced airway inflammation and hyperresponsiveness in mice. J Allergy Clin Immunol 2016;137(2):571–8.

[35] Johnston RA, Schwartzman IN, Flynt L, Shore SA. Role of interleukin-6 in murine airway responses to ozone. Am J Physiol Lung Cell Mol Physiol 2005;288(2):L390–7.

[36] Schmitz J, Owyang A, Oldham E, Song Y, Murphy E, McClanahan TK, et al. IL-33, an interleukin-1-like cytokine that signals via the IL-1 receptor-related protein ST2 and induces T helper type 2-associated cytokines. Immunity 2005;23(5):479–90.

[37] Barlow JL, Peel S, Fox J, Panova V, Hardman CS, Camelo A, et al. IL-33 is more potent than IL-25 in provoking IL-13-producing nuocytes (type 2 innate lymphoid cells) and airway contraction. J Allergy Clin Immunol 2013;132(4):933–41.

[38] Yagami A, Orihara K, Morita H, Futamura K, Hashimoto N, Matsumoto K, et al. IL-33 mediates inflammatory responses in human lung tissue cells. J Immunol 2010;185(10):5743–50.

[39] Yang Z, Grinchuk V, Urban Jr. JF, Bohl J, Sun R, Notari L, et al. Macrophages as IL-25/IL-33-responsive cells play an important role in the induction of type 2 immunity. PLoS One 2013;8(3):e59441.

[40] Kasahara DI, Kim HY, Mathews JA, Verbout NG, Williams AS, Wurmbrand AP, et al. Pivotal role of IL-6 in the hyperinflammatory responses to subacute ozone in adiponectin-deficient mice. Am J Physiol Lung Cell Mol Physiol 2014;306(6):L508–20.

[41] McCullough SD, Duncan KE, Swanton SM, Dailey LA, Diaz-Sanchez D, Devlin RB. Ozone induces a proinflammatory response in primary human bronchial epithelial cells through mitogen-activated protein kinase activation without nuclear factor-kappaB activation. Am J Respir Cell Mol Biol 2014;51(3):426–35.

[42] Marijsse GS, Seys SF, Schelpe AS, Dilissen E, Goeminne P, Dupont LJ, et al. Obese individuals with asthma preferentially have a high IL-5/IL-17A/IL-25 sputum inflammatory pattern. Am J Respir Crit Care Med 2014;189(10):1284–5.

[43] Pichavant M, Goya S, Meyer EH, Johnston RA, Kim HY, Matangkasombut P, et al. Ozone exposure in a mouse model induces airway hyperreactivity that requires the presence of natural killer T cells and IL-17. J Exp Med 2008;205(2):385–93.

[44] Iwakura Y, Ishigame H. The IL-23/IL-17 axis in inflammation. J Clin Invest 2006;116(5):1218–22.

[45] Laan M, Cui ZH, Hoshino H, Lotvall J, Sjostrand M, Gruenert DC, et al. Neutrophil recruitment by human IL-17 via C-X-C chemokine release in the airways. J Immunol 1999;162(4):2347–52.

[46] Shore SA, Williams ES, Zhu M. No effect of metformin on the innate airway hyperresponsiveness and increased responses to ozone observed in obese mice. J Appl Physiol (1985) 2008;105(4):1127–33.

[47] Johnson DE, Georgieff MK. Pulmonary neuroendocrine cells. Their secretory products and their potential roles in health and chronic lung disease in infancy. Am Rev Respir Dis 1989;140(6):1807–12.

[48] Busse WW, Holgate S, Kerwin E, Chon Y, Feng J, Lin J, et al. Randomized, double-blind, placebo-controlled study of brodalumab, a human anti-IL-17 receptor monoclonal antibody, in moderate to severe asthma. Am J Respir Crit Care Med 2013;188(11):1294–302.

[49] Tilg H, Moschen AR. Adipocytokines: mediators linking adipose tissue, inflammation and immunity. Nat Rev Immunol 2006;6(10):772–83.

[50] Williams AS, Chen L, Kasahara DI, Si H, Wurmbrand AP, Shore SA. Obesity and airway responsiveness: role of TNFR2. Pulm Pharmacol Ther 2013;26(4):444–54.

[51] Bouter B, Geary N, Langhans W, Asarian L. Diet-genotype interactions in the early development of obesity and insulin resistance in mice with a genetic deficiency in tumor necrosis factor-alpha. Metabolism 2010;59(7):1065–73.

[52] Liang H, Yin B, Zhang H, Zhang S, Zeng Q, Wang J, et al. Blockade of tumor necrosis factor (TNF) receptor type 1-mediated TNF-alpha signaling protected Wistar rats from diet-induced obesity and insulin resistance. Endocrinology 2008;149(6):2943–51.

[53] Park EJ, Lee JH, Yu GY, He G, Ali SR, Holzer RG, et al. Dietary and genetic obesity promote liver inflammation and tumorigenesis by enhancing IL-6 and TNF expression. Cell 2010;140(2):197–208.

[54] Uysal KT, Wiesbrock SM, Hotamisligil GS. Functional analysis of tumor necrosis factor (TNF) receptors in TNF-alpha-mediated insulin resistance in genetic obesity. Endocrinology 1998;139(12):4832–8.

[55] Uysal KT, Wiesbrock SM, Marino MW, Hotamisligil GS. Protection from obesity-induced insulin resistance in mice lacking TNF-alpha function. Nature 1997;389(6651):610–4.

[56] Castro-Giner F, Kogevinas M, Imboden M, de Cid R, Jarvis D, Machler M, et al. Joint effect of obesity and TNFA variability on asthma: two international cohort studies. Eur Respir J 2009;33(5):1003–9.

[57] Cho HY, Zhang LY, Kleeberger SR. Ozone-induced lung inflammation and hyperreactivity are mediated via tumor necrosis factor-alpha receptors. Am J Physiol Lung Cell Mol Physiol 2001;280(3):L537–46.

[58] Shore SA, Schwartzman IN, Le Blanc B, Murthy GG, Doerschuk CM. Tumor necrosis factor receptor 2 contributes to ozone-induced airway hyperresponsiveness in mice. Am J Respir Crit Care Med 2001;164(4):602–7.

[59] Mathews JA, Kasahara DI, Cho Y, Bell LN, Gunst PR, Karoly ED, et al. Effect of acute ozone exposure on the lung metabolomes of obese and lean mice. PLoS One. 2017;12(7):e0181017.

[60] Yang H, Magilnick N, Ou X, Lu SC. Tumour necrosis factor alpha induces co-ordinated activation of rat GSH synthetic enzymes via nuclear factor kappaB and activator protein-1. Biochem J 2005;391(Pt 2):399–408.

[61] van Luijn JC, Danz M, Bijlsma JW, Gribnau FW, Leufkens HG. Post-approval trials of new medicines: widening use or deepening knowledge? Analysis of 10 years of etanercept. Scand J Rheumatol 2011;40(3):183–91.

[62] Romanet-Manent S, Charpin D, Magnan A, Lanteaume A, Vervloet D, Group EC. Allergic vs nonallergic asthma: what makes the difference? Allergy 2002;57(7):607–13.

[63] Murdoch JR, Lloyd CM. Chronic inflammation and asthma. Mutat Res 2010;690(1–2):24–39.

[64] Monticelli LA, Sonnenberg GF, Abt MC, Alenghat T, Ziegler CG, Doering TA, et al. Innate lymphoid cells promote lung-tissue homeostasis after infection with influenza virus. Nat Immunol 2011;12(11):1045–54.

[65] Licona-Limon P, Kim LK, Palm NW, Flavell RA. TH2, allergy and group 2 innate lymphoid cells. Nat Immunol 2013;14(6):536–42.

[66] Shimbara A, Christodoulopoulos P, Soussi-Gounni A, Olivenstein R, Nakamura Y, Levitt RC, et al. IL-9 and its receptor in allergic and nonallergic lung disease: increased expression in asthma. J Allergy Clin Immunol 2000;105(1 Pt 1):108–15.

[67] Kotsimbos TC, Ernst P, Hamid QA. Interleukin-13 and interleukin-4 are coexpressed in atopic asthma. Proc Assoc Am Physicians 1996;108(5):368–73.

[68] Hamid Q, Azzawi M, Ying S, Moqbel R, Wardlaw AJ, Corrigan CJ, et al. Expression of mRNA for interleukin-5 in mucosal bronchial biopsies from asthma. J Clin Invest 1991;87(5):1541–6.

[69] Hamelmann E, Takeda K, Haczku A, Cieslewicz G, Shultz L, Hamid Q, et al. Interleukin (IL)-5 but not immunoglobulin E reconstitutes airway inflammation and airway hyperresponsiveness in IL-4-deficient mice. Am J Respir Cell Mol Biol 2000;23(3):327–34.

[70] Kasaian MT, Miller DK. IL-13 as a therapeutic target for respiratory disease. Biochem Pharmacol 2008;76(2):147–55.

[71] Kung TT, Luo B, Crawley Y, Garlisi CG, Devito K, Minnicozzi M, et al. Effect of anti-mIL-9 antibody on the development of pulmonary inflammation and airway hyper-responsiveness in allergic mice. Am J Respir Cell Mol Biol 2001;25(5):600–5.

[72] Johnston RA, Zhu M, Rivera-Sanchez YM, Lu FL, Theman TA, Flynt L, et al. Allergic airway responses in obese mice. Am J Respir Crit Care Med 2007;176(7):650–8.

[73] Lintomen L, Calixto MC, Schenka A, Antunes E. Allergen-induced bone marrow eosin-ophilopoiesis and airways eosinophilic inflammation in leptin-deficient Ob/Ob mice. Obesity (Silver Spring) 2012;20(10):1959–65.

[74] Dahm PH, Richards JB, Karmouty-Quintana H, Cromar KR, Sur S, Price RE, et al. Effect of antigen sensitization and challenge on oscillatory mechanics of the lung and pulmonary inflammation in obese carboxypeptidase E-deficient mice. Am J Physiol Regul Integr Comp Physiol 2014;307(6):R621–33.

[75] Kumar RK, Herbert C, Foster PS. The "classical" ovalbumin challenge model of asthma in mice. Curr Drug Targets 2008;9(6):485–94.

[76] Jose PJ, Griffiths-Johnson DA, Collins PD, Walsh DT, Moqbel R, Totty NF, et al. Eotaxin: a potent eosinophil chemoattractant cytokine detected in a Guinea pig model of allergic airways inflammation. J Exp Med 1994;179(3):881–7.

[77] Wardlaw AJ. Eosinophils in the 1990s: new perspectives on their role in health and disease. Postgrad Med J 1994;70(826):536–52.

[78] Brusselle GG, Kips JC, Tavernier JH, van der Heyden JG, Cuvelier CA, Pauwels RA, et al. Attenuation of allergic airway inflammation in IL-4 deficient mice. Clin Exp Allergy 1994;24(1):73–80.

[79] Grunig G, Warnock M, Wakil AE, Venkayya R, Brombacher F, Rennick DM, et al. Requirement for IL-13 independently of IL-4 in experimental asthma. Science 1998;282(5397):2261–3.

[80] Taube C, Wei X, Swasey CH, Joetham A, Zarini S, Lively T, et al. Mast cells, Fc epsilon RI, and IL-13 are required for development of airway hyperresponsiveness after aerosol-ized allergen exposure in the absence of adjuvant. J Immunol 2004;172(10):6398–406.

[81] Leibel RL, Chung WK, Chua Jr. SC. The molecular genetics of rodent single gene obe-sities. J Biol Chem 1997;272(51):31937–40.

[82] Kato H, Ueki S, Kamada R, Kihara J, Yamauchi Y, Suzuki T, et al. Leptin has a prim-ing effect on eotaxin-induced human eosinophil chemotaxis. Int Arch Allergy Immunol 2011;155(4):335–44.

[83] Wong CK, Cheung PF, Lam CW. Leptin-mediated cytokine release and migration of eosinophils: Implications for immunopathophysiology of allergic inflammation. Eur J Immunol 2007;37(8):2337–48.

[84] Shore SA, Schwartzman IN, Mellema MS, Flynt L, Imrich A, Johnston RA. Effect of leptin on allergic airway responses in mice. J Allergy Clin Immunol 2005;115(1):103–9.

[85] Whitehead GS, Walker JK, Berman KG, Foster WM, Schwartz DA. Allergen-induced airway disease is mouse strain dependent. Am J Physiol Lung Cell Mol Physiol 2003;285(1):L32–42.

[86] Bohm M, Apel M, Sugawara K, Brehler R, Jurk K, Luger TA, et al. Modulation of basophil activity: a novel function of the neuropeptide alpha-melanocyte-stimulating hormone. J Allergy Clin Immunol 2012;129(4):1085–93.

[87] Miyamoto S, Shikata K, Miyasaka K, Okada S, Sasaki M, Kodera R, et al. Cholecystokinin plays a novel protective role in diabetic kidney through anti-inflammatory actions on macrophage: anti-inflammatory effect of cholecystokinin. Diabetes 2012;61(4):897–907.

[88] Mito N, Kitada C, Hosoda T, Sato K. Effect of diet-induced obesity on ovalbumin-specific immune response in a murine asthma model. Metabolism 2002;51(10):1241–6.

[89] Calixto MC, Lintomen L, Schenka A, Saad MJ, Zanesco A, Antunes E. Obesity enhances eosinophilic inflammation in a murine model of allergic asthma. Br J Pharmacol 2010;159(3):617–25.

[90] Ge XN, Greenberg Y, Hosseinkhani MR, Long EK, Bahaie NS, Rao A, et al. High-fat diet promotes lung fibrosis and attenuates airway eosinophilia after exposure to cockroach allergen in mice. Exp Lung Res 2013;39(9):365–78.

[91] Calixto MC, Lintomen L, Andre DM, Leiria LO, Ferreira D, Lellis-Santos C, et al. Metformin attenuates the exacerbation of the allergic eosinophilic inflammation in high fat-diet-induced obesity in mice. PLoS One 2013;8(10):e76786.

[92] Kim JY, Sohn JH, Lee JH, Park JW. Obesity increases airway hyperresponsiveness via the TNF-alpha pathway and treating obesity induces recovery. PLoS One 2015;10(2):e0116540.

[93] Ather JL, Chung M, Hoyt LR, Randall MJ, Georgsdottir A, Daphtary NA, et al. Weight loss decreases inherent and allergic methacholine Hyperresponsiveness in mouse models of diet-induced obese asthma. Am J Respir Cell Mol Biol 2016;55(2):176–87.

[94] Everaere L, Ait-Yahia S, Molendi-Coste O, Vorng H, Quemener S, LeVu P, et al. Innate lymphoid cells contribute to allergic airway disease exacerbation by obesity. J Allergy Clin Immunol 2016;138(5):1309–18. e11.

[95] Han W, Li J, Tang H, Sun L. Treatment of obese asthma in a mouse model by simvastatin is associated with improving dyslipidemia and decreasing leptin level. Biochem Biophys Res Commun 2017;484(2):396–402.

[96] Saraiva SA, Silva AL, Xisto DG, Abreu SC, Silva JD, Silva PL, et al. Impact of obesity on airway and lung parenchyma remodeling in experimental chronic allergic asthma. Respir Physiol Neurobiol 2011;177(2):141–8.

[97] Dietze J, Bocking C, Heverhagen JT, Voelker MN, Renz H. Obesity lowers the threshold of allergic sensitization and augments airway eosinophilia in a mouse model of asthma. Allergy 2012;67(12):1519–29.

[98] Silva FMC, Oliveira EE, Gouveia ACC, Brugiolo ASS, Alves CC, Correa JOA, et al. Obesity promotes prolonged ovalbumin-induced airway inflammation modulating T helper type 1 (Th1), Th2 and Th17 immune responses in BALB/c mice. Clin Exp Immunol 2017;189(1):47–59.

[99] Guilleminault L, Williams EJ, Scott HA, Berthon BS, Jensen M, Wood LG. Diet and asthma: is it time to adapt our message? Nutrients 2017;9(11).

[100] Wang X, Cheng M, Zhao M, Ge A, Guo F, Zhang M, et al. Differential effects of high-fat-diet rich in lard oil or soybean oil on osteopontin expression and inflammation of adipose tissue in diet-induced obese rats. Eur J Nutr 2013;52(3):1181–9.

[101] van Veen IH, Ten Brinke A, Sterk PJ, Rabe KF, Bel EH. Airway inflammation in obese and nonobese patients with difficult-to-treat asthma. Allergy 2008;63(5):570–4.

[102] Lessard A, Turcotte H, Cormier Y, Boulet LP. Obesity and asthma: a specific phenotype? Chest 2008;134(2):317–23.

[103] Desai D, Newby C, Symon FA, Haldar P, Shah S, Gupta S, et al. Elevated sputum interleukin-5 and submucosal eosinophilia in obese individuals with severe asthma. Am J Respir Crit Care Med 2013;188(6):657–63.

[104] Sutherland TJ, Cowan JO, Young S, Goulding A, Grant AM, Williamson A, et al. The association between obesity and asthma: Interactions between systemic and airway inflammation. Am J Respir Crit Care Med 2008;178(5):469–75.

[105] Todd DC, Armstrong S, D'Silva L, Allen CJ, Hargreave FE, Parameswaran K. Effect of obesity on airway inflammation: a cross-sectional analysis of body mass index and sputum cell counts. Clin Exp Allergy 2007;37(7):1049–54.

[106] Cockcroft DW. Bronchoprovocation methods: direct challenges. Clin Rev Allergy Immunol 2003;24(1):19–26.

[107] Boulet LP, Turcotte H, Martin J, Poirier P. Effect of bariatric surgery on airway response and lung function in obese subjects with asthma. Respir Med 2012;106(5):651–60.

[108] Andre DM, Calixto MC, Sollon C, Alexandre EC, Leiria LO, Tobar N, et al. Therapy with resveratrol attenuates obesity-associated allergic airway inflammation in mice. Int Immunopharmacol 2016;38:298–305.

[109] Kanehiro A, Lahn M, Makela MJ, Dakhama A, Joetham A, Rha YH, et al. Requirement for the p75 TNF-alpha receptor 2 in the regulation of airway hyperresponsiveness by gamma delta T cells. J Immunol 2002;169(8):4190–7.

[110] Nakae S, Lunderius C, Ho LH, Schafer B, Tsai M, Galli SJ. TNF can contribute to multiple features of ovalbumin-induced allergic inflammation of the airways in mice. J Allergy Clin Immunol 2007;119(3):680–6.

[111] Garrison L, McDonnell ND. Etanercept: therapeutic use in patients with rheumatoid arthritis. Ann Rheum Dis 1999;58(Suppl 1):I65–9.

[112] Goldenberg MM. Etanercept, a novel drug for the treatment of patients with severe, active rheumatoid arthritis. Clin Ther 1999;21(1):75–87. [discussion 1-2].

[113] Berry MA, Hargadon B, Shelley M, Parker D, Shaw DE, Green RH, et al. Evidence of a role of tumor necrosis factor alpha in refractory asthma. N Engl J Med 2006;354(7):697–708.

[114] Barros LL, Souza-Machado A, Correa LB, Santos JS, Cruz C, Leite M, et al. Obesity and poor asthma control in patients with severe asthma. J Asthma 2011;48(2):171–6.

[115] Heaney LG, Brightling CE, Menzies-Gow A, Stevenson M, Niven RM. British Thoracic Society difficult asthma N. Refractory asthma in the UK: Cross-sectional findings from a UK multicentre registry. Thorax 2010;65(9):787–94.

[116] Kim HY, Lee HJ, Chang YJ, Pichavant M, Shore SA, Fitzgerald KA, et al. Interleukin-17-producing innate lymphoid cells and the NLRP3 inflammasome facilitate obesity-associated airway hyperreactivity. Nat Med 2014;20(1):54–61.

[117] Abramson M, Kutin JJ, Raven J, Lanigan A, Czarny D, Walters EH. Risk factors for asthma among young adults in Melbourne, Australia. Respirology 1996;1(4):291–7.

[118] Kuehr J, Frischer T, Meinert R, Barth R, Schraub S, Urbanek R, et al. Sensitization to mite allergens is a risk factor for early and late onset of asthma and for persistence of asthmatic signs in children. J Allergy Clin Immunol 1995;95(3):655–62.

[119] Sears MR, Burrows B, Flannery EM, Herbison GP, Holdaway MD. Atopy in childhood. I. Gender and allergen related risks for development of hay fever and asthma. Clin Exp Allergy 1993;23(11):941–8.

[120] Li J, Huang Y, Lin X, Zhao D, Tan G, Wu J, et al. Influence of degree of specific allergic sensitivity on severity of rhinitis and asthma in Chinese allergic patients. Respir Res 2011;12:95.

[121] Carroll WD, Lenney W, Child F, Strange RC, Jones PW, Whyte MK, et al. Asthma severity and atopy: how clear is the relationship? Arch Dis Child 2006;91(5):405–9.

[122] Fitzpatrick AM, Gaston BM, Erzurum SC, Teague WG, National Institutes of Health/National Heart L, Blood Institute Severe Asthma Research P. Features of severe asthma in school-age children: atopy and increased exhaled nitric oxide. J Allergy Clin Immunol 2006;118(6):1218–25.

[123] Kosaka S, Tazawa M. Change of serum IgE concentration in asthmatic attack. Tohoku J Exp Med 1976;120(4):313–7.

[124] Fitzpatrick S, Joks R, Silverberg JI. Obesity is associated with increased asthma severity and exacerbations, and increased serum immunoglobulin E in inner-city adults. Clin Exp Allergy 2012;42(5):747–59.

[125] Visness CM, London SJ, Daniels JL, Kaufman JS, Yeatts KB, Siega-Riz AM, et al. Association of obesity with IgE levels and allergy symptoms in children and adolescents: results from the National Health and Nutrition Examination Survey 2005–2006. J Allergy Clin Immunol 2009;123(5):1163–9. 9 e1–4.

[126] Vieira VJ, Ronan AM, Windt MR, Tagliaferro AR. Elevated atopy in healthy obese women. Am J Clin Nutr 2005;82(3):504–9.

[127] Huang SL, Shiao G, Chou P. Association between body mass index and allergy in teenage girls in Taiwan. Clin Exp Allergy 1999;29(3):323–9.

[128] Chen Y, Rennie D, Cormier Y, Dosman J. Association between obesity and atopy in adults. Int Arch Allergy Immunol 2010;153(4):372–7.

[129] Forno E, Acosta-Perez E, Brehm JM, Han YY, Alvarez M, Colon-Semidey A, et al. Obesity and adiposity indicators, asthma, and atopy in Puerto Rican children. J Allergy Clin Immunol 2014;133(5):1308–14. 14 e1-5.

[130] Schachter LM, Salome CM, Peat JK, Woolcock AJ. Obesity is a risk for asthma and wheeze but not airway hyperresponsiveness. Thorax 2001;56(1):4–8.

[131] Bildstrup L, Backer V, Thomsen SF. Increased body mass index predicts severity of asthma symptoms but not objective asthma traits in a large sample of asthmatics. J Asthma 2015;52(7):687–92.

[132] Tantisira KG, Litonjua AA, Weiss ST, Fuhlbrigge AL, Childhood Asthma Management Program Research G. Association of body mass with pulmonary function in the Childhood Asthma Management Program (CAMP). Thorax 2003;58(12):1036–41.

[133] Oshiba A, Hamelmann E, Takeda K, Bradley KL, Loader JE, Larsen GL, et al. Passive transfer of immediate hypersensitivity and airway hyperresponsiveness by allergen-specific immunoglobulin (Ig) E and IgG1 in mice. J Clin Invest 1996;97(6):1398–408.

[134] Mehlhop PD, van de Rijn M, Goldberg AB, Brewer JP, Kurup VP, Martin TR, et al. Allergen-induced bronchial hyperreactivity and eosinophilic inflammation occur in the absence of IgE in a mouse model of asthma. Proc Natl Acad Sci U S A 1997;94(4):1344–9.

[135] Oettgen HC, Martin TR, Wynshaw-Boris A, Deng C, Drazen JM, Leder P. Active anaphylaxis in IgE-deficient mice. Nature 1994;370(6488):367–70.

[136] Coyle AJ, Wagner K, Bertrand C, Tsuyuki S, Bews J, Heusser C. Central role of immunoglobulin (Ig) E in the induction of lung eosinophil infiltration and T helper 2 cell cytokine production: inhibition by a non-anaphylactogenic anti-IgE antibody. J Exp Med 1996;183(4):1303–10.

[137] Ouchi N, Kihara S, Arita Y, Okamoto Y, Maeda K, Kuriyama H, et al. Adiponectin, an adipocyte-derived plasma protein, inhibits endothelial NF-kappaB signaling through a cAMP-dependent pathway. Circulation 2000;102(11):1296–301.

[138] Arita Y, Kihara S, Ouchi N, Takahashi M, Maeda K, Miyagawa J, et al. Paradoxical decrease of an adipose-specific protein, adiponectin, in obesity. Biochem Biophys Res Commun 1999;257(1):79–83.

[139] Berg AH, Combs TP, Du X, Brownlee M, Scherer PE. The adipocyte-secreted protein Acrp30 enhances hepatic insulin action. Nat Med 2001;7(8):947–53.

[140] Wolf AM, Wolf D, Rumpold H, Enrich B, Tilg H. Adiponectin induces the anti-inflammatory cytokines IL-10 and IL-1RA in human leukocytes. Biochem Biophys Res Commun 2004;323(2):630–5.

[141] Yamamoto R, Ueki S, Moritoki Y, Kobayashi Y, Oyamada H, Konno Y, et al. Adiponectin attenuates human eosinophil adhesion and chemotaxis: implications in allergic inflammation. J Asthma 2013;50(8):828–35.

[142] Hug C, Wang J, Ahmad NS, Bogan JS, Tsao TS, Lodish HF. T-cadherin is a receptor for hexameric and high-molecular-weight forms of Acrp30/adiponectin. Proc Natl Acad Sci U S A 2004;101(28):10308–13.

[143] Williams AS, Kasahara DI, Verbout NG, Fedulov AV, Zhu M, Si H, et al. Role of the adiponectin binding protein, T-cadherin (Cdh13), in allergic airways responses in mice. PLoS One 2012;7(7):e41088.

[144] Shore SA, Terry RD, Flynt L, Xu A, Hug C. Adiponectin attenuates allergen-induced airway inflammation and hyperresponsiveness in mice. J Allergy Clin Immunol 2006;118(2):389–95.

[145] Medoff BD, Okamoto Y, Leyton P, Weng M, Sandall BP, Raher MJ, et al. Adiponectin deficiency increases allergic airway inflammation and pulmonary vascular remodeling. Am J Respir Cell Mol Biol 2009;41(4):397–406.

[146] Verbout NG, Benedito L, Williams AS, Kasahara DI, Wurmbrand AP, Si H, et al. Impact of adiponectin overexpression on allergic airways responses in mice. J Allergy (Cairo) 2013;2013:349520.

[147] Kattan M, Kumar R, Bloomberg GR, Mitchell HE, Calatroni A, Gergen PJ, et al. Asthma control, adiposity, and adipokines among inner-city adolescents. J Allergy Clin Immunol 2010;125(3):584–92.

[148] Sood A, Seagrave J, Herbert G, Harkins M, Alam Y, Chiavaroli A, et al. High sputum total adiponectin is associated with low odds for asthma. J Asthma 2014;51(5):459–66.

[149] Baek HS, Kim YD, Shin JH, Kim JH, Oh JW, Lee HB. Serum leptin and adiponectin levels correlate with exercise-induced bronchoconstriction in children with asthma. Ann Allergy Asthma Immunol 2011;107(1):14–21.

[150] Kim KW, Shin YH, Lee KE, Kim ES, Sohn MH, Kim KE. Relationship between adipokines and manifestations of childhood asthma. Pediatr Allergy Immunol 2008;19(6):535–40.

[151] Considine RV, Caro JF. Leptin and the regulation of body weight. Int J Biochem Cell Biol 1997;29(11):1255–72.

[152] Loffreda S, Yang SQ, Lin HZ, Karp CL, Brengman ML, Wang DJ, et al. Leptin regulates proinflammatory immune responses. FASEB J 1998;12(1):57–65.

[153] Tsaroucha A, Daniil Z, Malli F, Georgoulias P, Minas M, Kostikas K, et al. Leptin, adiponectin, and ghrelin levels in female patients with asthma during stable and exacerbation periods. J Asthma 2013;50(2):188–97.

[154] Nasiri Kalmarzi R, Ataee P, Mansori M, Moradi G, Ahmadi S, Kaviani Z, et al. Serum levels of adiponectin and leptin in asthmatic patients and its relation with asthma severity, lung function and BMI. Allergol Immunopathol (Madr) 2017;45(3):258–64.

[155] Guler N, Kirerleri E, Ones U, Tamay Z, Salmayenli N, Darendeliler F. Leptin: does it have any role in childhood asthma? J Allergy Clin Immunol 2004;114(2):254–9.

[156] Nagel G, Koenig W, Rapp K, Wabitsch M, Zoellner I, Weiland SK. Associations of adipokines with asthma, rhinoconjunctivitis, and eczema in German schoolchildren. Pediatr Allergy Immunol 2009;20(1):81–8.

[157] Zhang L, Yin Y, Zhang H, Zhong W, Zhang J. Association of asthma diagnosis with leptin and adiponectin: a systematic review and meta-analysis. J Invest Med 2017;65(1):57–64.

[158] Sideleva O, Suratt BT, Black KE, Tharp WG, Pratley RE, Forgione P, et al. Obesity and asthma: an inflammatory disease of adipose tissue not the airway. Am J Respir Crit Care Med 2012;186(7):598–605.

Obesity, mitochondrial dysfunction, and obstructive lung disease

7

Rituparna Chaudhuri*, **Michael A. Thompson†**, **Christina Pabelick†**, **Anurag Agrawal***, **Y.S. Prakash†**

CSIR Institute of Genomics & Integrative Biology, Delhi University, New Delhi, India* Department of Anesthesiology and Perioperative Medicine, Mayo Clinic, Rochester, MN, United States†

ABBREVIATIONS

ASM	airway smooth muscle cell
CR	calorie restriction
Drp	dynamin related protein
eNOS	endothelial nitric oxide synthase
FADH$_2$	dihydro flavine adenine dinucleotide
FEV1	forced expiratory volume in 1 second
Fis	fission protein
FVC	forced vital capacity
HETE	hydroxyeicosatetraenoic acid
HFD	high fat diet
HIF	hypoxia inducible factor
HODE	hydroxyoctadecadienoic acid
IL-6	interleukin 6
IMM	inner mitochondrial membrane
LOX	lipoxygenase
MAM	mitochondria-associated ER membrane
MAPK	mitogen activated protein kinase
MetS	metabolic syndrome
Mfn	mitofusin
mTOR	mammalian target of rapamycin
NADH	nicotinamide adenine dinucleotide
ND2	NADH dehydrogenase subunit 2
NEFA	nonesterified fatty acid
NLRP	Nod-like receptor protein
NO	nitric oxide

Mechanisms and Manifestations of Obesity in Lung Disease. https://doi.org/10.1016/B978-0-12-813553-2.00007-5

Opa1	optic atrophy type
PGC1alpha	peroxisome proliferator-activated receptor gamma coactivator 1-alpha
SIRT	sirtuin
SOD	superoxide dismutase
TCA	tricarboxylic acid cycle
TNFR	tumor necrosis factor receptor
UCP	uncoupling protein
WAT	white adipose tissue

INTRODUCTION

Obesity is characterized by excessive lipid storage in visceral adipose tissue as well as deposition in nonadipose tissues such as skeletal muscle and liver. Etiological components of obesity include genetic, environmental, and lifestyle factors. Defective cell metabolism, an imbalance between nutrient intake and its utilization, leads to dysfunctional fatty acid oxidation (FAO). This results in increased free fatty acids and fatty acyl coenzyme As (CoAs) in adipocytes, skeletal muscle, and liver, leading to a cascade of events, resulting in compromised insulin signaling and storage of excess fat. Because most metabolic pathways converge at the level of the mitochondrion, compromised mitochondrial function is a logical mechanism that has indeed been demonstrated in obesity [1].

Obstructive lung disease is a class of respiratory diseases including asthma, chronic obstructive pulmonary disease (COPD), bronchiectasis, obstructive sleep apnea and cystic fibrosis that involve varying degrees of airway inflammation, airway hyperreactivity, reversible vs. irreversible airway thickening and remodeling, and altered airway collapsibility. Severe cases may require hospitalization, and these diseases overall contribute to reduced quality of life. Interestingly, there is now recognition in a number of studies that body mass index (BMI) correlates to severity of obstructive lung diseases [2]. Separately, mitochondrial dysfunction is now thought to be involved in such lung disorders [3, 4]. Accordingly, given mitochondrial involvement in both obesity and obstructive lung disease, we explore possible molecular mechanistic links that may lead to the pathogenesis of either condition. Potential therapeutics that target mitochondria for prevention of either development or escalation of both these conditions or of lung disease in obese patients represents an entirely novel concept worthy of further exploration.

MITOCHONDRIA AND THEIR MULTIFACETED ROLES

Mitochondria originated as prokaryotic cells capable of performing their own oxidative mechanisms not possible in eukaryotic cells and became endosymbionts hosted by eukaryotes [5]. Thus they have many features common with ancient aerobic bacteria and possess their own individual genome, consisting of double-stranded circular DNA. The mitochondrial chromosome contains genes coding for enzymes of the tricarboxylic acid (TCA) cycle, and other genes encoding for some ribosomal RNAs and transfer RNAs (tRNAs) necessary for translation of messenger RNA into protein. They possess a double membranous structure with its outer membrane that contains enzymes such as NADH cytochrome c reductase, monoamine oxidase, etc. and can associate with the endoplasmic reticulum (ER) membrane in structures called mitochondria-associated ER membranes (MAMs). The intermembrane space is the space between the outer and inner membranes, which contains an important enzyme, cytochrome c [6]. The inner mitochondrial membrane (IMM) contains a number of important proteins, including those required for oxidative phosphorylation, ATP synthesis, and mitochondrial fission and fusion [7]. There is a membrane potential across the IMM formed by the action of enzymes of the electron transport chain (ETC). The mitochondrial matrix contains hundreds of enzymes, tRNA, and several copies of the mitochondrial DNA genome [7]. The MAM is enriched in enzymes for lipid synthesis; this is particularly useful as mitochondria are dynamic organelles that continuously undergo fission and fusion, and need phospholipids to maintain membrane integrity [8, 9]. Another important function of MAMs is in regulating calcium signaling along with the ER [10, 11]. Close physical contact between mitochondria and ER at certain points facilitates optimum Ca^{2+} transmission from ER to mitochondria. This signaling is especially critical as a certain level of Ca^{2+} maintains mitochondria, and thus cellular homeostasis. This Ca^{2+} is required to stimulate metabolism by activating dehydrogenase enzymes critical to carry out the citric acid cycle. However, if mitochondria become overloaded with Ca^{2+}, the intrinsic pathway of apoptosis is activated, and membrane potential required for metabolism collapses.

A predominant role for mitochondria is to produce ATP by oxidizing pyruvate and nicotinamide adenine dehydrogenase (NADH) through aerobic respiration [12]. Pyruvate molecules produced by glycolysis in the cytoplasm are actively transported into the matrix where they can be either oxidized to form acetyl-CoA and NADH, or be carboxylated to form oxaloacetate (OAA). The latter reaction builds up the OAA store of the cycle such that acetyl-CoA production can be increased based on tissue demands [13]. Acetyl-CoA on the other hand, derived from the β-oxidation of fatty acids is the only fuel that can enter the TCA cycle. The oxidation of acetyl-CoA derived from carbohydrates, fats, and proteins to carbon dioxide produces reduced intermediates—three molecules of NADH and one molecule of dihydro flavine adenine dinucleotide ($FADH_2$), which can act as source of electrons for the ETC. The ETC is a series of complexes in the IMM that facilitates the passage of electrons from NADH and $FADH_2$ through an electron donor to a more electronegative acceptor. This releases energy, which is used to generate protons and create a proton gradient across the mitochondrial membrane by actively pumping them across the IMM into the intermembrane space. This electrochemical proton gradient drives ATP synthesis through ATP synthase (i.e., oxidative phosphorylation (OxPhos)). Although ETC and OxPhos are vital parts of metabolism, some electrons leak from complex I and III in ETC and are prematurely leaked to oxygen, the final electron acceptor. This partial reduction of oxygen generates superoxide, which is dismutated to hydrogen peroxide by superoxide dismutases (SOD). Superoxide, hydrogen peroxide, and hydroxyl radical together constitute mitochondrial ROS (mtROS). If the generation of such mtROS increases or the natural antioxidants of the cell (reducing agents such as glutathione, catalase, peroxidase, etc.) decrease, a state of imbalance between the manifestation of ROS and the body's detoxification ability results in oxidative stress. These free radicals are highly reactive and can cause major damage in cells, oxidizing proteins, mutating DNA, inducing inflammation, and overall contributing to disease pathophysiology. As ROS production increases at high membrane potentials, mitochondria maintain their membrane potential at an optimum level to balance ATP production against oxidant generation. Another important checkpoint in place to regulate excessive oxidant production are uncoupling proteins (UCP), which can dissipate the proton gradient generated by proton pumping from the matrix to the intermembrane space. The energy lost in dissipating the proton gradient is not used for ATP generation but is released as heat. Thus this process is also linked to thermogenesis [14]. Interestingly, the expression of UCPs is increased by leptin, thus connecting mitochondria to satiety signaling [15]. This promotes mitochondrial uncoupling, which allows protons to bypass ATP synthase, reduces amount of mtROS significantly, thus decreasing inflammatory responses.

Apart from their major function in cellular bioenergetics, mitochondria have now been shown to perform other nonenergetic roles such as regulation of metabolism, apoptosis, and inflammatory responses. Mitochondria exist as a dynamic network with regulated fusion and fission. At the molecular level, Dynamin-1-like (Drp1) and fission 1 (Fis1) regulate fission, whereas optic atrophy 1 (Opa1) and mitofusin 1 and 2 (Mfn 1/2) control mitochondrial fusion. Although the phenomenon of continuous fission-fusion as a contributor to good mitochondrial health has been established, the reasons underlying the dynamic nature of mitochondrial structure is not as clear [16]. A reasonable model postulates that fusion is especially useful as it allows mixing of contents of both mitochondrial membranes and the matrix. This serves the purpose of reducing heterogeneity amongst the several hundred mitochondria in a typical cell. This is probably why mammalian mitochondria can tolerate a surprisingly high amount of mtDNA mutations without compromising respiratory capacity. It has been shown that mtDNA undergoes exponential increase in mice skeletal muscle in the first 2 months of life. This developmental increase does not occur in mice with skeletal muscle-targeted deletion mutants of Mfn1/2, leading to less than 10% of the amount of mtDNA compared with wild-type muscle, giving rise to severe respiratory deficiency [17]. Although disruption of mitochondrial fusion factor *fuzzy onions* (Fzo1) has been shown to result in yeast colonies lacking mtDNA, Mfn2 mutant lines were found to have normal amounts of mtDNA. Mice with error-prone mtDNA polymerase along with Mfn1 mutation have been shown to have severely compromised respiratory function, in contrast to those carrying single mutations [17]. However, mitochondria with loss of membrane potential are segregated from healthy ones and marked for mitophagy. Fission is mostly seen as crucial in apoptosis wherein mitochondrial size is reduced so as to be suitable substrates for engulfment by autophagosomes. Mitochondrial fission also helps the cell to adapt to stress. It is thus interesting to note that cells with moderate dysfunction in their mitochondria can be rescued by fusion, but severely damaged mitochondria will be degraded [16]. Failure of mitophagy leads to the build-up of dysfunctional mitochondria in the cell, resulting in increased mitoROS, inflammation, and eventual apoptosis.

Stimulation of inflammatory responses leads to the formation of an inflammasome complex, which in turn activates an inflammatory cascade. Under severe mitochondrial stress, enhanced mtROS can trigger inflammation via various pathways such as MAPK, TNFR, and NLRP3 inflammasome signals [18, 19]. Multiple studies have shown that decreased mtROS reduces the release of proinflammatory mediators such as IL-6 or TNF [17]. When mtDNA is damaged beyond repair from oxidant stress or other factors, inflammasome formation is activated followed by increased cytokine release from immune cells as well as resident tissues such as epithelium and smooth muscle. Otherwise, damaged mitochondria can also release mtDNA into the cytoplasm or circulation, which act as mitochondrial damage-associated molecular patterns (DAMPs) and perpetuate a noninfectious inflammatory response [20, 21]. Further, a persistent increase in extracellular ATP will also act as a DAMP and mobilize Ca^{2+}, activate the inflammasome, release mtDNA into the cytosol, and increase mitoROS. It may be noted that inflammatory cytokines such as IL-17, TNFα, and IFNγ can cause the opening of mitochondrial permeability transition pore (MPTP), which leads to abolition of the proton gradient, often resulting in cell death [3].

MITOCHONDRIAL DYSFUNCTION IN OBESITY: EXPLORING THE POTENTIAL MECHANISMS

Understanding the major functions of mitochondria, as described in the previous section, allows for assessment of mitochondrial dysfunction. Defects in ETC and OxPhos decrease mitochondrial activity, loss of membrane potential, reduced mitochondrial numbers (decreased biogenesis), imbalance in fission-fusion, excessive ROS generation, inflammation, and/or decrease in mitochondrial substrate (carbohydrate/lipid) oxidation resulting from a general decrease in OxPhos, and insufficient mitophagy, overall leading to accumulation of damaged mitochondria with compromised function.

Obesity is characterized by excessive adipose tissue accumulation arising from an imbalance in energy intake versus consumption. Obesity is marked by mitochondrial dysfunction, inflammation, and inhibition of antioxidant defenses. There are numerous instances of mitochondrial dysfunction in obesity, some of which are discussed later (Figure 7.1).

In both diet-induced and genetically induced obesity mouse models, mitochondrial function is compromised in liver and skeletal muscle, as represented by increased fission (enhanced Drp1 and Fis1) and decreased fusion (reduction in Mfn2) [22]. Obese Zucker rats show decreased mitochondrial fusion, and less branched and smaller mitochondria, observations corroborated in obese type II diabetic humans [23].

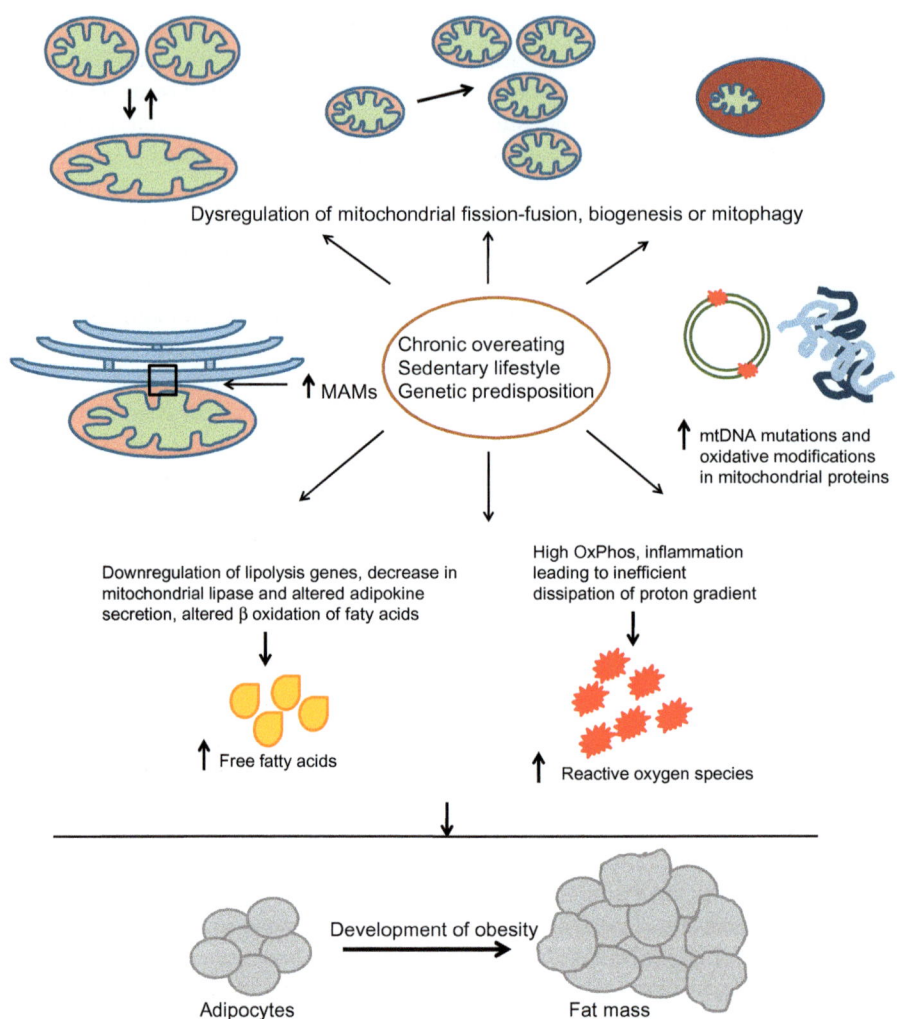

FIGURE 7.1

Mitochondrial dysfunction in obesity. Chronic overnutrition along with physical inactivity act as external stressors that lead to mitochondrial dysfunction. Multiple pathways directly or indirectly related to mitochondria, including misregulation of mitochondrial bioenergetics, excess production of ROS, abnormal fission-fusion, biogenesis, and mitophagy, may contribute to obesity. An increase in ER-mitochondria associated membranes (MAMs) has been implicated in the altered calcium homeostasis in the mitochondria, leading to compromised energetics. Changes in expression of lipid metabolism genes, such as adiponectin, and a higher number of mitochondrial mutations also have been shown to lead to excess fat deposition.

A shift in mitochondrial dynamics toward fission is also seen in liver and skeletal muscle of genetically induced db/db, ob/ob, as well as mice with diet-induced obesity. Saturated fatty acids have been shown to increase fission processes in immortalized mouse myoblast C2C12 cells associated with mitochondrial dysfunction [24]. Such increased fragmentation has been suggested to increase total surface area to enhance accessibility of metabolic substrates to carrier proteins. This could represent an adaptive response to increased nutrition thus increasing oxidation of surplus dietary fatty acids, which would result in elevated ROS production (reviewed in Ref. [25]). In comparison to lean individuals, mitochondria of obese people are also structurally and functionally distinct. They have less energy generating capacity, less well-defined IMM, and reduced FAO.

Overnutrition tends to overload mitochondria due to increased substrates leading to more exhaustive bioenergetics. Normally, when mitochondrial respiration is uncoupled from ATP production, protons can leak across IMM thus releasing energy in the form of heat and facilitating ROS disposal. However, proton dissipation is decreased with chronic high food intake and resulting need for higher rates of metabolism, leading to higher amounts of ROS. Obesity or feeding mice a high-fat diet (HFD) increases the H_2O_2-emitting potential of mitochondria, and the cellular environment becomes more oxidized and decreases redox-buffering capacity [26]. Increased mtROS production is seen in genetically obese db/db mice as well as mice with diet-induced obesity. This causes posttranslational oxidative modifications to mitochondrial proteins and leads to dysfunctional oxidative respiration and ATP production [27]. HFD and hyperglycemia enhance ROS production in mouse adipocytes. Thus there is increased oxidative stress in obese individuals and in genetically obese mice, accompanied by abnormal adipokine production. In accordance, several human studies show that obese and insulin-resistant individuals have lower oxidative enzyme activities and reduced lipid metabolism in skeletal muscle compared with lean subjects [28]. $NADH:O_2$ oxidoreductase activity and mitochondrial size is smaller in skeletal muscle of obese patients with type II diabetes [29]. Undernutrition conditions such as calorie restriction (CR) and starvation have been found to be associated with less oxidative injury related to changes in mitochondrial dynamics.

A number of enzymes associated with mitochondrial bioenergetics and biogenesis are found to be affected by obesity. Expression of PGC1α, the master regulator of mitochondrial biogenesis, is decreased in obese individuals [30], as well as in genetically obese mice and mice with diet-induced obesity [31], thus correlating with mitochondrial functional impairment and lipid accumulation. Adipocytes respond to metabolic challenges by altering mitochondrial number or distribution of mtDNA content [32]. mRNA and protein levels of carnitine palmitoyltransferase-1 (CPT-1, a mitochondrial enzyme responsible for fatty acid β-oxidation), and mitochondrial biogenesis-related factors, such as mitochondrial transcription factor A (TFAM) and nuclear respiratory factor-1 (NRF-1), as well as citrate synthase (a critical enzyme of the Krebs cycle) in liver tissue are markedly decreased in mice fed a HFD compared with mice fed a control standard diet [33]. Mfn2 levels positively correlate with insulin sensitivity in morbidly obese patients after bariatric surgery [34]. In another study, sustained exposure of mice to a HFD modified the liver mitochondrial proteome, with significant changes in proteins related to oxidative phosphorylation and lipid metabolism [35]. Twenty-two mitochondrial proteins were found to be significantly altered; activity of cytochrome c oxidase, and TCA cycle enzymes malate and pyruvate dehydrogenase were found to be decreased, whereas there was increase in ATP synthase subunits (F1α and β) as well as increased sensitivity to the nitric oxide (NO)-dependent inhibition of mitochondrial respiration. These changes are indicative of compromised mitochondrial respiration in HFD-fed mice [36]. In HFD-induced obesity, genes related to lipolysis, mitochondrial bioenergetics, and fatty acid metabolism are downregulated, whereas inflammation-associated genes are up-regulated. Interestingly, protein arginine methyl transferase (PRMT1) is decreased in liver of HFD-fed mice. This leads to hypomethylation of PGC-1α, which, along with the finding that overfeeding impairs AMPK-dependent phosphorylation of PGC-1α, suggests an epigenetic deregulation of mitochondrial metabolism in contributing to obesity development [37]. Prolonged HFD in mice alters mtDNA copy number and increases mitochondrial damage in liver and skeletal muscle. HFD also significantly increases levels of DNA repair protein 8-oxoguanine DNA glycosylase (OGG1) in skeletal muscle [38].

During overnutrition, an early event seems to be reduction and reorganization of the total ER membrane content in liver cells, along with increased number of ER-mitochondrial contact points (MAMs). Experimental induction of ER stress with tunicamycin has shown a similar pattern of ER rearrangement, indicating that this may be an adaptive response to boost mitochondrial function. However, with chronic overnutrition, these connections are persistently maintained. This enhances calcium transfer from ER to mitochondria, leading to calcium accumulation in mitochondria, consequently leading to increased oxidative stress, compromised OxPhos and respiratory inhibition, higher ROS levels, and opening of MPTPs. Opening of permeability transition pores results in a drop in mitochondrial membrane potential and increased mitochondrial swelling with loss of nucleotides and cytochrome c, and subsequently apoptotic cell death. Increase in mitochondrial calcium has also been shown to induce mitophagy. Increased MAM formation has been observed in livers of both ob/ob and HFD mice [39].

Normally, when there is a negative energy balance (i.e., energy requirement outweighs food intake), lipolysis in white adipose tissue (WAT) is enhanced using nonesterified fatty acids (NEFA) as a substrate for FAO in liver and skeletal muscle with associated insulin sensitization. On the other hand, in chronic presence of excess nutrients, NEFA accumulates, leading to mitochondrial dysfunction and fat storage. WAT produces adipokines that often regulate fat storage in the body through mitochondrial pathways, such as activin B, which has been shown to block breakdown of lipids and increase triglyceride accumulation by downregulating mitochondrial lipase expression [32]. In the continued presence of excessive caloric intake along with a sedentary lifestyle that requires less energy expenditure, effective uncoupling and ROS dissipation does not take place in mitochondria. Excess ROS reduces oxygen consumption in adipocytes, and FAO is blocked eventually, leading to lipid accumulation. In HFD-fed obese mice, mitochondria of metabolically active tissues are impaired and mitochondrial biogenesis is reduced in white adipocytes [40]. Mitochondrial dysfunction, characterized by increased lipid peroxidation in the brain, mtROS, and lower mitochondrial membrane potential, has been observed in Sprague Dawley rats with diet-induced obesity [41]. Excess food intake has been shown to impair mitochondrial ability for respiration through mTOR pathways and increase susceptibility to apoptosis [37, 42]. The sirtuins (silent mating type information regulation 2 homolog) are nicotinamide adenine dinucleotide-dependent protein deacetylases, which are involved in various processes of cell metabolism. SIRT3–5 are localized in mitochondria and deacetylate various critical enzymes to regulate mitochondrial function. Deacetylation of key enzymes by SIRT3 increases mitochondrial FAO in liver, whereas SIRT3 deficiency leads to metabolic syndrome (MetS)-like features by mitochondrial protein hyperacetylation. HFD increases hyperacetylation of mitochondrial proteins in liver and reduces SIRT3 [43]. PGC-1α, which regulates SIRT3 gene expression [44], is also reduced in MetS. This suggests that CR or fasting could reverse MetS features by activating both SIRT3 and PGC-1α thus increasing FAO and mitochondrial biogenesis, respectively [45].

Whereas the studies previously cited provide various mechanistic links between mitochondrial dysfunction and obesity, most of them report effects induced by genetic or diet-induced obesity and cannot be taken to assess whether mitochondrial dysfunction is a cause or consequence of obesity. In the next section, we review more direct evidence demonstrating mitochondrial dysfunction as the cause for development of obesity.

MITOCHONDRIAL DYSFUNCTION LEADS TO OBESITY
FUNCTIONAL STUDIES

Overexpression of IMM UCP, UCP1, decreases weight gain in obesity-prone mice due to increased energy expenditure and reduced fatty acid synthesis. UCP1 expression can directly regulate fat deposition and contribute to obesity [46]. With age, PGC1α knockout (KO) mice show higher body weight and fat deposition in the liver, along with decreased mitochondrial oxidative capacity, whereas mice with even slightly elevated levels of PGC1α are resistant to age-related obesity [28, 47]. Mice with muscle-specific deletion of cytochrome c oxidase 10 (*COX10*) or mice expressing mutant *Peo1* (a mitochondrial helicase) leads to development of a starvation-like response, along with an increase in fibroblast growth factor (FGF) 21, a metabolic hormone. This indicates that there can be noncell-autonomous effects of mitochondrial function on systemic metabolism [48]. Targeted disruption of *Atg7*, an autophagy gene in skeletal muscle, protects mice from diet-induced obesity and insulin resistance [49]. Interestingly, these effects of mitochondria are mediated in a noncell-autonomous manner through the mitokine FGF21 [50]. Removal of mitochondria through enhanced mitophagy could reduce mitochondrial numbers, resulting in decreased substrate oxidation, further enhancing lipid accumulation. Interestingly, it has been shown that treatment of rats with mitochondria-targeted antioxidant or overexpression of catalase to scavenge ROS completely preserves insulin sensitivity in spite of an HFD [26]. Mitochondrial complex IV activity is inhibited in white adipocytes of middle-aged mice, elucidating a possible molecular mechanism for age-dependent obesity. Moreover, this decrease in activity is enough to impair FAO and lead to obesity during aging. It is well established now that WAT expansion results in hypoxia inside the tissue due to poor oxygenation. This promotes HIF1α accumulation, which in turn exacerbates WAT growth through complex IV inhibition in a feed-forward loop. Accumulation of HIF1α could be promoted not only by hypoxia but also by intracellular ROS and lipid accumulation [51]. Dysfunction of mitochondria results in increased production of ROS, inflammation, and upregulated inflammatory cytokines, which mediate the reduction of endothelial nitric oxide synthase (eNOS) expression in adipocytes. This decreases NO production through inhibition of PGC1α and hampers beta oxidation of fatty acids in mitochondria, leading to lipid accumulation in adipocytes, eventually precipitating as abdominal obesity. Resultant obesity leads to release of free fatty acids, which damage various organs such as skeletal muscle, pancreas, and liver, leading to a further decrease in metabolic function of mitochondria, affecting glucose uptake, and decreasing insulin sensitivity. In addition, 12/15-lipoxygenase (LOX) may cause mitochondrial dysfunction through its metabolites such as 13-S-hydroxyoctadecadienoic acid (13-S-HODE) and 12-S-hydroxyeicosatetraenoic acid (12-S-HETE), which can cause mitochondrial degradation [45]. Further, fat-specific deletion of 12/15-LOX improves glucose metabolism and protects against obesity-mediated complications [52]. Overall, these findings emphasize that mitochondrial function often lies upstream of metabolic consequences.

MITOCHONDRIAL GENOME VARIATIONS IN OBESITY PATIENTS

Numerous studies have attempted to identify causal mutations that could explain the rising obesity epidemic. However, obesity is a very complex disorder, with a greater contribution of lifestyle and environment, so the search for such mutations has proved to be challenging. Amongst monogenic mutations, the most common ones were found in the leptin and melanocortin pathway in the central nervous system [53]. Although these mutations do cause obesity, hyperphagia, hyperglycemia, and hyperinsulinemia, the rates of occurrence of these in humans is very low and cannot account for the exponential rise in obesity over the past 2 decades. Human obesity usually results from a combined effect of multiple genes, environmental factors, and lifestyle. Offspring have been found to have higher BMI correlations with their mothers than with fathers [54–56]. This points to the mitochondrial genome's contribution toward body weight as being at least a partially inherited trait from the mother. The location of the mitochondrial genome inside the mitochondrial matrix without a protective histone covering and limited repair mechanisms makes it 3–10 times more susceptible than nuclear DNA to oxidative modifications that accumulate over time [57].

An interesting finding has been that maternal diet-induced obesity in a mouse model is associated with alterations in mitochondrial activity such as increased membrane potential and ROS, changed mtDNA content and biogenesis, and decreased glutathione, resulting in a more oxidized environment in zygotes [58]. This increased state of oxidative stress in oocytes and zygotes during the periconceptual period was found to be detrimental in embryogenesis [59]. Mitochondria are involved in oocyte maturation, fertilization, and initiation/progression of embryo implantation [60]. Given mitochondria's primary role as the cell's "powerhouse," they produce ATP by coupling nutrient oxidation with reduction of NADPH and $FADH_2$. Consumption of a high energy-rich maternal diet has been shown to place mitochondria in a "positive energy balance" (high substrate with low ATP demand), thus perturbing mitochondrial metabolism in the oocyte and zygote.

Mitochondrial genome polymorphisms such as mtSNP, 8684C>T (T53I) in the mitochondrial ATP synthase subunit 6 gene (ATP6), 3497C>T (A64V) in the NADH dehydrogenase subunit 1 gene (ND1), and 1119T>C (472U>C) in the 12S rRNA gene have been found to be involved in the development of obesity syndrome. Several research groups have studied the associations of different mtSNPs with obesity in various human ethnic groups during the course of evolution. Mitochondrial SNPs—ND2, COX2, and ATP8—have been observed within genes encoding proteins of oxidative phosphorylation and electron transport in subhaplogroups of the Pima Indians. These were adaptations acquired for a more energy-efficient metabolism due to an adoption of a more restricted caloric intake when this population migrated to the desert. However, with time, migrated humans adopted a more sedentary lifestyle thus acquiring a positive energy balance with high calorie intake and low ATP demand. Today, these same mutations may potentially contribute to obesity. For example, *UCP* is a mitochondrial inner membrane protein that is a regulated proton channel. *UCP* can dissipate the proton gradient generated by NADH-powered pumping of protons from the mitochondrial matrix to the inner mitochondrial space. Polymorphisms in the *UCP*2 gene (rs660339 and rs659366) have been found to be associated with the risk of abdominal obesity and abnormal body fat distribution in the Spanish population. The *UCP*2 A55V variant was also found to be associated with obesity and related phenotypes in an aboriginal community in Taiwan. *UCP*1 variants, g.IVS4-208T>G SNP, was associated with obesity in Southern Italy in the severely obese population. MtDNA haplogroup X and two mtSNPs (mt4823 and mt8873) have been observed to be significant markers associated with reduced body fat mass. *UCP*3 gene shows an association with obesity phenotypes in Caucasian families. A common polymorphism in the promoter of the *UCP*2 gene is associated with obesity and an allele associated with obesity and hyperinsulinemia in north Indians. In another study, overexpression of the transgene hOGG1 in human 8-oxoguanine-DNA glycosylase 1 (hOGG1) transgenic (TG) mice led to increased oxidative damage of mtDNA and manifested in obesity and hepatosteatosis. MtDNA variant (cytosine to adenine) in NADH dehydrogenase subunit 2 was found to directly affect ROS production from complexes I and III, which will have implications in inflammation and weight gain [61]. Thus mitochondrial mutations, in conjunction with environmental cues and lifestyle choices, often result in obesity.

A recent genome-wide association study (GWAS) meta-analysis identified an SNP in the IP3R2 (ER-membrane receptor for inositol 1,4,5 trisphosphate) locus to be associated with alterations in the waist-to-hip ratio adjusted for BMI [62]. Protein expression as well as phosphorylation status of IP3R were increased in a genetic model of obesity (*db/db* mice), resulting in higher cytosolic calcium. In addition, the metabolic effects of mitochondrial dysfunction usually differ depending on the site of mitochondrial dysfunction, its severity, or type, thus providing an explanation for distinct metabolic phenotypes of patients with mitochondrial DNA mutations [50]. Multiple symmetrical lipomatosis, which is characterized by abnormal body fat storage and development of lipomas in adulthood, is a primary mitochondrial disorder showing mtDNA mutations and characteristic mitochondrial pathology in muscle histological sections [63]. mtDNA sequence changes can alter mitochondrial function thus changing the balance between interlinked processes of ATP production, oxidant generation, and heat generation. These factors will therefore influence the cross-talk with nuclear DNA and affect individual susceptibility to disorders associated with metabolism, such as obesity. Table 7.1 illustrates the various mutations and polymorphisms in mitochondrial DNA that have been shown to affect obesity.

Table 7.1 List of Single Nucleotide Polymorphisms Implicated in Obesity

Mitochondrial Gene Implicated in Obesity	Polymorphism Detected	Population Studied	Reference
Mitochondrial ATP synthase subunit 6	m.8614T>C, m.8994G>A	Turkish children	[61]
NADH dehydrogenase subunit 1	rs1064794438	–	[61]
NADH dehydrogenase subunit 2	m.5460	Pima Indians	[61]
12S rRNA	rs143000400	–	[61]
COX2	m.7859	Pima Indians	[61]
ATP8	m.8541	Pima Indians	[61]
UCP1	rs1494808; m.8344A>G	Southern Italy	[61, 63]
UCP2	rs660339, rs659366	Spanish population; aboriginal community in Taiwan	[61]
UCP3	rs1800849	Caucasian	[103]
MtDNA haplogroup X	m.4823; m.8873	Caucasian	[61]

EXPLORING THE LINK BETWEEN OBESITY AND OBSTRUCTIVE LUNG DISEASES

An initial observation linking obesity to obstructive airway disease was that leptin-deficient (ob/ob) mice show higher airway resistance and smaller lung volumes compared with their lean counterparts [64]. Obese mice also develop innate airway hyperresponsiveness even without allergen immunization [65]. In humans, more severe bronchospasm, greater bronchodilator use, and prolonged recovery time have been reported in obese asthmatic children [66]. Although a number of factors may contribute to this link, such as systemic inflammation, genetic or epigenetic changes, altered nutrient levels, or changes in gut microbiome, the previously presented data also suggest that obesity poses a risk factor for development of obstructive lung diseases.

MITOCHONDRIAL DYSFUNCTION: THE OVERLAPPING ELEMENT BETWEEN OBESITY AND OBSTRUCTIVE LUNG DISEASES

In previous chapters of this textbook, the epidemiological data linking obesity with asthma, COPD, and obstructive sleep apnea has been thoroughly discussed. The strong epidemiological links between obesity and obstructive lung disease form a basis to explore whether there are unifying molecular mechanisms that play a role in these two seemingly unrelated disorders. Some of the strongest overlaps were found to be a variety of proteins and metabolites involved in dysregulation of mitochondrial function (illustrated in Figure 7.2).

12/15 LOX, a nonheme iron dioxygenase, catalyzes hydroperoxidation of polyunsaturated fatty acids in adipocytes [67]. Fat-specific deletion of 12/15 LOX protected mice from obesity-related complications [52]. Absence of 12/15 LOX led to resolution of inflammation through reduction of macrophage infiltration in adipose tissue [67]. Interestingly, genetic deletion of 12/15 LOX was found to also alleviate asthmatic features [68]. 12/15 LOX is known to depolarize mitochondria in both *in vitro* and *in vivo* models [45]. In addition, 12/15 LOX may cause mitochondrial degradation and dysfunction through metabolites such as 13-S-hydroxyoctadecadienoic acid (13-S-HODE) and 12-S-hydroxyeicosatetraenoic acid (12-S-HETE) [69, 70]. Another common pathophysiological feature in both asthma and obesity is reduced bioavailability of L-arginine [71], partly due to an increase in methylated arginines, such as asymmetric dimethyl arginine (ADMA), that competitively binds and uncouples eNOS [72]. eNOS has a protective effect in both asthma and obesity via NO formation, whereas uncoupled eNOS forms superoxide without generating NO. Reduction in bioavailability of NO inhibits mitochondrial biogenesis and decreases FAO. Supplementation with a high dose of L-arginine or overexpression of eNOS has been shown to alleviate symptoms of both asthma and MetS [45]. IL-4, a key pro-inflammatory cytokine in asthma, has been shown to promote intracellular ADMA accumulation, leading to mitochondrial loss through oxo-nitrative stress and hypoxic response. This provides a novel understanding of how obesity with high ADMA

Obesity   **Involvement of mitochondria**   Lung disease

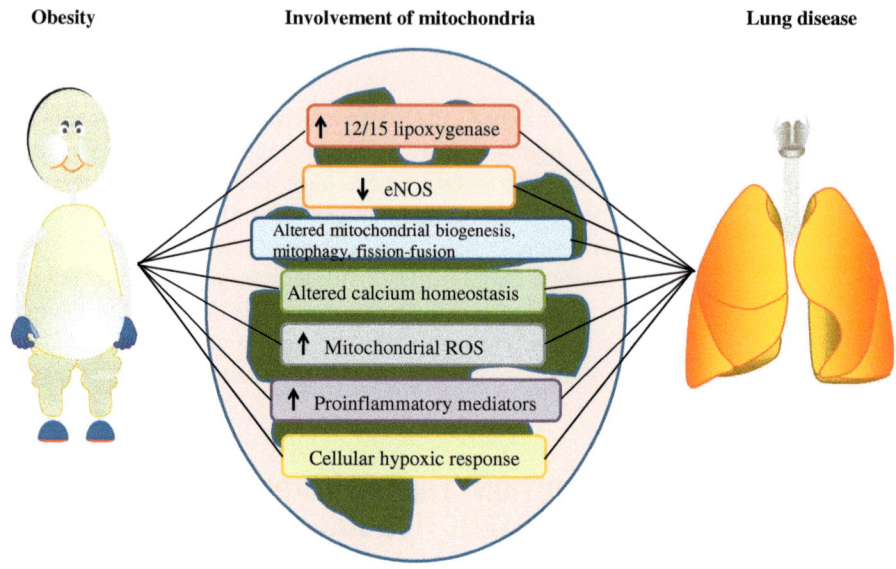

↑ 12/15 lipoxygenase

↓ eNOS

Altered mitochondrial biogenesis, mitophagy, fission-fusion

Altered calcium homeostasis

↑ Mitochondrial ROS

↑ Proinflammatory mediators

Cellular hypoxic response

FIGURE 7.2

Involvement of mitochondria in the pathogenesis of obesity and lung disorders, highlighting overlapping mechanistic links through which one may contribute to the other. An increase in 12/15 LOX in the adipocytes due to nutrient overload, or in airway epithelial cells in obstructive lung disease, leads to inflammation and reduced eNOS expression. This decreases fatty acid oxidation in the mitochondria and increases lipid accumulation in adipocytes. The resultant release in free fatty acids causes mitochondrial dysfunction. This causes reduction in mitochondrial biogenesis via the PGC-1α pathway. 12/15 LOX metabolites also cause mitochondrial dysfunction through an increase in calcium overload. Cellular hypoxic response mounted in both obesity and obstructive lung diseases also leads to reduction in mitochondrial genesis. NOS, Nitric oxide synthase; ROS, reactive oxygen species; COPD, chronic obstructive pulmonary disorder.

levels and asthma with high IL-4 levels can together potentiate each other and cause mitochondrial dysfunction [73]. Amongst the MetS components, abdominal obesity has the strongest known link to deteriorating lung function. Hypoxic response is a phenomenon that has been reported in both obesity and lung diseases. In obesity, due to large masses of adipose tissue, there is often a lack of adequate tissue oxygen supply leading to chronic hypoxia. On the other hand, airway epithelium in mice with allergic asthma and COPD airways have also been shown to have marked cellular hypoxic response, even in normoxia. This leads to a decrease in mitochondrial biogenesis [74]. An increase in certain proinflammatory mediators such as leptin, IL-6, TNFα, and C-reactive protein, and decrease of adiponectin has been found in obese asthmatics [75].

MITOCHONDRIAL DYSFUNCTION AND OBSTRUCTIVE LUNG DISEASES

EXPERIMENTAL LINKS BETWEEN MITOCHONDRIAL DYSFUNCTION AND LUNG DISEASE

Mitochondrial Dysfunction in COPD—Airway smooth muscle (ASM) cells from COPD patients show mitochondrial dysfunction represented by reduced membrane potential, ATP content, and decreased respiration and complex activity, in addition to increased mtROS. This is reversed by mitochondria-targeted antioxidant therapy [76]. Mitochondrial membrane potential in thrombocytes derived from COPD is found to be reduced compared with healthy subjects. Acute cigarette smoke (CS) exposure in A/J mice for 4 weeks alters glucose metabolism in the lung, leading to a decrease in glycolytic rate and an increase in the pentose phosphate pathway. These changes were associated with a change in cellular and mitochondrial redox status. CS exposure for 8 weeks results in changes in genes involved in ETC, OxPhos, metabolism, mitochondrial dynamics, and redox metabolism [77]. Patients with COPD show impaired β-oxidation of fatty acids, ETC, and citric acid cycle in airway epithelial and smooth muscle cells. Mitochondrial biogenesis is decreased in COPD airways, although both increased mitochondrial fission and hyperfusion have been observed [78]. COPD and associated hypoxia can detrimentally affect mitochondria. Bitter taste receptor TAS2R agonists, chloroquine and quinine, were found to induce autophagy and subsequent ASM cell death through decreasing mitochondrial membrane potential and cellular ATP levels, and increasing mtROS and fragmentation [79]. These diverse findings show that the cellular mechanisms utilized by the bitter taste receptors to alleviate obstructive airway diseases involves mitochondrial pathway. In mice, genetic deficiency of FAM13A, a well-replicated COPD GWAS gene, results in resistance to CS-induced emphysema. FAM13A is known to induce expression of carnitine palmitoyl transferase 1A (CPT1A), a key mitochondrial protein involved in the FAO pathway, which ultimately increases FAO. CS induction produces a metabolic shift in cells from glucose utilization to greater lipid utilization in ASM. Chronic exposure to CS may thus lead to sustained increase in FAO followed by mtROS accumulation, mitochondrial damage, and cell death [80]. Another study showed mice deficient in the COPD susceptibility gene iron responsive element binding protein 2 (*IRP2*) are resistant to CS-induced experimental COPD. IRP2 regulates mitochondrial heme biosynthesis in the lung. CS-induced increase in IRP2 levels increases iron load in the mitochondria and COX levels. Excessive iron deposition in alveolar macrophages has been associated with poorer clinical outcomes in COPD patients [57]. Accelerated cell senescence in airway and alveolar epithelial cells is a pathological characteristic of COPD. Mitophagy via the PTEN-induced kinase 1 (PINK1) Parkin (PARK2) pathway has been identified as a primary pathway by which CS-induced damaged mitochondria are removed. However, studies show that PARK2 is decreased in COPD lungs [81], and Parkin translocation to damaged mitochondria is hindered [82]. This impairs mitophagy leading to accumulation of unhealthy mitochondria and higher ROS, as seen in COPD. Mice with genetic deficiency of PINK1 are protected against CS-induced cell death, mitochondrial dysfunction, altered mucociliary clearance, and airspace enlargement (emphysema) [83]. In patients with COPD, tissue injury and cell death have been related to increased necrosis, rather than the noninflammatory process of apoptosis. This leads to inflammation and damage to airway cells. Mitochondria were found to be responsible for this switch, as CS exposure inhibits complex I and II activities thus decreasing production of ATP, which is crucial for apoptosis-mediated cell death. Thus, targeting mitochondria along with CS cessation may retard the damage to airway tissue in lung diseases [84].

Mitochondria of bronchial epithelial cells are more prone to fragmentation in the lungs of COPD patients [85]. Increased mitochondrial fission is associated with enhanced mtROS and cellular senescence in CS-exposed human bronchial epithelial cells, and thus may potentially contribute to the pathogenesis of COPD. UCP3 levels have been found to increase with overeating in healthy subjects to protect mitochondria from dysregulation of fatty acid metabolism. However, UCP3 levels are significantly lower in limb muscles of COPD patients, indicating failure to upregulate UCP3. Exercise training proves useful in such patients, increasing UCP3 levels and improving lipid peroxidation status [86].

One of the major pathways involved in the pathogenesis of COPD is an imbalance between oxidant vs. antioxidant levels, with increased oxidative stress being a hallmark [87]. Although generation of ROS and reactive nitrogen species (RNS) are part of normal mitochondrial functioning, disturbances in antioxidant defenses play an important role in such imbalances. MtROS drives metalloprotease MMP2 activation, which results in a negative feedback cycle and degrades mitochondrial membrane potential and impairs mitochondrial function. MtROS can also act as signal transducers to release proinflammatory mediators [88]. The relevance of mitochondrial respiration in lung disease lies in the increasing recognition that factors such as inflammation in asthma and CS exposure in COPD influence bioenergetics and ROS balance and vice versa. ATP levels are increased in BAL with COPD, which then acts as DAMPs to activate the inflammasome, increase mtROS, and further perpetuate mitochondrial dysfunction. Inflammation can also affect other aspects of mitochondrial function; for example, TNFα and IL-13 increase ROS generation and mitochondrial membrane permeabilization, promoting loss of mitochondrial Ca^{2+} and ATP production, finally initiating apoptosis [88].

Mitochondrial Dysfunction in Asthma—Mitochondrial dysfunction was shown to exacerbate allergic airway inflammation [89]. Mice treated with a low dose of inhaled rotenone, an inhibitor of ETC complex I (complex I is the site within the mitochondria where NADH is reduced and a proton gradient is generated, thus driving the ETC), develop features of airway hyperresponsiveness and airway remodeling, whereas withdrawal of rotenone leads to reversal of such features [89]. Dysmorphic mitochondria have been observed in epithelium and ASM of asthmatic patients and in mouse models of allergic asthma. Cultured airway epithelial cells from asthma patients and in experimental models show mitochondrial fragmentation and swelling, and activation of apoptotic pathways, as well as enhanced oxidative damage [4]. Because mitochondrial respiration capacity is compromised, there is an increase in the number of abnormal mitochondria with higher ROS production [72]. A two-fold increase in mitochondrial numbers has been reported in ASM of patients with severe asthma [90].

The role of mitochondria in regulation of Ca^{2+} in ASM to protect from hypercontractility is found to be abnormal in asthmatic ASM in the presence of inflammatory cytokines such as TNFα. The Ca^{2+} channel blocker, gallopamil, reduces ASM proliferation in asthmatics as well as decreases patient exacerbations [90]. Apart from inflammatory cytokines, a number of lipid metabolites, such as 13-S HODE and 12/15 lipoxygenase, induce severe asthma in mouse models, along with robust mitochondrial dysfunction. These lipid intermediates may be generated in the body due to inflammation or be produced as a result of other disorders that precipitate asthma. Key inflammatory Th2 cytokines such as IL-4 and IL-13 have been shown to oxidize 13S-HODE to contribute to asthma by inducing mitochondrial dysfunction [70]. An increase in ADMA, a methyl arginine that uncouples NO synthase generates higher ROS and induces mitochondrial dysfunction in a model of allergic asthma. Interestingly, ADMA is also upregulated in obesity due to increased protein turnover [72]. IL-4-induced mitotoxicity is higher in the presence of exogenous ADMA [73]. IL-4 promotes intracellular ADMA accumulation, resulting in loss of mitochondria through hypoxia and oxo-nitrative stress [45]. Exposure to allergens, such as ragweed pollen, damages mitochondrial ETC components [91]. Platelets from asthma patients are less reliant on glycolysis and show increased TCA cycle activity, consistent with previous reports that OxPhos is more efficient in asthmatic individuals. This potential metabolic shift may be due to hypoxic adaptation in asthma patients [92]. Mice with experimental allergic asthma show reduction in expression of cytochrome *c* and complex I in bronchial epithelium, loss of mitochondrial cristae, and a decrease in ATP levels in the lung [4]. Decrease in levels of an antioxidant enzyme, superoxide dismutase, is seen in asthmatic airways. Preexisting mitochondrial dysfunction in airway epithelia worsen asthma symptoms [45]. Interestingly, mitochondrial biogenesis is increased in asthmatic ASM cells, probably necessary for airway remodeling in asthma [76]. Thus the status of mitochondria may be different in different cell types in asthmatic lungs.

GENETIC ASSOCIATION BETWEEN MITOCHONDRIA AND LUNG DISEASE

Although there are limited data on inherited risk in COPD, asthma has a stronger hereditary component. Maternal inheritance is considered to be one of the strongest risk factors for asthma and other atopic diseases [93–96]. Hence, mitochondria could be involved in vertical transmission of asthma. Mitochondrial haplogroup U has been shown to be associated with increased serum IgE levels in the European population. Various mutations in mitochondrial genes encoding mitochondrial tRNAs are associated with asthma. In addition, ATP synthase mitochondrial F1 complex assembly factor 1 gene is associated with asthma in Caucasian European children [97]. A3243G tRNA Leu (UUR) functional mutation is present in some rare forms of asthma, which is also associated with ischemic heart disease, hypertension, and age-related maculopathy. These genetic associations may indicate a further causal effect of mitochondria in asthma [45]. Pathogenic defects in mtDNA have been found to be present in asthma patients [92]. Ratio of mtDNA/nDNA is higher in exhaled breath condensate (EBC) of 13 COPD, 14 asthmatic, and 23 asthma-COPD (ACOS) patients [98]. COPD muscle has been shown to harbor higher levels of oxidatively damaged DNA and mtDNA deletion mutations. Long range PCR revealed that mtDNA from COPD limb muscle is more prone to deletion mutations, with a 72% mutation frequency as opposed to a 16% in the age-matched control group [99]. Mitochondrial gene expression is changed in the placenta of asthmatic mothers [90]. A genome-wide study including a sample of 372 asthmatic children and 395 healthy children detected polymorphisms in cytochrome B (in males) and NADH dehydrogenase 2/16S RNA (in females). These genes are strongly associated with ROS production [90]. Causal role for mitochondrial dysfunction has been shown by mouse studies with genetic deficiency of mitochondrial ubiquinol cytochrome *c* reductase core II protein in airway epithelium. These mice have mitochondrial dysfunction in the airways and show greater inflammation and airway remodeling than normal mice sensitized to and challenged with antigen [100].

THERAPIES TARGETING MITOCHONDRIA

Preceding sections highlight the role of mitochondria as intricately linked to the pathobiology of two seemingly disparate diseases, obesity and obstructive lung disorders. Although the occurrence of lung disease in obese patients, or vice versa, may be associative with some shared comorbidities, there is a plethora of evidence now to believe that mitochondrial dysfunction in obesity can be directly responsible for development of lung diseases. Considering the significant involvement of mitochondrial dysfunction in such disease scenarios, targeting mitochondria for alleviating disease or relief of symptoms is an attractive option, irrespective of it being the cause or the consequence. Currently, there is no specific approved therapy for the distinct group of patients with both obesity and obstructive lung disease. However, it is increasingly being recognized that patients with obese-asthma or COPD may not be very effectively treated with antiinflammatory therapy [90]. Weight loss is the first line of treatment in these cases; both exercise and CR are beneficial to decrease body weight as well as stimulate mitochondrial biogenesis. However, these options may not always be feasible for such patients; thus more precise therapies to target mitochondria are needed.

General antioxidants vitamin E (α-tocopherol) or vitamin C have been used, but they have not shown any benefits in clinical studies [90]. One way to negate the harmful effects of mitochondrial dysfunction is to target mtROS. MitoQ, MitoTempo, and Tiron are powerful mitochondrial antioxidants that have shown efficacy in preventing and reversing mitochondrial oxidative damage [90]. An alternate approach to optimize oxidant-antioxidant levels in the cell is to increase levels of Nrf-2, a master antioxidant transcription factor. Nrf-2-stimulating molecule, sulforaphane, has been shown to be protective against COPD. The caveat with this therapeutic approach to reduce ROS is that low levels of ROS are beneficial in the cell, and ROS production is heavily dependent on the cell type and disease conditions.

Another reliable strategy is to administer metformin, which has shown antiinflammatory effects in the allergic model of asthma as well as in obese mice. Also, sirtuin activator (resveratrol) inhibitors of 12/15 LOX (baicalein and esculetin) have also been shown to be efficacious in mouse models through targeting mitochondrial biogenesis. Supplementation of L-arginine was shown to target the NO pathway and improve mitochondrial function.

The most recent advance in mitochondrial therapy is replacement of damaged mitochondria through transfer of healthy mitochondria from donor cells. There are numerous studies now to substantiate the benefits of such a strategy [101, 102]. Mesenchymal stem cells have been found to be very effective donors to transfer mitochondria to recipient cells with mitochondrial dysfunction. Although a number of factors such as optimization of dose, source of MSC, donor efficiency, etc. would impact the final outcome, this direction looks promising. This is currently in clinical trial phases I and II for some diseases [90].

SUMMARY

Mitochondrial dysfunction in adipose tissue, skeletal muscle, and liver has long been known in obesity. On the other hand, the involvement of mitochondria in airway diseases such as asthma and COPD are also well demonstrated. There appears to be a bidirectional relationship. We highlight the contribution of mitochondria as a unifying link between these two diseases. Although there are lacunae in our understanding of the field, therapeutic targeting of mitochondria may be a viable strategy to treat asthma or COPD in obesity.

REFERENCES

[1] Bugger H, Abel ED. Molecular mechanisms for myocardial mitochondrial dysfunction in the metabolic syndrome. Clin Sci 2008;114:195–210.
[2] Ho WC, Lin YS, Caffrey JL, Lin MH, Hsu HT, Myers L, et al. Higher body mass index may induce asthma among adolescents with pre-asthmatic symptoms: a prospective cohort study. BMC Public Health 2011;11(1):542.
[3] Yue L, Yao H. Mitochondrial dysfunction in inflammatory responses and cellular senescence: pathogenesis and pharmacological targets for chronic lung diseases. Br J Pharmacol 2016;173:2305–18.
[4] Mabalirajan U, Dinda AK, Kumar S, Roshan R, Gupta P, Sharma SK, et al. Mitochondrial structural changes and dysfunction are associated with experimental allergic asthma. J Immunol 2008;181(5):3540–8.
[5] Margulis L, Sagan D. Origins of sex. Three billion years of genetic recombination. vol. 87. Yale University Press; 1986. p. 69–71.

[6] Chipuk JE. Mitochondrial outer membrane permeabilization during apoptosis: the innocent bystander scenario. Cell Death Differ 2006;13:1396–402.

[7] Alberts B, Johnson A, Lewis J, Raff M, Roberts K, Walter P. Molecular biology of the cell; 1994.

[8] Osman C, Voelker DR, Langer T. Making heads or tails of phospholipids in mitochondria. J Cell Biol 2011;192(1):7–16.

[9] Twig G, Elorza A, Molina AJA, Mohamed H, Wikstrom JD, Walzer G, et al. Fission and selective fusion govern mitochondrial segregation and elimination by autophagy. EMBO J 2008;27(2):433–46.

[10] De Brito OM, Scorrano L. Endoplasmic reticulum—mitochondria relationship. EMBO J 2010;29(16):2715–23.

[11] Rizzuto R, Marchi S, Bonora M, Aguiari P, Bononi A, Stefani DD, et al. Ca^{2+} transfer from the ER to mitochondria: when, how and why. Biochim Biophys Acta 2009;1787(11):1342–51.

[12] Donald Voet, Judith G. Voet, Charlotte W. Pratt. Fundamentals of biochemistry, 2nd ed. 2006; 547–556.

[13] Stryer L. Citric acid cycle. Biochemistry. 4th ed; 1995. 509–527, 569–579, 614–616, 638–641, 732–735, 739–748, 770–773.

[14] Rousset S, Mozo J, Miroux B, Ricquier D. The biology of mitochondrial uncoupling proteins. Diabetes 2004;53.

[15] Gong D, He Y, Karas M, Reitman M. Uncoupling protein-3 is a mediator of thermogenesis regulated by thyroid hormone, 3-adrenergic agonists. JBC 1997;272(39):24129–33.

[16] Chan DC. Fusion and fission: interlinked processes critical for mitochondrial health. Annu Rev Genet 2012;46:265–87.

[17] Chen H, Vermulst M, Wang YE, Chomyn A, Prolla TA, McCaffrey JM, et al. Mitochondrial fusion is required for mtDNA stability in skeletal muscle and tolerance of mtDNA mutations. Cell 2010;141(2):280–9.

[18] Zhou R, Yazdi A, Menu P, Tschopp J. A role for mitochondria in NLRP3 inflammasome activation. Nature 2011;469:221–5.

[19] Emre Y, Hurtaud C, Ubel T, Criscuolo F, Ricquier D, Cassard-Doulcier A-M. Mitochondria contribute to LPS-induced MAPK activation via uncoupling protein UCP2 in macrophages. Biochem J 2007;402:271–8.

[20] Shimada K, Crother TR, Karlin J, Dagvadorj J, Chen S, Ramanujan VK, et al. Oxidized mitochondrial DNA activates the NLRP3 inflammasome during apoptosis. Immunity 2012;36(3):401–14.

[21] Oka T, Hikoso S, Yamaguchi O, Taneike M, Toshihiro T, Tamai T, et al. Mitochondrial DNA that escapes from autophagy causes inflammation and heart failure. Nature 2012;485:251–5.

[22] Lionetti L, Mollica MP, Donizzetti I, Gifuni G, Sica R, Pignalosa A, et al. High-lard and high-fish-oil diets differ in their effects on function and dynamic behaviour of rat hepatic mitochondria. PLoS One 2014;9(3).

[23] Bach D, Pich S, Soriano FX, Vega N, Baumgartner B, Oriola J, et al. Mitofusin-2 determines mitochondrial network architecture and mitochondrial metabolism: a novel regulatory mechanism altered in obesity. J Biol Chem 2003;278(19):17190–7.

[24] Jheng H-F, Tsai P-J, Guo S-M, Kuo L-H, Chang C-RC-S, Su I-J, et al. Mitochondrial fission contributes to mitochondrial dysfunction and insulin resistance in skeletal muscle. Mol Cell Biol 2012;32:309–19.

[25] Putti R, Sica R, Migliaccio V, Lionetti L. Diet impact on mitochondrial bioenergetics and dynamics. Front Physiol 2015;6:1–7.

[26] Anderson EJ, Lustig ME, Boyle KE, Woodlief TL, Kane DA, Lin C, et al. Mitochondrial H_2O_2 emission and cellular redox state link excess fat intake to insulin resistance in both rodents and humans. J Clin Investig 2009;119(3):573–81.

[27] Xu XJ, Babo E, Qin F, Croteau D, Colucci WS. Short-term caloric restriction in db/db mice improves myocardial function and increases high molecular weight (HMW) adiponectin. IJC Metab Endocr 2016;13:28–34.

[28] Montgomery MK, Turner N. Mitochondrial dysfunction and insulin resistance: an update. Endocr Connect 2015;4:1–15.

[29] Kelley DE, Goodpaster B, Wing RR, Simoneau J, Maclean PS, Bergouignan A, et al. Skeletal muscle fatty acid metabolism in association with insulin resistance, obesity, and weight loss. Am J Physiol 1999;277:1130–41.

[30] Semple RK, Crowley VC, Sewter CP, Laudes M, Christodoulides C, Considine RV, et al. Expression of the thermogenic nuclear hormone receptor coactivator PGC-1alpha is reduced in the adipose tissue of morbidly obese subjects. Int J Obes Relat Metab Disord 2004;28(1):176–9.

[31] Crunkhorn S, Dearie F, Mantzoros C, Gami H, Silva WS, Espinoza D, et al. Peroxisome proliferator activator receptor ϒ Coactivator-1 expression is reduced in obesity. JBC 2007;282(21):15439–50.

[32] Bournat JC, Brown CW. Mitochondrial dysfunction in obesity. Curr Opin Endocrinol Diabetes Obes 2010;17(5):446–52.

[33] Chen C, Lee T, Kwok C, Hsu Y, Shih K, Lin Y, et al. Cannabinoid receptor type 1 mediates high-fat diet-induced insulin resistance by increasing forkhead box O1 activity in a mouse model of obesity. Int J Mol Med 2016;743–54.

[34] Bach D, Naon D, Pich S, Soriano FX, Vega N, Rieusset J, et al. Expression of Mfn2, the Charcot-Marie-tooth neuropathy type 2A gene, in human skeletal muscle. Diabetes 2005;54:2685–93.

[35] Eccleston HB, Andringa KK, Betancourt AM, King AL, Mantena SK, Swain TM, et al. Chronic exposure to a high-fat diet induces hepatic steatosis, impairs nitric oxide bioavailability. Antioxid Redox Signal 2011;15(2).

[36] Nuño-lámbarri N, Barbero-becerra VJ, Uribe M. Mitochondrial molecular pathophysiology of nonalcoholic fatty liver disease: a proteomics approach. Int J Mol Sci 2016;17:281.

[37] Engin A. Eat and death: chronic over-eating. Obesity and lipotoxicity. Adv Exp Med Biol 2017;960.

[38] Yuzefovych LV, Musiyenko SI, Wilson GL, Rachek LI. Mitochondrial DNA damage and dysfunction, and oxidative stress are associated with endoplasmic reticulum stress, protein degradation and apoptosis in high fat diet-induced insulin resistance mice. PLoS One 2013;8(1).

[39] Arruda AP, Pers BM, Parlakgul G, Guney E, Inouye K, Hotamisligil GS. Chronic enrichment of hepatic ER-mitochondria contact sites leads to calcium dependent mitochondrial dysfunction in obesity. Nat Med 2014;20(12):1427–35.

[40] Tedesco L, Valerio A, Dossena M, Cardile A, Ragni M, Pagano C, et al. Cannabinoid receptor stimulation impairs mitochondrial biogenesis in mouse white adipose tissue, muscle, and liver: the role of eNOS, p38 MAPK, and AMPK pathways. Diabetes 2010;59(11):2826–36.

[41] Ma W, Yuan L, Yu H, Xi Y, Xiao R. Mitochondrial dysfunction and oxidative damage in the brain of diet-induced obese rats but not in diet-resistant rats. Life Sci 2014;110(2):53–60.

[42] Hernández-aguilera A, Rull A, Rodríguez-gallego E, Riera-borrull M, Luciano-mateo F, Camps J, et al. Mitochondrial dysfunction: a basic mechanism in inflammation-related non-communicable diseases and therapeutic opportunities. Mediators Inflamm 2013;.

[43] Hirschey MD, Shimazu T, Goetzman E, Jing E, Lombard DB, Grueter CA, et al. SIRT3 regulates fatty acid oxidation via reversible enzyme deacetylation. Nature 2010;464(7285):121–5.

[44] Kong X, Wang R, Xue Y, Liu X, Zhang H, Chen Y, et al. Sirtuin 3, a new target of PGC-1α, plays an important role in the suppression of ROS and mitochondrial biogenesis. PLoS One 2010;5(7).

[45] Mabalirajan U, Ghosh B. Mitochondrial dysfunction in metabolic syndrome and asthma. J Allergy 2013;.

[46] Rossmeisl M, Syrovy I, Baumruk F, Flachs P, Janovska P, Kopecky J. Decreased fatty acid synthesis due to mitochondrial uncoupling in adipose tissue. FASEB J 2000;14(2):1793–800.

[47] Leone TC, Lehman JJ, Finck BN, Schaeffer PJ, Wende AR, Boudina S, et al. PGC-1 α deficiency causes multi-system energy metabolic derangements: muscle dysfunction, abnormal weight control and hepatic steatosis. PLoS Biol 2005;3(4).

[48] Tyynismaa H, Caroll CJ, Raimundo N, Ahola-Erkkila S, Wenz T, Ruhanen H, et al. Mitochondrial myopathy induces a starvation-like response. Hum Mol Genet 2010;19(20):3948–58.

[49] Kim KH, Jeong YT, Oh H, Seong HK, Jae MC, Yo-Na K, et al. Autophagy deficiency leads to protection from obesity and insulin resistance by inducing Fgf21 as a mitokine. Nat Med 2013;19:83–92.

[50] Lee M. Effect of mitochondrial stress on systemic metabolism. Ann N Y Acad Sci 2015;1350:61–5.

[51] Soro-Arnaiz I, Li QOY, Torres-Capelli M, Meléndez-Rodríguez F, Veiga S, Veys K, et al. Role of mitochondrial complex IV in age-dependent obesity. Cell Rep 2016;16(11):2991–3002.

[52] Cole BK, Morris MA, Grzesik WJ, Leone KA, Nadler JL. Adipose tissue-specific deletion of 12/15-lipoxygenase protects mice from the consequences of a high-fat diet. Mediators Inflamm 2012;.

[53] Dunham-snary KJ, Ballinger SW. Mitochondrial genetics & obesity: evolutionary adaptation & contemporary disease susceptibility. Free Radic Biol Med 2013;65:.

[54] Price RA, Gottesman II. Body fat in identical twins reared apart: roles for genes and environment. Behav Genet 1991;21(1):1–7.

[55] Sorensen T, Holst C, Stunkard AJ. Adoption study of environmental modifications of the genetic influences on obesity. Int J Obes Relat Metab Disord 1998;22(1):73–81.

[56] Zonta LA, Javakar SD, Bosisio M, Galante A, Penneti V. Genetic analysis of human obesity in an Italian sample. Hum Hered 1987;37(3):129–39.

[57] Cloonan SM, Choi AMK. Mitochondria in lung disease. J Clin Invest 2016;126(3):809–20.

[58] Samuelsson A, Matthews PA, Argenton M, Christie MR, Mcconnell JM, Jansen EHJM, et al. Diet-induced obesity in female mice leads to offspring and insulin resistance a novel murine model of developmental programming. Hypertension 2008;51:383–92.

[59] Igosheva N, Abramov AY, Poston L, Eckert JJ, Fleming TP, Michael R, et al. Maternal diet-induced obesity alters mitochondrial activity and redox status in mouse oocytes and zygotes. PLoS One 2010;5(4):1–8.

[60] Dumollard R, Duchen M, Carroll J. The role of mitochondrial function in the oocyte and embryo. Curr Top Dev Biol 2007;77:21–49.

[61] Rao KR, Lal N, Giridharan NV. Genetic and epigenetic approach to human obesity. Indian J Med Res 2014;140(5):589–603.

[62] Shungin D, Winkler TW, Croteau-Chonka DC, Ferreira T, Locke AE, Reedik M. New genetic loci link adipose and insulin biology to body fat distribution. Nature 2015;518(7538):187–96.

[63] Plummer C, Spring PJ, Marotta R, Chin J, Taylor G, Sharpe D, et al. Multiple symmetrical lipomatosis—a mitochondrial disorder of brown fat. Mitochondrion 2013;13(4):269–76.

[64] Fiorino EK, Obesity BLJ. Respiratory diseases in childhood. Clin Chest Med 2009;30(3):601–8.

[65] Johnston RA, Zhu M, Rivera-Sanchez YM, Lu FL, Theman TA, Flynt L. Allergic airway responses in obese mice. Am J Respir Crit Care Med 2007;176(7).

[66] Galicia-negrete G, Falfán-valencia R. Mediators of inflammatory response in asthma and its association with obesity. Rev Alerg Mex 2017;64(2):198–205.

[67] Nunemaker CS, Chen M, Pei H, Kimble SD, Keller SR, Carter JD, et al. 12-Lipoxygenase knockout mice are resistant to inflammatory effects of obesity induced by western diet. Am J Physiol Endocrinol Metab 2008;295(5):1065–75.

[68] Andersson CK, Claesson HE, Rydell-Törmänen K, Swedmark S, Hällgren A, Erjefält JS. Mice lacking 12/15-lipoxygenase have attenuated airway allergic inflammation and remodeling. Am J Respir Cell Mol Biol 2008;39(6):648–56.

[69] Mabalirajan U, Rehman R, Ahmad T, Kumar S, Leishangthem GD, Singh S, et al. 12/15-lipoxygenase expressed in non-epithelial cells causes airway epithelial injury in asthma. Sci Rep 2013;3:1540.

[70] Mabalirajan U, Rehman R, Ahmad T, Kumar S, Singh S, Leishangthem GD, et al. Linoleic acid metabolite drives severe asthma by causing airway epithelial injury. Sci Rep 2013;3:1349.

[71] Mabalirajan U, Ahmad T, Leishangthem GD, Joseph DA, Dinda AK, Agrawal A, et al. Beneficial effects of high dose of L-arginine on airway hyperresponsiveness and airway inflammation in a murine model of asthma. J Allergy Clin Immunol 2010;125(3):626–35.

[72] Ahmad T, Mabalirajan U, Ghosh B, Agrawal A. Altered asymmetric dimethyl arginine metabolism in allergically inflamed mouse lungs. Am J Respir Cell Mol Biol 2009;42(1).

[73] Pattnaik B, Bodas M, Bhatraju K, Ahmad T, Pant R. IL-4 promotes asymmetric dimethylarginine accumulation, oxo-nitrative stress, and hypoxic response—induced mitochondrial loss in airway epithelial cells. J Allergy Clin Immunol 2016;.

[74] Ahmad T, Kumar M, Mabalirajan U, Pattnaik B, Aggarwal S, Singh R, et al. Hypoxia response in asthma: differential modulation on inflammation and epithelial injury. Am J Respir Cell Mol Biol 2012;47(1):1–10.

[75] Sears DD, Miles PD, Chapman J, Ofrecio JM, Almazan F, Thapar D, et al. 12/15-lipoxygenase is required for the early onset of high fat diet-induced adipose tissue inflammation and insulin resistance in mice. PLoS One 2009;4(9):7250.

[76] Wiegman CH, Michaeloudes C, Haji G, Narang P. Oxidative stress—induced mitochondrial dysfunction drives inflammation and airway smooth muscle remodeling in patients with chronic obstructive pulmonary disease. J Allergy Clin Immunol 2015;136:769–80.

[77] Agarwal AR, Yin F, Cadenas E. Short-term cigarette smoke exposure leads to metabolic alterations in lung alveolar cell. Am J Respir Cell Mol Biol 2014;51(2):284–93.

[78] Mirrakhimov AE. Chronic obstructive pulmonary disease and glucose metabolism: a bitter sweet symphony. Cardiovasc Diabetol 2012;11(1):1.

[79] Pan S, Sharma P, Shah SD, Deshpande DA. Bitter taste receptor agonists alter mitochondrial function and induce autophagy in airway smooth muscle cells. Am J Physiol Lung Cell Mol Physiol 2017;154–65.

[80] Jiang Z, Knudsen NH, Wang G, Qiu W, Zar Z, Naing C, et al. Genetic control of fatty acid β-oxidation in chronic obstructive pulmonary disease. Am J Respir Cell Mol Biol 2017;1–54.

[81] Ito S, Araya J, Kurita Y, Kobayashi K, Takasaka N, Yoshida M, et al. PARK2-mediated mitophagy is involved in regulation of HBEC senescence in COPD pathogenesis. Clin Res Paper 2015;547–59.

[82] Ahmad T, Sundar IK, Lerner CA, Gerloff J, Tormos AM, Yao H, et al. Impaired mitophagy leads to cigarette smoke stress- induced cellular senescence: implications for chronic obstructive pulmonary disease. FASEB J 2015;1–18.

[83] Mizumura K, Cloonan SM, Nakahira K, Bhashyam AR, Cervo M, Kitada T, et al. Mitophagy-dependent necroptosis contributes to the pathogenesis of COPD. J Clin Invest 2014;124(9):3987–4003.

[84] Van Der TM, Slebos D, De BHG, Leuvenink HG, Bakker SJL, Gans ROB, et al. Cigarette smoke-induced blockade of the mitochondrial respiratory chain switches lung epithelial cell apoptosis into necrosis. Am J Physiol Lung Cell Mol Physiol 2007;1211–8.

[85] Hara H, Araya J, Ito S, Kobayashi K, Takasaka N, Yoshii Y, et al. Mitochondrial fragmentation in cigarette smoke-induced bronchial epithelial cell senescence. Am J Physiol Lung Cell Mol Physiol 2013;.

[86] Gosker HR, Schrauwen P, Broekhuizen R, Hesselink MKC, Moonen-kornips E, Ward KA, et al. Exercise training restores uncoupling protein-3 content in limb muscles of patients with chronic obstructive pulmonary disease. AJP Endocrinol Metab 2006;3:976–81.

[87] Jerzy A, Bia B, Sitarek PB, B JM, Piotrowski WJ, Górski PB. The role of mitochondria and oxidative/antioxidative imbalance in pathobiology of chronic obstructive pulmonary disease. Oxid Med Cell Longev 2016;.

[88] Yoon CM, Nam M, Oh YM, Cruz CSD, Kang MJ. Mitochondrial regulation of inflammosome activation in chronic obstructive pulmonary disease. J Innate Immun 2016;8(2):121–8.

[89] Aguilera-Aguirre L, Bacsi A, Saavedra-Molina A, Kurosky A, Sur S, Boldogh I. Mitochondrial dysfunction increases allergic airway inflammation. J Immunol 2009;183(8):5379–87.

[90] Agrawal A, Mabalirajan U. Rejuvenating cellular respiration for optimizing respiratory function: targeting mitochondria. Am J Physiol Lung Cell Mol Physiol 2016;310:103–13.

[91] Aravamudan B, Thompson MA, Pabelick C, Prakash YS. Mitochondria in lung diseases. Expert Rev Respir Med 2013;7(6):631–46.

[92] Xu W, Cardenes N, Corey C, Erzurum SC, Shiva S. Platelets from asthmatic individuals show less reliance on glycolysis. PLoS One 2015;10(7):1–15.

[93] Oliveti JF, Kercsmar CM, Redline S. Pre- and perinatal risk factors for asthma in inner city African-American children. Am J Epidemiol 1996;143(6):5.

[94] Litonjua AA, Carey VJ, Burge HA, Weiss ST, Gold DR. Parental history and the risk for childhood asthma. Am J Respir Crit Care Med 1998;158(1).

[95] Soto-Quiros ME, Silverman EK, Hanson LA, Weiss MD, Celedon JC. Maternal history, sensitization to allergens, and current wheezing, rhinitis, and eczema among children in Costa Rica. Pediatr Pulmonol 2002;33(4):237–43.

[96] Lim RH, Kobzik L, Dahl M. Risk for asthma in offspring of asthmatic mothers versus fathers: a meta analysis. PLoS One 2010;5(4):e10134.

[97] Schauberger EM, Ewart SL, Arshad SH, Huebner M, Karmaus W, Holloway JW, et al. Identification of *ATPAF1* as a novel candidate gene for asthma in children. J Allergy Clin Immunol 2011;128(4):753–60.

[98] Carpagnano GE, Lacedonia D, Carone M, Soccio P, Cotugno G, Palmiotti GA, et al. Study of mitochondrial DNA alteration in the exhaled breath condensate of patients affected by obstructive lung diseases. J Breath Res 2016;10(2).

[99] Konokhova Y, Spendiff S, Jagoe RT, Aare S, Kapchinsky S, Macmillan NJ, et al. Failed upregulation of TFAM protein and mitochondrial DNA in oxidatively deficient fibers of chronic obstructive pulmonary disease locomotor muscle. Skelet Muscle 2016;1–16.

[100] Agrawal A, Prakash YS. Obesity, metabolic syndrome, and airway disease: a bioenergetic problem? Immunol Allergy Clin N Am 2014;34(4):785–96.

[101] Ahmad T, Mukherjee S, Pattnaik B, Kumar M, Singh S, Rehman R, et al. Miro1 regulates intercellular mitochondrial transport & enhances mesenchymal stem cell rescue efficacy. EMBO J 2014;33(9):994–1010.

[102] Islam MN, Das SR, Emin MT, Wei M, Sun L, Westphalen K, et al. Mitochondrial transfer from bone-marrow-derived stromal cells to pulmonary alveoli protects against acute lung injury. Nat Med 2012;18(5):759–65.

[103] Qian L, Xu K, Xu X, Gu R, Liu X, Shan S, et al. UCP2-866G/A, Ala55Val and UCP3-55C/T polymorphisms in association with obesity susceptibility—a meta-analysis study. PLoS ONE 2013;8(4):e58939.

Metabolic syndrome and sleep apnea: A bidirectional relationship

Haris Younas*, Chenjuan Gu†, Aman Rathore*, Jonathan C. Jun*, Vsevolod Y. Polotsky*

Division of Pulmonary and Critical Care, Department of Medicine, Johns Hopkins University, Baltimore, MD, United States* Department of Pulmonary and Critical Care Medicine, Ruijin Hospital, Shanghai Jiao Tong University School of Medicine, Shanghai, China†

ABBREVIATIONS

OSA	obstructive sleep apnea
PSG	polysomnography
AHI	apnea hypopnea index
IH	intermittent hypoxia
CPAP	continuous positive airway pressure
BMI	body mass index
T2DM	type 2 diabetes mellitus
MHO	metabolically healthy obese
MONW	metabolically obese normal weight
SF	subcutaneous fat
VF	visceral fat
NC	neck circumference
MetS	metabolic syndrome
LDL	low density lipoproteins
HDL	high density lipoproteins
TG	triglyceride
SNS	sympathetic nervous system
IR	insulin resistance
RAS	renin-angiotensin system
GWAS	genome-wide association studies
EE	energy expenditure
GH	growth hormone
RCTs	randomized control trials
miRNAs	micro-ribonucleic acids
LPL	lipoprotein lipase
SREBP-1	sterol regulatory element binding protein 1
SCD-1	stearoyl coenzyme A desaturase 1

Mechanisms and Manifestations of Obesity in Lung Disease. https://doi.org/10.1016/B978-0-12-813553-2.00008-7

OBSTRUCTIVE SLEEP APNEA
DEFINITION

Obstructive sleep apnea (OSA) is a disorder of repeated collapse of the pharyngeal airway during sleep, resulting in arousals and oxyhemoglobin desaturations. Symptoms include snoring, sleepiness, gasping, dry mouth, headache, and nocturia. Patients may also be asymptomatic and present with snoring or witnessed pauses in breathing. Risk factors for OSA include obesity, male gender, smoking, craniofacial abnormalities, enlarged tonsils, and family history. During sleep, upper airway muscles relax and may occlude breathing. Interruptions in breathing are detected with polysomnography (PSG) in a laboratory or using newer portable monitors. Breathing pauses lasting at least 10 seconds are termed "apneas," whereas reduced periods of airflow are termed "hypopneas" when followed by oxygen desaturation or arousal. The sum of apneas and hypopneas divided by hours of sleep time yields the apnea hypopnea index (AHI). By most guidelines, adult OSA is defined as an AHI $\geq$ 5 events/hour. OSA severity can be characterized as mild (AHI 5–15 events/hour), moderate (AHI 15–30 events/hour), or severe (AHI $\geq$ 30 events/hour) [1]. Recent prevalence estimates of moderate to severe sleep apnea (AHI $\geq$ 15) are 10% among 30- to 49-year-old men; 17% among 50- to 70-year-old men; 3% among 30- to 49-year-old women; and 9% among 50- to 70-year-old women [2].

PATHOPHYSIOLOGY

Anatomical, neuromuscular, and breathing control defects during sleep contribute to OSA. Anatomical factors include bony abnormalities, adenotonsillar hypertrophy, pharyngeal lymphoid tissue, and adipose tissue in the neck or tongue [3]. Inadequate airway muscle tone during sleep is another factor. Chemo-reflexes to hypercapnia and hypoxia, as well as how easily one is awakened can destabilize sleep and breathing. Various phenotypes of OSA are characterized by different contributions of these factors [4].

TREATMENT

Therapy for OSA was limited to tracheostomy and weight loss until the advent of continuous positive airway pressure (CPAP) in the 1980s [5]. CPAP is now the first-line treatment for OSA and is highly effective but often difficult for patients to tolerate. Other treatment options include weight loss, oral appliances, hypoglossal nerve stimulation, and surgery.

CONSEQUENCES

A classic complaint of OSA patients is unrefreshing sleep and daytime sleepiness. Untreated OSA increases risk of occupational and motor vehicle accidents [6].

Children with OSA may present with hyperactivity, irritability, or inattention [7]. In addition, OSA is associated with cardiovascular, neurocognitive, and metabolic dysfunction. It remains unclear whether OSA is a mediator or marker of cardiometabolic risk.

METABOLIC SYNDROME AND VISCERAL OBESITY
DEFINITION OF OBESITY

Obesity is defined as a body mass index (BMI) of ≥ 30 kg/m^2. More than 2 billion adults and children worldwide are obese [8]. Obesity increases risks of chronic diseases such as type 2 diabetes (T2DM) and hypertension. However, the relation between BMI and these outcomes is not straightforward. For example, it is estimated that 30% of the obese population do not have dyslipidemia, hypertension, or insulin resistance (IR) [9]. These persons are classified as "metabolically healthy obese" (MHO) [10]. Conversely, 5%–45% of the nonobese population manifest metabolic dysfunction and are described as "metabolically obese normal weight" (MONW) [11].

IMPORTANCE OF FAT DISTRIBUTION

BMI does not quantify regional adiposity. This is an important limitation, because distribution of fat strongly predicts its metabolic impact. In 1956, French physician Jean Vague recognized that patients with central abdominal and upper-body "android" fat were susceptible to cardiometabolic disease [12]. He distinguished these patients from those with lower-body "gynoid" fat. "Android" and "gynoid" are often equated with "apple-shaped" male-pattern obesity and "pear-shaped" female-pattern obesity, respectively. Adipose tissue can also be characterized as subcutaneous fat (SF) or visceral fat (VF). SF comprises ~80% of body fat and functions as energy storage and insulation. VF comprises ~15% of body fat and includes the mesenteric and omental fat depots, which surround the abdominal organs. Although smaller in mass than SF, VF is associated with greater detrimental impacts. McLaughlin *et al.* performed CT scans in subjects classified as insulin sensitive or insulin resistant. At the same BMI, VF mass was larger in the insulin-resistant group, whereas increased SF mass was *protective* against IR [13]. Similarly, in a German study, insulin-sensitive and insulin-resistant participants were indistinguishable by total body fat or SF but differed by VF mass [14]. VF is associated with increased mortality, whereas SF is associated with decreased all-cause mortality [15].

METABOLIC SYNDROME

Abdominal VF tends to enlarge waist circumference, whereas SF accumulates more below the waist, increasing the hip circumference. Waist circumference is one of the criteria for metabolic syndrome (MetS), a constellation of cardiovascular disease (CVD) risk factors that cluster together in the same individual. Gerald Reaven was among the first to describe clustering of IR with atherosclerosis, hyperglycemia, increased low-density lipoproteins (LDL), decreased high-density lipoproteins (HDL), and hypertension. He coined the term "Syndrome X" and proposed that IR lay at the root of this disorder [16]. Many other criteria for MetS have since been proposed, some of which consider waist circumference a prerequisite, and others that do not [17]. Of relevance to OSA, fat can also deposit in the neck or tongue, and its presence reflects MetS, in some cases more so than waist circumference or BMI [18–20]. In morbidly obese subjects, neck circumference correlated with VF ($R^2 = 0.67$), more strongly than waist circumference ($R^2 = 0.35$) [21]. In the Framingham cohort [22], neck circumference was associated with MetS even after adjustment for VF.

FUNCTIONAL AND ANATOMICAL FEATURES OF METS

The *functional* basis for MetS is resistance to the effects of insulin. Skeletal muscle IR leads to hyperglycemia and hyperinsulinemia; liver IR increases glucose output and promotes lipid synthesis; adipose IR leads to high-plasma free fatty acids (FFA) that, in turn, impair muscle glucose uptake and increase liver triglyceride (TG) synthesis. Hyperinsulinemia causes vascular smooth muscle proliferation, activates the sympathetic nervous system (SNS), and causes renal sodium retention [23]. Activation of the renin-angiotensin system (RAS) also contributes to hypertension. Adipose tissue is a major site of angiotensinogen synthesis. Massiera *et al.* showed a significant contribution of adipose-derived angiotensinogen to plasma levels in mice [24]. RAS activation also promotes lipogenesis and fat hypertrophy, particularly of the abdominal VF [25], and impairs adiponectin secretion [26]. *In vivo*, angiotensin 2 promotes secretion of inflammatory cytokines from human adipocytes [27]. Thus adipose activation of the RAS may induce a vicious cycle of fat expansion and inflammation. The *anatomical* correlate of MetS is VF accumulation. It is unclear whether VF is a source or passive marker of metabolic dysfunction [28]. High FFA causes fat accumulation in skeletal muscle and liver leading to IR and inflammation. This ectopic accumulation of fat in nonadipose tissues has been referred to as "lipotoxicity". Mesenteric fat delivers FFA to the liver via the portal circulation and may contribute to liver IR. However, omentectomy did not confer additional metabolic benefits over gastric bypass [29]. Visceral adipocytes also secrete more inflammatory cytokines than SF and are more resistant to antilipolytic effects of insulin [30, 31]. By contrast, SF is protective against MetS, perhaps by sequestering fatty acids against upper body fat expansion. This function of SF is powerfully illustrated by Gavrilova *et al.* who transplanted SF into the mice with lipodystrophic diabetes—a form of diabetes associated with decreased total body fat mass but increased fat in visceral and nonadipose depots. SF transplant reversed metabolic defects in the recipient animal [32].

CAUSES OF VISCERAL ADIPOSITY AND METS

Genetic and environmental factors contribute to MetS. Body adiposity is 5%–30% heritable, with VF more heritable than SF mass or BMI [33]. From twin studies, waist-hip ratio heritability was estimated at 31% [34]. Genome-wide association studies (GWAS) screen for common genetic variants among large numbers of people associated with a particular trait. GWAS identified dozens of loci associated with obesity, dyslipidemia, and IR [35], many of which are polymorphisms in insulin-signaling pathways. Yaghootkar *et al.* found a genetic basis for MHO and MONW phenotypes by examining how loci associated with IR clustered with adiposity and metabolic traits. Interestingly, 11 risk alleles clustered with increased TG, low HDL, increased VF, and *lower* SF—consistent with a MONW phenotype [36]. Genetic variation in isolated "founder" populations also provide clues to the genetic basis of IR. Exome sequencing in a population from Greenland found a 17% prevalence of an allele carrying a nonsense mutation in the gene expressing TBC1D4, resulting in in lower muscle GLUT4 and IR [37]. Second, the environment can play a role in promoting MetS. Indisputably, the primary cause of obesity is excess energy consumption and insufficient physical activity. However, other factors may influence where fat accumulates. For example, better physical fitness is associated with less VF accumulation and MetS features [38]. Stress may promote central obesity or "Cushing's disease of the omentum" [39] via chronic cortisol elevations. Visceral adipocytes contain higher levels of 11β-hydroxysteroid dehydrogenase-1 (which converts inert cortisone to cortisol) compared with SF. In primate studies, social subordination was related to greater abdominal obesity [40]. In rodents, stress and high-fat diet stimulated VF expansion, in association with neuropeptide Y; inhibition of NPY receptors prevented fat accumulation and MetS [41].

OBESITY, VISCERAL OBESITY, AND OSA
SLEEP APNEA AND OBESITY

The role of obesity in the pathogenesis of OSA is well established; at least half of adult OSA is directly attributable to excess weight [42]. In patients with BMI ≥ 30, OSA prevalence is more than 50% [43]. Among patients referred for bariatric surgery, prevalence exceeded 70% [44], and OSA was nearly ubiquitous in a study of patients with BMI ≥ 40 [45]. In a seminal study, 690 middle-aged Wisconsin residents were followed 4 years for development of OSA [46]. A 10% increase or decrease in weight predicted an increase or decrease of AHI by 32% or 26%, respectively. In the Cleveland Family Study, 1149 participants underwent two sleep studies, 5 years apart [47]. AHI was associated with BMI (OR 1.14 per 1-unit increase) and with waist-hip ratio (OR 1.61 per 0.1 unit increase). At the genetic level, the correlation between AHI and adiposity metrics was ~0.6, revealing shared genes in the pathogenesis of obesity and OSA [48].

SLEEP APNEA AND VF

OSA is more closely associated with upper body fat than with general obesity. This includes abdominal VF and extrathoracic subcutaneous depots such as neck fat. Neck circumference correlated better with nocturnal oxygen desaturation (a surrogate of OSA) than BMI, low set hyoid bone, or soft palate length [49, 50]. In a study of patients with suspected OSA, increased cross-sectional area of abdominal VF measured by CT scan was highly correlated with AHI. A cut-off point of 106.2 cm^2 area of cross-sectional abdominal fat on CT scan provided highest sensitivity and specificity [51]. In a study of 149 Chinese subjects, AHI correlated with mesenteric ($r = 0.43$) but not SF thickness. After adjustment for BMI and neck circumference, each 1-cm increase in mesenteric fat thickness was associated with a sevenfold increased risk for OSA [52]. In some studies, the association of VF and neck fat with OSA was less consistent or was modified by sex. In Japanese patients with suspected OSA, neck circumference was associated with OSA severity, independent of VF mass [53]. In a German study, AHI correlated with intra-abdominal fat but not neck fat [54]. In a Turkish study, each SD increment in neck circumference increased the likelihood of OSA by 13% in men, yet waist circumference was more predictive of OSA in women [55]. Thus OSA can be considered a manifestation of visceral and/or neck fat [56].

IMPACT OF WEIGHT LOSS ON OSA

In several studies, weight loss improves OSA [57]. A metaanalysis of lifestyle modification for OSA [58] showed a reduction of 13.7 kg and a corresponding reduction of the AHI by 16. The Sleep Action for Health in Diabetes (AHEAD) cohort [59] studied diabetic patients from four U.S. centers enrolled in lifestyle intervention versus usual care. Lifestyle changes reduced weight by 10.7 kg and AHI by ~10 after 1 year, versus negligible changes in the control group [57]. A metaanalysis of 342 OSA patients undergoing a variety of bariatric surgeries showed a decrease in AHI from 54.7 to 15.8 after weight loss [60]. However, only a minority of patients achieve complete resolution of OSA with medical or surgical weight loss [61].

ASSOCIATION OF METS WITH OSA

OSA is closely associated with MetS. In fact, some have labeled the coexistence of OSA and MetS "Syndrome Z" [62]. This overlap between obesity, OSA, and metabolic impairment is shown in Figure 8.1. A recent study showed that, in OSA patients, the prevalence of MetS was 35.6% [63]. In a Brazilian study of 50 consecutive patients presenting with MetS, 68% were found to have OSA [64]. The pooled OR for MetS in those with OSA was 2.87 in a metaanalysis [65]. However, interpretation of these associations is complex. As mentioned previously, OSA is a manifestation of visceral, more so than total body adiposity. Studies seeking to quantify the "isolated" impact of OSA have not consistently adjusted for metrics of VF. The importance of this distinction is illustrated by two studies that arrive at dissimilar conclusions: (1) Coughlin *et al.* compared OSA to controls and showed that those with OSA were more obese and had a markedly higher prevalence of MetS (87% versus 35%) [66]. After adjustment for BMI, MetS was nine times more likely to be present in OSA

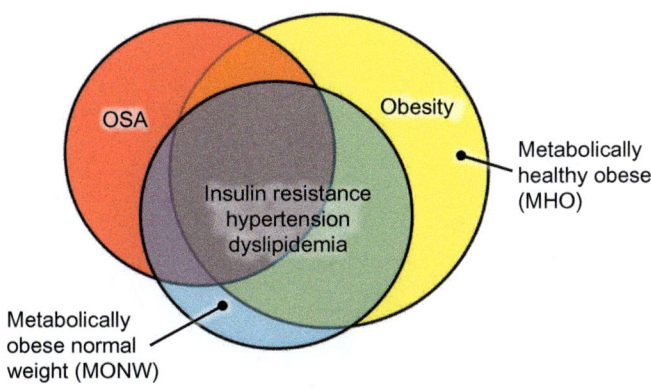

FIGURE 8.1

Overlap of obesity, obstructive sleep apnea, and metabolic dysfunction. *Obesity-OSA overlap*: Among OSA patients, most are obese; among obese persons, about 50% have OSA. *Obesity-MetS overlap*: Among obese persons, prevalence of MetS is 30%–75%. The prevalence of metabolically healthy obese may be 30%. The prevalence of metabolically obese normal weight is 5%–45%. *MetS-OSA overlap*: Among OSA patients, one-third have MetS; among patients with MetS, two-thirds have OSA.

versus controls. (2) In a Japanese study, men with OSA were compared with controls, matched for VF, BMI, and age. The two groups did not differ in terms of TG, HDL, or diastolic BP, but the OSA patients had higher fasting blood glucose and prevalence of hypertension [67]. Taken together, OSA may have an independent detrimental effect on metabolism, but the presence of VF is an important confounder.

HOW VF CONTRIBUTES TO OSA

Mechanical factors likely explain associations between VF and OSA. First, neck or tongue fat, shown to correlate with VF, reduces airway caliber. OSA patients have thicker necks and narrower pharyngeal airways [68] compared with controls, even with the same waist circumference [69]. Fat accumulates posterolateral to the oropharyngeal airspace in obese OSA patients [70]. Schwab *et al.* showed that the upper airway soft tissues are enlarged in OSA patients [71]. Similarly, Mortimore *et al.* compared neck circumference and neck fat in BMI-matched controls and OSA patients [72]. OSA patients exhibited 10% greater neck tissue volume and 27% neck fat volume than controls. Second, abdominal fat exerts caudal pressure on the lungs, reducing expiratory reserve volume [73]. Stadler *et al.* showed that inflating a cuff around the abdomen reduced end-expiratory lung volume and increased upper airway collapsibility [74]. Schwartz *et al.* also manipulated the end-expiratory lung volume with a "body box" in healthy subjects and showed that end-expiratory lung volume was inversely related to pharyngeal collapsibility [75]. Third, obesity activates the RAS [76], which increases plasma fluid volume. During sleep, recumbent position causes fluid to shift from the lower body into the neck region, reducing airway caliber [77]. Figure 8.2 highlights visceral and ectopic fat accumulation as a shared basis for MetS and OSA.

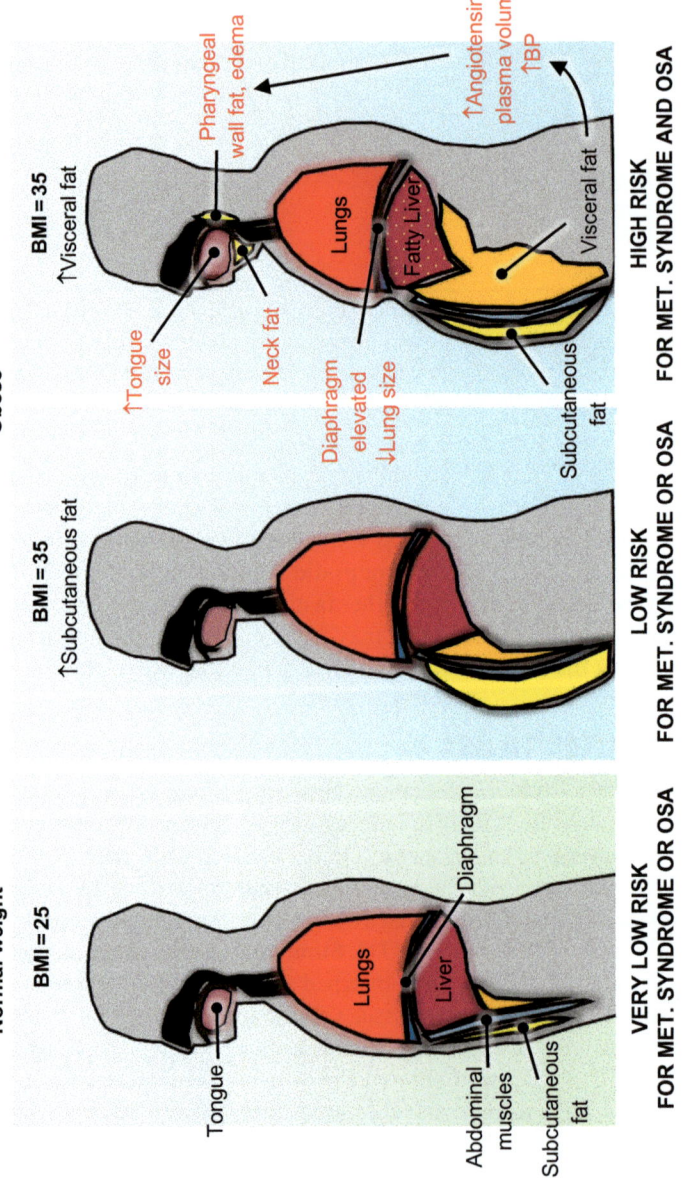

FIGURE 8.2

Comparison of body composition among persons with normal weight, subcutaneous, or visceral adiposity. Even at the same BMI, individuals with a visceral pattern of obesity are at greater risk for metabolic syndrome and obstructive sleep apnea (OSA). Factors that contribute to OSA in visceral obesity are shown in red text.

THE EFFECT OF OSA ON METABOLIC SYNDROME
IMPACT OF OSA ON BODY COMPOSITION
Can OSA induce weight gain or visceral adiposity?

Some have questioned whether OSA is the "chicken" rather than the "egg" of abdominal fat [78]. First, sleep loss is known to stimulate appetite, and cross-sectional studies show an association between short sleep and increased weight [79]. This is at odds with the observation that OSA and short sleep time increases energy expenditure (EE), particularly during the nocturnal period [80]. To engender obesity, calorie intake would need to overcompensate for this heightened EE. However, this positive energy balance has not been empirically observed in OSA [81, 82]. Second, sleepiness from OSA may impair physical activity. In one study, CPAP therapy increased physical activity of OSA patients [83] but did not increase activity in other studies [84, 85]. Exercise capacity may also be improved with short-term [86] and long-term [87] CPAP therapy. Third, OSA reduces growth hormone (GH) levels due to loss of slow-wave sleep; CPAP can restore GH and downstream insulin-like growth factor-1 [88]. Reduced activity of the GH axis predisposes to visceral adiposity [89]. However, studies have yet to show that OSA affects weight via this pathway. A related question is whether OSA hinders attempts at weight loss. In a study observing the response of men with visceral obesity to a 1-year lifestyle intervention program, OSA subjects achieved a smaller reduction in weight and less metabolic improvement than controls [90]. A caveat of this study is that OSA patients had higher BMI, waist circumference, and TG at baseline than controls.

Effect of CPAP on body weight and composition

Most studies indicate that treatment of OSA with CPAP actually increases weight, presumably by reducing EE [85, 91]. In terms of body composition, the effects of CPAP are inconsistent. Munzer et al. showed that CPAP for 8 months increased lean mass and abdominal SF [92]. Chin et al. examined fat distribution in 22 OSA patients pre/post 6 months of CPAP by CT scan [93]. They showed a 22% reduction in VF and a 12% reduction in SF area. Interestingly, VF decreased in nine patients who showed no change in weight, yet SF decreased in those with an overall reduction in weight. In a Spanish study, Catala et al. compared 50 OSA patients treated with CPAP with 35 untreated patients over 2 years. The CPAP group decreased SF, whereas the untreated group increased intra-abdominal fat [94]. CPAP in 29 obese patients for ~12 weeks reduced VF among adherent subjects [95]. Similarly, CPAP reduced liver fat in Israeli patients with nonalcoholic fatty-liver disease after 2 years but had no effect in noncompliant patients [96]. However, several CPAP studies failed to show improvements in body composition: Vgonztas et al. showed that CPAP for 3 months had no impact on VF, adipokines, or glucose metabolism [97]. In randomized studies, 8–12 weeks of CPAP did not affect abdominal or liver fat compared with sham CPAP [98–101].

THE IMPACT OF OSA ON INSULIN RESISTANCE

Cross-sectional studies

Several epidemiologic and clinical studies reveal close links between OSA and altered glucose metabolism. A recent systematic review summarizes many of these studies [102], so we will focus on a few representative studies in this article. The prevalence of T2DM in OSA is 30%–40% [103, 104], whereas the prevalence of OSA in T2DM is 58%–86% [105]. Data from the Sleep Heart Health Study, which enrolled 2656 subjects, demonstrated that subjects with mild or moderate OSA had an OR of 1.27 and 1.46, respectively, for fasting glucose intolerance, after adjustment for parameters including BMI and waist circumference [106]. In another multisite community-based study, moderate-to-severe OSA was associated with abnormal fasting glucose in African Americans (OR = 2.14) and white participants (OR = 2.85) relative to subjects without OSA, after adjusting for age, sex, WC, and sleep duration [107]. Data from Chinese subjects also revealed an independent association between OSA, abnormal fasting glucose, and impaired glucose tolerance after adjustment for BMI [108]. In mildly obese individuals without known diabetes, an AHI $\geq$ 5 was associated with an increased risk of impaired glucose tolerance (OR = 2.15) after adjusting for BMI and percent body fat [109]. For T2DM patients, concomitant OSA was associated with higher HbA$_1$c levels and diabetes complications after controlling for BMI or VF [110–112]. OSA may also affect insulin sensitivity before the onset of T2DM. In a cohort of 270 subjects without diabetes, those with OSA were more insulin resistant. Although obesity was the major determinant of IR, AHI and minimum oxygen saturation were also independent determinants [113]. In severely obese subjects without diabetes, oxygen desaturation $\geq$4.6% was associated with a 1.5-fold increase in IR (estimated by HOMA-IR) [114]. Data from a large community-based study in Japan showed that HOMA-IR was independently associated with oxygen desaturation index after adjusting for BMI [115]. Insulin sensitivity was measured in another study of OSA patients using a frequently sampled intravenous glucose tolerance test. Authors observed a graded decrease in insulin sensitivity as a function of OSA severity, independent of age, sex, race, and percent body fat [116]. Furthermore, even in healthy lean men, the presence of mild OSA was associated with a 27% decrease in insulin sensitivity [117]. Interestingly, in a study of nonobese OSA/control Chinese patients with similar BMI ~22 and WC ~80 cm, OSA patients exhibited higher HOMA-IR and TG [118]. In patients with T2DM, the prevalence of OSA is much higher than that in the general population; in the Sleep AHEAD cohort, 86.3% obese T2DM patients had OSA [119]. In summary, these cross-sectional studies demonstrate a close relationship of OSA with different stages of diabetes, which need to be validated by prospective and interventional studies.

Prospective studies

Observational studies also demonstrate that OSA is associated with incident T2DM [120–122] or hyperglycemia. In a metaanalysis of six prospective cohort studies including a total of 5953 participants, moderate-severe OSA was associated with a 63% greater adjusted risk of developing T2DM [123]. In a Swedish community-based prospective study, OSA was associated with deterioration of insulin sensitivity (HOMA-IR) and > fourfold risk of developing T2DM after 11 years of follow-up, after adjusting for BMI and interval weight gain [124]. Among patients with established T2DM, glycemic control may be compromised by OSA. For example, in diabetic patients with moderate-severe OSA, continuous glucose monitoring demonstrated a glucose peak following the lowest oxygen level during sleep [125]. No correlation between hypoxia and glucose was apparent in nondiabetic OSA patients [126].

Interventional studies

CPAP can improve the glycemic health of OSA patients, especially in those with pre- or established T2DM [127]. The strongest evidence derives from T2DM patients undergoing serial glucose sampling during sleep or over a 24-hour period, in which glucose levels were decreased by 8%–14% by CPAP [128–130]. Similarly, in prediabetic OSA patients, 2 weeks of CPAP improved morning glucose tolerance [131]. In U.S. veterans with OSA, CPAP reduced fasting blood glucose and decreased incident T2DM but only among those with excellent CPAP adherence [132]. Other studies show no glycemic benefit to CPAP therapy or indicate that benefits pertain only to certain outcomes or subgroups. For example, 6–12 weeks of CPAP versus sham therapy had no effect on fasting glucose or IR despite a robust improvement in BP [133] in nondiabetic patients [101]. CPAP withdrawal acutely increased nocturnal glucose by 17 mg/dL in T2DM but had a minimal effect on glucose in nondiabetics. Compensatory hyperinsulinemia in nondiabetics likely prevented glucose excursions [134]. In a randomized double-blind crossover study of CPAP vs. sham CPAP, Weinstock *et al.* found no change in morning oral glucose tolerance except in those with severe OSA [100]. Similarly, glucose tolerance, but not HOMA-IR, was improved after 12 weeks of CPAP in morbidly obese patients with severe OSA [135]. In T2DM patients, 3 months of CPAP (versus sham CPAP) did not improve HbA_1c [136]. The same null effect was observed in a 6-month CPAP trial of patients with well-controlled diabetes (HbA_1c 6.5%–8.5%) [137]. In contrast, 6 months of CPAP reduced HbA_1c and improved IR in diabetic patients with poorly controlled T2DM [138]. Despite the heterogeneity, metaanalyses [139, 140] show an overall decrease of ~0.5 in HOMA-IR with CPAP. In summary, interventional studies clearly show that OSA affects glucose metabolism, particularly in those with T2DM, severe OSA, and good CPAP adherence.

Mechanisms linking OSA to impaired glucose metabolism

Several mechanisms could explain how OSA impairs glucose metabolism. Intermittent hypoxia (IH) is recognized as a cornerstone of OSA pathogenesis, which can promote SNS activation, hypothalamic-pituitary-adrenal (HPA) axis activation, reduced adiponectin, and inflammation (Figure 8.3).

Intermittent hypoxia

IH refers to the cyclical pattern of hypoxia and reoxygenation during sleep in OSA. Clinical data demonstrates a relationship between IH severity and IR in OSA patients

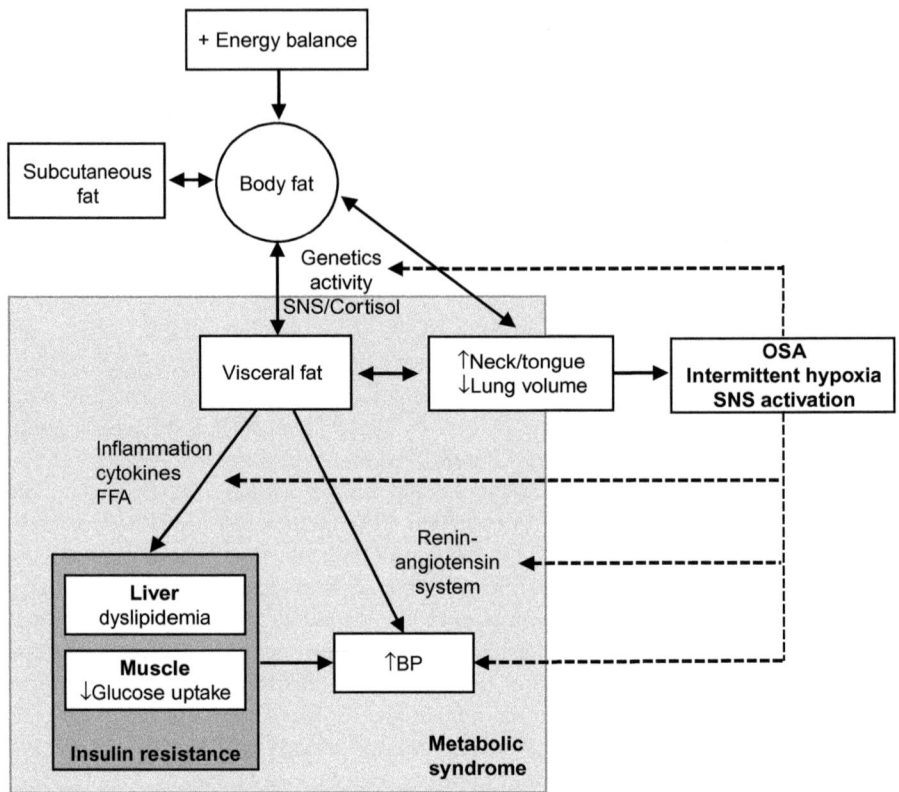

FIGURE 8.3

Relation between fat distribution, MetS, and OSA. Total body fat is increased by positive energy balance. Genetic and environmental factors determine the distribution of adipose tissue. VF is a mediator or marker of IR and MetS. OSA, in turn, may aggravate MetS by affecting fat distribution, stimulating SNS and HPA axes, inflammation and lipolysis, and the RAS.

[106, 141]. Animal models of OSA have been developed, many of which deliver IH to rodents [142]. Iiyori *et al.* exposed mice to IH for 9 hours and conducted hyperinsulinemic euglycemic clamps to measure insulin sensitivity. Mice developed IR and decreased glucose utilization in oxidative muscle fibers [143]. Other animal studies show that longer periods of IH also induce IR [144, 145]. In the sole human study, healthy volunteers were subjected to 5 hours of IH during wakefulness. Their insulin sensitivity (intravenous glucose tolerance test) worsened after exposure [146]. These studies confirm the major contributing role of IH in OSA-induced glucose metabolic disorders. The potential mechanisms are further discussed later.

Activation of the sympathetic nervous system

SNS activation strongly affects glucose metabolism [147]. For example, acute psychological stress increases plasma glucose in patients with T2DM [148]. Stressful stimuli of OSA, such as IH and arousals, also activate the SNS. IH exposure decreased insulin sensitivity in healthy volunteers, accompanied by increased SNS activity [146]. In another human study, acute hypoxia-induced IR was attenuated by sympathetic inhibition with clonidine [149]. During CPAP withdrawal, nocturnal glucose increased dynamically with elevations of the heart rate suggesting automatic influence [134]. A collection of hypoxia-sensitive cells in the carotid artery, called the carotid body, is responsible for SNS activation when blood oxygenation decreases. Shin *et al.* performed carotid sinus nerve denervation on mice before exposing them to chronic IH. Denervation prevented IH-stimulated hepatic glucose output and liver expression of phosphoeno/pyruvate carboxykinase, a rate-limiting hepatic enzyme of gluconeogenesis. [150]. However, other rodent studies reported that IH decreased insulin sensitivity independently of sympathetic activation. For example, adrenal medullectomy improved glucose tolerance but did not prevent IR and fasting hyperglycemia stimulated by IH [151]; α-adrenergic blockade, β-adrenergic blockade, and adrenal medullectomy mitigated hyperglycemia and lipolysis during IH but did not prevent IR [152].

Hypothalamic-pituitary-adrenal (HPA) axis

HPA axis activation inhibits insulin-mediated glucose uptake [153]. Some studies demonstrate higher cortisol in OSA subjects [134, 154], yet others do not [155]. CPAP withdrawal increased cortisol in proportion to sleep fragmentation and awakenings [134]. Similarly, in a mouse study, IH reversed the diurnal blood glucose rhythm and augmented diurnal peak corticosterone [156].

Hypoadiponectinemia

Adiponectin is an adipocyte-derived hormone with insulin-sensitizing and antiinflammatory properties. Plasma adiponectin levels are reduced in obesity-linked IR [157]. In patients with severe OSA, serum adiponectin was lower compared with those with milder OSA, independent of obesity or abdominal VF [158]. In patients with OSA and T2DM, 6-month CPAP treatment decreased IR, accompanied by an increase in adiponectin [138]. Adiponectin was negatively correlated with HOMA-IR and urine catecholamines after adjustment for adiposity. Furthermore, 5-week exposure of rats to IH increased HOMA-IR, decreased serum adiponectin, and expression of adipocyte adiponectin and Hypoxia Inducible Factor-1α [159]. In cultured 3T3-L1 adipocytes, 48-hour exposure to IH decreased secretion of adiponectin [160].

Inflammation

Inflammation plays an important role in linking obesity, IR and diabetes [161]. OSA/IH has been shown to activate inflammatory pathways [162–164]. Exposure of mice to chronic IH increased nuclear factor-kappa B (NF-κB, a major transcriptional regulator of inflammation) binding activity in aorta, lungs, and heart. Interestingly, 1 month of CPAP in five severe OSA patients reduced NF-κB activity from peripheral blood mononuclear cells to that of five controls [165]. The severity of IH was associated with increased levels of tumor necrosis factor-α and interleukin (IL)-8 (proinflammatory mediators known to inhibit insulin signaling in tissues such as liver and adipose tissue) [166], and both were reduced by CPAP therapy [167]. Murphy *et al.* exposed mice and 3T3-L1 adipocytes to IH, causing IR in mice and decreasing insulin-mediated glucose uptake in cultured 3T3-L1 adipocytes. IH induced proinflammatory M1 macrophage polarization in visceral adipose tissue. In the same study, these findings were further investigated in a clinical cohort in which OSA severity was found to correlate with M1 macrophage polarization in adipose tissue [168]. Several other IH animal studies and *in vitro* studies demonstrate IH-induced inflammation, manifested by adipose macrophage recruitment [169], a shift of macrophages toward a M1 proinflammatory subtype [170], activation of NF-κB [171], and increased cytokine release [169, 171, 172]. Some of these inflammatory responses to IH were accompanied by whole-body or adipose IR [169, 170]. However, recent clinical studies show conflicting findings with regard to OSA-induced inflammation [173]. It is also unclear whether IH-induced adipose inflammation and IR can account for defects in whole-body glucose uptake, which typically originates from skeletal muscle.

Sleep loss and fragmentation

Decreased sleep efficiency and frequent arousals are also common features of OSA, both of which may contribute to defects in glucose metabolism. In a rat model, 3 months of chronic sleep deprivation caused IR (HOMA-IR and insulin tolerance test) [174]. In a human study, 11 healthy volunteers had their sleep experimentally fragmented via auditory and mechanical stimuli for two nights. After fragmented sleep, morning insulin sensitivity was decreased in association with HPA and SNS activation [175].

THE EFFECT OF OSA ON HYPERTENSION

Cross-sectional studies

The Sleep Heart Health Study enrolled 6132 participants in the 1990s to evaluate the association between OSA and hypertension. The OR for hypertension was 1.37, comparing the highest AHI category (≥30) with the lowest (<1.5), after adjustment for BMI, neck circumference, and other demographics and anthropometric variables [176]. Data from the European Sleep Apnea Database cohort study based on 11,911 adults from 24 sleep centers also demonstrated an independent association between prevalent OSA/nocturnal hypoxemia and hypertension [177]. Among Brazilian patients with poorly controlled grade 2 or 3 hypertension, 55% of the participants had concomitant OSA, which was independently associated with a nocturnal nondipping pattern of BP and arterial stiffness, even after at least 30 days of standard antihypertensive treatment [178]. Patients with hypertension and concomitant moderate to severe OSA showed a relation between untreated severe OSA (AHI ≥ 30) and elevated BP, compared with that of moderate OSA despite intensive antihypertensive pharmacotherapy [179].

Prospective studies

The Wisconsin Sleep Cohort Study enrolled 709 participants from the general population and showed a dose-dependent association between OSA severity at baseline and the presence of hypertension 4 years later. The adjusted OR for incident hypertension was 2.03 and 2.89 for mild and more severe OSA, respectively [180]. Further analysis of the same cohort demonstrated that even sleep apnea, primarily in REM sleep, was associated with hypertension [181].

Interventional studies

Several RCTs support a beneficial role of CPAP in lowering BP. In a metaanalysis that included 31 RCTs comparing CPAP with passive (sham CPAP, placebo drugs, etc.) or active (oral appliances and antihypertensive drugs) treatments, CPAP achieved a significantly higher net decrease in systolic (2.6 mmHg) and diastolic BP (2.0 mmHg) compared with passive treatments [182]. Patients with a higher baseline AHI and resistant hypertension [183–185] may exhibit a greater BP-lowering effect. However, in a Brazilian RCT of OSA patients with resistant hypertension, 6 months of CPAP reduced nocturnal BP only in patients with uncontrolled ambulatory BP [186]. Conversely, in patients with minimally symptomatic OSA, CPAP had no impact on BP [187] or lowered BP only in highly adherent patients [188]. Hence, the CPAP response to hypertension is influenced by baseline BP, OSA severity, sleepiness, and adherence. Moreover, some specific microribonucleic acids (miRNAs) functionally associated with the cardiovascular system appeared to be able to predict BP responses to CPAP treatment in patients with OSA and resistant hypertension [189].

Mechanisms linking OSA to hypertension

In the 1990s, Fletcher *et al.* exposed rats to chronic IH, aiming to study the impact of OSA on BP. After 35 days, IH induced a 13.7 mmHg increase in mean artery pressure [190]. An increase in BP was prevented by both surgical denervation of peripheral chemoreceptors and adrenal demedullation in this model [191]. In a canine model of OSA, intermittent airway occlusion increased nighttime BP and caused sustained daytime hypertension [192]. In a study of healthy humans, 14-day nocturnal IH increased daytime BP, which was accompanied by increased muscle sympathetic nerve activity [193]. Overactivation of the SNS and renin-angiotensin-aldosterone system (RAS) activation are currently considered major pathways in OSA-induced hypertension.

Roles of sympathetic nervous system

OSA and IH exposure sensitize chemoreflexes, increasing sympathetic outflow even during wakefulness [194–196]. Underlying mechanisms may include reactive oxygen species (ROS) production in the carotid body, NADPH oxidase activation, and inflammation [197–202]. One of the mechanisms underlying IH-induced increased ROS generation in the carotid body is an imbalance between HIF-1 and HIF-2, favoring transcription of prooxidant over antioxidant enzymes [203]. IH also affects central sites of sympathetic regulation. In animal studies, IH increased expression of FosB/ΔFosB (indicating repeated neuron activation) in autonomic nuclei of several central regions [204]. Inhibition of ΔFosB in the median preoptic nucleus significantly reduced the BP response to IH [205]. Decreased release of oxytocin from hypothalamic paraventricular nucleus neurons, which inhibited the activation of cardiac vagal neurons in the dorsal motor nucleus of the vagus, also contributed to increased BP in rats exposed to chronic IH and hypercapnia [206]. Another site of SNS stimulation in OSA originates from renal sympathetic nerves. In a pig model of OSA, breathing against tracheal occlusion led to reductions in renal blood flow, followed by a surge in systemic BP. This postapneic BP rise was inhibited by renal sympathetic denervation but not by angiotensin receptor blockade [207].

Roles of renin-angiotensin-aldosterone system

The RAS is activated in OSA [208], especially in patients with hypertension [209], whereas CPAP therapy has been shown to attenuate RAS activity [210]. In a chronic IH rat model of OSA, animals developed sustained BP elevation, accompanied by a fourfold increase in plasma renin activity, which was prevented by losartan treatment [211]. Targets of angiotensin have been identified in the brain and carotid body of mice. Circulating angiotensin II binds to angiotensin II type Ia (AT1aR) receptors in the subfornical region of the brain. Saxena *et al.* injected small-hairpin RNA against either AT1aR or a scrambled control sequence in this region of the rat brain before exposure to chronic IH. Central AT1aR inhibition prevented chronic IH induced during the waking hours [212]. Moreover, chronic IH exposure stimulated expression of RAS components in the carotid body, including angiotensinogen and AT1 receptors. Pretreatment with losartan attenuated carotid body oxidative stress and macrophage infiltration [201, 202]. Hence, OSA/IH upregulates both angiotensin levels and sites of receptor activation. However, a randomized crossover study compared the effect of CPAP versus the angiotensin receptor blocker, valsartan, on BP in hypertensive patients with OSA. Both CPAP and valsartan reduced BP, although the drug had a much stronger effect [213]. However, in another study, CPAP added to losartan reduced nocturnal systolic BP by 4.7 mmHg beyond the effect of losartan alone [214]. Hence, OSA-induced hypertension cannot be solely ascribed to RAS activation.

THE EFFECT OF OSA ON HYPERLIPIDEMIA

Cross-sectional studies

OSA is associated with atherogenic dyslipidemia. A metaanalysis demonstrated that patients with OSA had higher total cholesterol, LDL, TG, and lower HDL compared with those without OSA, and in metaregression, AHI was associated with LDL and TG, independent of BMI [215]. In a large cross-sectional study from China, the prevalence of dyslipidemia was significantly higher in OSA versus non-OSA patients; LDL was associated with OSA adjusted for BMI and waist-hip ratio [216]. However, another study showed that lipoprotein levels in OSA patients may be associated with underlying IR, rather than OSA severity *per se* [217].

Interventional studies

Some studies support a helpful role of CPAP in dyslipidemia, whereas others do not. In a metaanalysis from 29 studies with 1958 subjects, CPAP treatment decreased total cholesterol and LDL, and increased HDL. However, most of the included studies did not include an untreated control group [218]. In a randomized placebo-controlled crossover trial, 2 months of CPAP improved the 24-hour postprandial TG profile, and decreased mean 24-hour cholesterol by 0.19 mmol/L [219]. In another randomized trial of a more obese OSA cohort, CPAP decreased TG only when combined with weight loss [220]. CPAP withdrawal for three nights did not affect the nocturnal TG profile, nor morning lipoprotein levels [134].

Mechanisms

Several animal studies show that IH increases cholesterol and TG levels. In mice, 5-day IH exposure increased serum TG, TC, and HDL [221], and 4-week IH elevated TC and LDL [222]. Four- to 12-week IH exacerbated dyslipidemia in atherosclerosis-prone ApoE$^{-/-}$ mice [223]. In general, TG and lipoprotein levels in plasma depend upon lipolysis of adipose tissue TG, uptake of TG-fatty acids from VLDL and chylomicron remnants, and hepatic *de novo* lipogenesis. All of these pathways are affected by IR [224], and some of these pathways have been examined in OSA or IH:

Adipose tissue lipolysis: Delivery of FFA to the liver and subsequent re-esterification is a major determinant in TG synthesis and lipoprotein secretion [225]. In a Spanish study, OSA patients had higher fasting FFA levels when compared with BMI-matched controls [226]. OSA increased nocturnal plasma FFA in heart failure patients in a hypoxia-dependent manner [227]. Furthermore, OSA elicited by CPAP withdrawal increased nocturnal plasma FFA in nonheart failure patients [134]. The elevation of plasma FFA in response to OSA has also been recapitulated in animal models of IH [152, 228, 229]. The increase in FFA induced by IH was abolished by acipimox, a lipolysis inhibitor, in one study [229] and by propranolol in another [152].

Reduced clearance of circulating lipoproteins: Lipoprotein lipase (LpL) hydrolyzes circulating lipoproteins, liberating fatty acids for uptake into tissues. Chronic IH in mice reduced LpL activity and increased the expression of angiopoietin-like protein 4 (Angptl4), a potent inhibitor of LpL, in the epididymal fat [230]. Mice with partial global deficiency of HIF-1α did not present increased plasma fasting TG and adipose Angptl4 after IH exposure, suggesting that HIF-1α may increase adipose Angptl4, thereby reducing LpL activity [223]. Jun *et al.* showed that acute sustained hypoxia for 6 hours impaired lipid clearance, yet lowered hepatic VLDL secretion [231]. A major site of hypoxia-induced suppression of lipid uptake was brown adipose tissue. In mice housed at thermoneutrality (30°C), hypoxia had no effect on plasma TG clearance [232]. Hence, ambient temperature is a critical determinant of the effect of hypoxia on lipid kinetics.

Upregulation of hepatic lipid biosynthesis: IH upregulated a key hepatic transcription factor of *de novo* lipid biosynthesis, sterol regulatory element binding protein 1 (SREBP-1), as well as a SREBP-1-regulated enzyme, stearoyl coenzyme

A desaturase 1 (SCD-1) [221], whereas interruption of liver SREBP-1 signaling in transgenic mice [233] and depletion of SCD-1 with antisense oligonucleotides abolished hyperlipidemia [234]. It is not known whether IH directly induced these transcriptional changes or if they reflect liver IR [224].

CONCLUSION AND CLINICAL PERSPECTIVE

OSA is a respiratory manifestation of VF and a likely aggravator of metabolic dysfunction. Thus patients presenting with MetS and OSA may be caught in a vicious cycle. In terms of improving metabolism, clinical studies show that our attention should be foremost directed at underlying obesity. Improvements in insulin sensitivity with CPAP are more apparent in lean than in obese patients [235], suggesting a "ceiling effect" of obesity. Similarly, in obese OSA patients, a lifestyle intervention resulted in ~7 kg weight loss over 24 weeks and yielded improvements in inflammation, IR, and TG levels, whereas CPAP alone had comparatively little effect other than a synergistic reduction in BP [220]. Thus the importance of weight loss and attenuation of traditional CVD risk factors should be emphasized. On the other hand, CPAP can improve OSA symptoms and potentially improve metabolic function, particularly in vulnerable "metabolic responders" such as T2DM patients or those exhibiting signs of nocturnal stress [134]. Excellent adherence is necessary to realize the full benefits of CPAP. Further studies are required to determine mechanisms and predictors of cardiometabolic dysfunction in OSA to discover targeted therapies for at-risk patients.

REFERENCES

[1] Sleep-related breathing disorders in adults: recommendations for syndrome definition and measurement techniques in clinical research. The report of an American Academy of Sleep Medicine Task Force. Sleep 1999;22(5):667–89.

[2] Peppard PE, Young T, Barnet JH, Palta M, Hagen EW, Hla KM. Increased prevalence of sleep-disordered breathing in adults. Am J Epidemiol 2013;177(9):1006–14.

[3] Schwab RJ, Kim C, Bagchi S, Keenan BT, Comyn FL, Wang S, et al. Understanding the anatomic basis for obstructive sleep apnea syndrome in adolescents. Am J Respir Crit Care Med 2015;191(11):1295–309.

[4] Edwards BA, Wellman A, Sands SA, Owens RL, Eckert DJ, White DP, et al. Obstructive sleep apnea in older adults is a distinctly different physiological phenotype. Sleep 2014;37(7):1227–36.

[5] Sullivan CE, Issa FG, Berthon-Jones M, Eves L. Reversal of obstructive sleep apnoea by continuous positive airway pressure applied through the nares. Lancet (London, England) 1981;1(8225):862–5.

[6] Garbarino S, Guglielmi O, Sanna A, Mancardi GL, Magnavita N. Risk of occupational accidents in workers with obstructive sleep apnea: systematic review and meta-analysis. Sleep 2016;39(6):1211–8.

[7] Marcus CL, Brooks LJ, Draper KA, Gozal D, Halbower AC, Jones J, et al. Diagnosis and management of childhood obstructive sleep apnea syndrome. Pediatrics 2012;130(3):e714–55.

[8] Ng M, Fleming T, Robinson M, Thomson B, Graetz N, Margono C, et al. Global, regional, and national prevalence of overweight and obesity in children and adults during 1980–2013: a systematic analysis for the Global Burden of Disease Study 2013. Lancet (London, England) 2014;384(9945):766–81.

[9] Primeau V, Coderre L, Karelis AD, Brochu M, Lavoie ME, Messier V, et al. Characterizing the profile of obese patients who are metabolically healthy. Int J Obes (2005) 2011;35(7):971–81.

[10] Roberson LL, Aneni EC, Maziak W, Agatston A, Feldman T, Rouseff M, et al. Beyond BMI: the "Metabolically healthy obese" phenotype & its association with clinical/subclinical cardiovascular disease and all-cause mortality—a systematic review. BMC Public Health 2014;14:14.

[11] Ding C, Chan Z, Magkos F. Lean, but not healthy: the "metabolically obese, normal-weight" phenotype. Curr Opin Clin Nutr Metab Care 2016;19(6):408–17.

[12] Vague J. The degree of masculine differentiation of obesities: a factor determining predisposition to diabetes, atherosclerosis, gout, and uric calculous disease. Am J Clin Nutr 1956;4(1):20–34.

[13] McLaughlin T, Lamendola C, Liu A, Abbasi F. Preferential fat deposition in subcutaneous versus visceral depots is associated with insulin sensitivity. J Clin Endocrinol Metab 2011;96(11):E1756–60.

[14] Stefan N, Kantartzis K, Machann J, Schick F, Thamer C, Rittig K, et al. Identification and characterization of metabolically benign obesity in humans. Arch Intern Med 2008;168(15):1609–16.

[15] Lee SW, Son JY, Kim JM, Hwang S-s, Han JS, Heo NJ. Body fat distribution is more predictive of all-cause mortality than overall adiposity. Diabetes Obes Metab.

[16] Reaven GM. Banting lecture 1988. Role of insulin resistance in human disease. Diabetes 1988;37(12):1595–607.

[17] Huang PL. A comprehensive definition for metabolic syndrome. Dis Model Mech 2009;2(5–6):231–7.

[18] Laakso M, Matilainen V, Keinanen-Kiukaanniemi S. Association of neck circumference with insulin resistance-related factors. Int J Obes Rel Metab Disorders 2002;26(6):873–5.

[19] Stabe C, Vasques AC, Lima MM, Tambascia MA, Pareja JC, Yamanaka A, et al. Neck circumference as a simple tool for identifying the metabolic syndrome and insulin resistance: results from the Brazilian Metabolic Syndrome Study. Clin Endocrinol 2013;78(6):874–81.

[20] Kim AM, Keenan BT, Jackson N, Chan EL, Staley B, Poptani H, et al. Tongue fat and its relationship to obstructive sleep apnea. Sleep 2014;37(10):1639–48.

[21] Yang L, Samarasinghe YP, Kane P, Amiel SA, Aylwin SJ. Visceral adiposity is closely correlated with neck circumference and represents a significant indicator of insulin resistance in WHO grade III obesity. Clin Endocrinol 2010;73(2):197–200.

[22] Preis SR, Massaro JM, Hoffmann U, D'Agostino Sr. RB, Levy D, Robins SJ, et al. Neck circumference as a novel measure of cardiometabolic risk: the Framingham Heart study. J Clin Endocrinol Metab 2010;95(8):3701–10.

[23] Rocchini AP. Insulin resistance, obesity and hypertension. J Nutr 1995;125(6 Suppl):1718s–24s.

[24] Massiera F, Bloch-Faure M, Ceiler D, Murakami K, Fukamizu A, Gasc JM, et al. Adipose angiotensinogen is involved in adipose tissue growth and blood pressure regulation. FASEB J 2001;15(14):2727–9.

[25] Giacchetti G, Faloia E, Mariniello B, Sardu C, Gatti C, Camilloni MA, et al. Overexpression of the renin-angiotensin system in human visceral adipose tissue in normal and overweight subjects. Am J Hypertens 2002;15(5):381–8.

[26] Furuhashi M, Ura N, Higashiura K, Murakami H, Tanaka M, Moniwa N, et al. Blockade of the renin-angiotensin system increases adiponectin concentrations in patients with essential hypertension. Hypertension (Dallas, TX: 1979) 2003;42(1):76–81.

[27] Skurk T, van Harmelen V, Hauner H. Angiotensin II stimulates the release of interleukin-6 and interleukin-8 from cultured human adipocytes by activation of NF-kappaB. Arterioscler Thromb Vasc Biol 2004;24(7):1199–203.

[28] Hardy OT, Czech MP, Corvera S. What causes the insulin resistance underlying obesity? Curr Opin Endocrinol Diabetes Obes 2012;19(2):81–7.

[29] Andersson DP, Thorell A, Lofgren P, Wiren M, Toft E, Qvisth V, et al. Omentectomy in addition to gastric bypass surgery and influence on insulin sensitivity: a randomized double blind controlled trial. Clin Nutr (Edinburgh, Scotland) 2014;33(6):991–6.

[30] Albu JB, Curi M, Shur M, Murphy L, Matthews DE, Pi-Sunyer FX. Systemic resistance to the antilipolytic effect of insulin in black and white women with visceral obesity. Am J Physiol 1999;277(3 Pt 1):E551–60.

[31] Foster MT, Pagliassotti MJ. Metabolic alterations following visceral fat removal and expansion: beyond anatomic location. Adipocytes 2012;1(4):192–9.

[32] Gavrilova O, Marcus-Samuels B, Graham D, Kim JK, Shulman GI, Castle AL, et al. Surgical implantation of adipose tissue reverses diabetes in lipoatrophic mice. J Clin Invest 2000;105(3):271–8.

[33] Bouchard C, Perusse L, Leblanc C, Tremblay A, Theriault G. Inheritance of the amount and distribution of human body fat. Int J Obes 1988;12(3):205–15.

[34] Selby JV, Newman B, Quesenberry Jr. CP, Fabsitz RR, Carmelli D, Meaney FJ, et al. Genetic and behavioral influences on body fat distribution. Int J Obes 1990;14(7):593–602.

[35] Brown AE, Walker M. genetics of insulin resistance and the metabolic syndrome. Curr Cardiol Rep 2016;18(8):75.

[36] Yaghootkar H, Scott RA, White CC, Zhang W, Speliotes E, Munroe PB, et al. Genetic evidence for a normal-weight "metabolically obese" phenotype linking insulin resistance, hypertension, coronary artery disease, and type 2 diabetes. Diabetes 2014;63(12):4369–77.

[37] Moltke I, Grarup N, Jorgensen ME, Bjerregaard P, Treebak JT, Fumagalli M, et al. A common Greenlandic TBC1D4 variant confers muscle insulin resistance and type 2 diabetes. Nature 2014;512(7513):190–3.

[38] Arsenault BJ, Lachance D, Lemieux I, et al. Visceral adipose tissue accumulation, cardiorespiratory fitness, and features of the metabolic syndrome. Arch Intern Med 2007;167(14):1518–25.

[39] Bujalska IJ, Kumar S, Stewart PM. Does central obesity reflect "Cushing's disease of the omentum". Lancet (London, England) 1997;349(9060):1210–3.

[40] Tamashiro KL, Sakai RR, Shively CA, Karatsoreos IN, Reagan LP. Chronic stress, metabolism, and metabolic syndrome. Stress (Amsterdam, Netherlands) 2011;14(5):468–74.

[41] Kuo LE, Czarnecka M, Kitlinska JB, Tilan JU, Kvetnansky R, Zukowska Z. Chronic stress, combined with a high-fat/high-sugar diet, shifts sympathetic signaling toward

neuropeptide Y and leads to obesity and the metabolic syndrome. Ann N Y Acad Sci 2008;1148:232–7.

[42] Young T, Peppard PE, Taheri S. Excess weight and sleep-disordered breathing. J Appl Physiol (Bethesda, MD: 1985) 2005;99(4):1592–9.

[43] Resta O, Foschino-Barbaro MP, Legari G, Talamo S, Bonfitto P, Palumbo A, et al. Sleep-related breathing disorders, loud snoring and excessive daytime sleepiness in obese subjects. Int J Obes Rel Metab Disorders 2001;25(5):669–75.

[44] Lopez PP, Stefan B, Schulman CI, Byers PM. Prevalence of sleep apnea in morbidly obese patients who presented for weight loss surgery evaluation: more evidence for routine screening for obstructive sleep apnea before weight loss surgery. Am Surg 2008;74(9):834–8.

[45] Valencia-Flores M, Orea A, Castano VA, Resendiz M, Rosales M, Rebollar V, et al. Prevalence of sleep apnea and electrocardiographic disturbances in morbidly obese patients. Obes Res 2000;8(3):262–9.

[46] Peppard PE, Young T, Palta M, Dempsey J, Skatrud J. Longitudinal study of moderate weight change and sleep-disordered breathing. JAMA 2000;284(23):3015–21.

[47] Tishler PV, Larkin EK, Schluchter MD, Redline S. Incidence of sleep-disordered breathing in an urban adult population: the relative importance of risk factors in the development of sleep-disordered breathing. JAMA 2003;289(17):2230–7.

[48] Patel SR, Larkin EK, Redline S. Shared genetic basis for obstructive sleep apnea and adiposity measures. Int J Obes (2005) 2008;32(5):795–800.

[49] Stradling JR, Crosby JH. Predictors and prevalence of obstructive sleep apnoea and snoring in 1001 middle aged men. Thorax 1991;46(2):85–90.

[50] Davies RJ, Stradling JR. The relationship between neck circumference, radiographic pharyngeal anatomy, and the obstructive sleep apnoea syndrome. Eur Respir J 1990;3(5):509–14.

[51] Ogretmenoglu O, Suslu AE, Yucel OT, Onerci TM, Sahin A. Body fat composition: a predictive factor for obstructive sleep apnea. Laryngoscope 2005;115(8):1493–8.

[52] Liu KH, Chu WC, To KW, Ko FW, Ng SS, Ngai JC, et al. Mesenteric fat thickness is associated with increased risk of obstructive sleep apnoea. Respirology (Carlton, VIC) 2014;19(1):92–7.

[53] Kawaguchi Y, Fukumoto S, Inaba M, Koyama H, Shoji T, Shoji S, et al. Different impacts of neck circumference and visceral obesity on the severity of obstructive sleep apnea syndrome. Obesity (Silver Spring, MD) 2011;19(2):276–82.

[54] Schafer H, Pauleit D, Sudhop T, Gouni-Berthold I, Ewig S, Berthold HK. Body fat distribution, serum leptin, and cardiovascular risk factors in men with obstructive sleep apnea. Chest 2002;122(3):829–39.

[55] Onat A, Hergenc G, Yuksel H, Can G, Ayhan E, Kaya Z, et al. Neck circumference as a measure of central obesity: associations with metabolic syndrome and obstructive sleep apnea syndrome beyond waist circumference. Clin Nutr (Edinburgh, Scotland) 2009;28(1):46–51.

[56] Vgontzas AN, Bixler EO, Chrousos GP. Sleep apnea is a manifestation of the metabolic syndrome. Sleep Med Rev 2005;9(3):211–24.

[57] Joosten SA, Hamilton GS, Naughton MT. Impact of weight loss management in OSA. Chest 2017;152(1):194–203.

[58] Mitchell LJ, Davidson ZE, Bonham M, O'Driscoll DM, Hamilton GS, Truby H. Weight loss from lifestyle interventions and severity of sleep apnoea: a systematic review and meta-analysis. Sleep Med 2014;15(10):1173–83.

[59] Kuna ST, Reboussin DM, Borradaile KE, Sanders MH, Millman RP, Zammit G, et al. Long-term effect of weight loss on obstructive sleep apnea severity in obese patients with type 2 diabetes. Sleep 2013;36(5):641–649a.

[60] Greenburg DL, Lettieri CJ, Eliasson AH. Effects of surgical weight loss on measures of obstructive sleep apnea: a meta-analysis. Am J Med 2009;122(6):535–42.

[61] Dixon JB, Schachter LM, O'Brien PE, Jones K, Grima M, Lambert G, et al. Surgical vs conventional therapy for weight loss treatment of obstructive sleep apnea: a randomized controlled trial. JAMA 2012;308(11):1142–9.

[62] Wilcox I, McNamara SG, Collins FL, Grunstein RR, Sullivan CE, Syndrome Z. the interaction of sleep apnoea, vascular risk factors and heart disease. Thorax 1998;53 (Suppl 3):S25–8.

[63] Bozkurt NC, Beysel S, Karbek B, Unsal IO, Cakir E, Delibasi T. Visceral obesity mediates the association between metabolic syndrome and obstructive sleep apnea syndrome. Metab Syndr Relat Disord 2016;14(4):217–21.

[64] Drager LF, Queiroz EL, Lopes HF, Genta PR, Krieger EM, Lorenzi-Filho G. Obstructive sleep apnea is highly prevalent and correlates with impaired glycemic control in consecutive patients with the metabolic syndrome. J Cardiometab Syndrome 2009;4(2):89–95.

[65] Xu S, Wan Y, Xu M, Ming J, Xing Y, An F, et al. The association between obstructive sleep apnea and metabolic syndrome: a systematic review and meta-analysis. BMC Pulmonary Med 2015;15:105.

[66] Coughlin SR, Mawdsley L, Mugarza JA, Calverley PM, Wilding JP. Obstructive sleep apnoea is independently associated with an increased prevalence of metabolic syndrome. Eur Heart J 2004;25(9):735–41.

[67] Kono M, Tatsumi K, Saibara T, Nakamura A, Tanabe N, Takiguchi Y, et al. Obstructive sleep apnea syndrome is associated with some components of metabolic syndrome. Chest 2007;131(5):1387–92.

[68] Katz I, Stradling J, Slutsky AS, Zamel N, Hoffstein V. Do patients with obstructive sleep apnea have thick necks? Am Rev Respir Dis 1990;141(5 Pt 1):1228–31.

[69] Hoffstein V, Mateika S. Differences in abdominal and neck circumferences in patients with and without obstructive sleep apnoea. Eur Respir J 1992;5(4):377–81.

[70] Horner RL, Mohiaddin RH, Lowell DG, Shea SA, Burman ED, Longmore DB, et al. Sites and sizes of fat deposits around the pharynx in obese patients with obstructive sleep apnoea and weight matched controls. Eur Respir J 1989;2(7):613–22.

[71] Schwab RJ, Pasirstein M, Pierson R, Mackley A, Hachadoorian R, Arens R, et al. Identification of upper airway anatomic risk factors for obstructive sleep apnea with volumetric magnetic resonance imaging. Am J Respir Crit Care Med 2003;168(5):522–30.

[72] Mortimore IL, Marshall I, Wraith PK, Sellar RJ, Douglas NJ. Neck and total body fat deposition in nonobese and obese patients with sleep apnea compared with that in control subjects. Am J Respir Crit Care Med 1998;157(1):280–3.

[73] Jones RL, Nzekwu MM. The effects of body mass index on lung volumes. Chest 2006;130(3):827–33.

[74] Stadler DL, McEvoy RD, Sprecher KE, Thomson KJ, Ryan MK, Thompson CC, et al. Abdominal compression increases upper airway collapsibility during sleep in obese male obstructive sleep apnea patients. Sleep 2009;32(12):1579–87.

[75] Squier SB, Patil SP, Schneider H, Kirkness JP, Smith PL, Schwartz AR. Effect of end-expiratory lung volume on upper airway collapsibility in sleeping men and women. J Appl Physiol 2010;109(4):977–85.

[76] Pantanetti P, Garrapa GG, Mantero F, Boscaro M, Faloia E, Venarucci D. Adipose tissue as an endocrine organ? A review of recent data related to cardiovascular complications of endocrine dysfunctions. Clin Exp Hypertens (New York, NY: 1993) 2004;26(4):387–98.

[77] White LH, Bradley TD. Role of nocturnal rostral fluid shift in the pathogenesis of obstructive and central sleep apnoea. J Physiol 2013;591(5):1179–93.

[78] Pillar G, Shehadeh N. Abdominal fat and sleep apnea: the chicken or the egg? Diabetes Care 2008;31(Suppl 2):S303–9.

[79] Van Cauter E, Knutson KL. Sleep and the epidemic of obesity in children and adults. Eur J Endocrinol 2008;159(Suppl 1):S59–66.

[80] Bamberga M, Rizzi M, Gadaleta F, Grechi A, Baiardini R, Fanfulla F. Relationship between energy expenditure, physical activity and weight loss during CPAP treatment in obese OSA subjects. Respir Med 2015;109(4):540–5.

[81] Shechter A. Obstructive sleep apnea and energy balance regulation: a systematic review. Sleep Med Rev 2017;34:59–69.

[82] Jun JC, Polotsky VY. Sleep and sleep loss: an energy paradox? Sleep 2012;35(11):1447–8.

[83] Jean RE, Duttuluri M, Gibson CD, Mir S, Fuhrmann K, Eden E, et al. Improvement in physical activity in persons with obstructive sleep apnea treated with continuous positive airway pressure. J Phys Act Health 2017;14(3):176–82.

[84] Batool-Anwar S, Goodwin JL, Drescher AA, Baldwin CM, Simon RD, Smith TW, et al. Impact of CPAP on activity patterns and diet in patients with obstructive sleep apnea (OSA). J Clin Sleep Med 2014;10(5):465–72.

[85] Tachikawa R, Ikeda K, Minami T, Matsumoto T, Hamada S, Murase K, et al. Changes in energy metabolism after continuous positive airway pressure for obstructive sleep apnea. Am J Respir Crit Care Med 2016;194(6):729–38.

[86] Edward Shifflett D, Walker EW, Gregg JM, Zedalis D, Herbert WG. Effects of short-term PAP treatment on endurance exercise performance in obstructive sleep apnea patients. Sleep Med 2001;2(2):145–51.

[87] Maeder MT, Ammann P, Munzer T, Schoch OD, Korte W, Hurny C, et al. Continuous positive airway pressure improves exercise capacity and heart rate recovery in obstructive sleep apnea. Int J Cardiol 2009;132(1):75–83.

[88] Hoyos CM, Killick R, Keenan DM, Baxter RC, Veldhuis JD, Liu PY. Continuous positive airway pressure increases pulsatile growth hormone secretion and circulating insulin-like growth factor-1 in a time-dependent manner in men with obstructive sleep apnea: a randomized sham-controlled study. Sleep 2014;37(4):733–41.

[89] Berryman DE, Glad CA, List EO, Johannsson G. The GH/IGF-1 axis in obesity: pathophysiology and therapeutic considerations. Nat Rev Endocrinol 2013;9(6):346–56.

[90] Borel AL, Leblanc X, Almeras N, Tremblay A, Bergeron J, Poirier P, et al. Sleep apnoea attenuates the effects of a lifestyle intervention programme in men with visceral obesity. Thorax 2012;67(8):735–41.

[91] Drager LF, Brunoni AR, Jenner R, Lorenzi-Filho G, Bensenor IM, Lotufo PA. Effects of CPAP on body weight in patients with obstructive sleep apnoea: a meta-analysis of randomised trials. Thorax 2015;70(3):258–64.

[92] Munzer T, Hegglin A, Stannek T, Schoch OD, Korte W, Buche D, et al. Effects of long-term continuous positive airway pressure on body composition and IGF1. Eur J Endocrinol 2010;162(4):695–704.

[93] Chin K, Shimizu K, Nakamura T, Narai N, Masuzaki H, Ogawa Y, et al. Changes in intra-abdominal visceral fat and serum leptin levels in patients with obstructive sleep apnea

syndrome following nasal continuous positive airway pressure therapy. Circulation 1999;100(7):706–12.

[94] Catala R, Ferre R, Sangenis S, Cabre A, Hernandez-Flix S, Masana L. Intraabdominal fat redistribution in long-term continuous positive airway pressure treatment in obstructive sleep apnea patients. Med Clin (Barc) 2016;146(11):484–7.

[95] Trenell MI, Ward JA, Yee BJ, Phillips CL, Kemp GJ, Grunstein RR, et al. Influence of constant positive airway pressure therapy on lipid storage, muscle metabolism and insulin action in obese patients with severe obstructive sleep apnoea syndrome. Diabetes Obes Metab 2007;9(5):679–87.

[96] Shpirer I, Copel L, Broide E, Elizur A. Continuous positive airway pressure improves sleep apnea associated fatty liver. Lung 2010;.

[97] Vgontzas AN, Zoumakis E, Bixler EO, Lin HM, Collins B, Basta M, et al. Selective effects of CPAP on sleep apnoea-associated manifestations. Eur J Clin Investig 2008;38(8):585–95.

[98] Kritikou I, Basta M, Tappouni R, Pejovic S, Fernandez-Mendoza J, Nazir R, et al. Sleep apnoea and visceral adiposity in middle-aged male and female subjects. Eur Respir J 2013;41(3):601–9.

[99] Sivam S, Phillips CL, Trenell MI, Yee BJ, Liu PY, Wong KK, et al. Effects of 8 weeks of continuous positive airway pressure on abdominal adiposity in obstructive sleep apnea. Eur Respir J 2012;40(4):913–8.

[100] Weinstock TG, Wang X, Rueschman M, Ismail-Beigi F, Aylor J, Babineau DC, et al. A controlled trial of CPAP therapy on metabolic control in individuals with impaired glucose tolerance and sleep apnea. Sleep 2012;35(5):617–625B.

[101] Hoyos CM, Killick R, Yee BJ, Phillips CL, Grunstein RR, Liu PY. Cardiometabolic changes after continuous positive airway pressure for obstructive sleep apnoea: a randomised sham-controlled study. Thorax 2012;67(12):1081–9.

[102] Tasali E, Mokhlesi B, Van Cauter E. Obstructive sleep apnea and type 2 diabetes: interacting epidemics. Chest 2008;133(2):496–506.

[103] Clarenbach CF, West SD, Kohler M. Is obstructive sleep apnea a risk factor for diabetes? Discov Med 2011;12(62):17–24.

[104] Pandey A, Demede M, Zizi F, Al Haija'a OA, Nwamaghinna F, Jean-Louis G, et al. Sleep apnea and diabetes: insights into the emerging epidemic. Curr Diab Rep 2011;11(1):35–40.

[105] Kaur A, Mokhlesi B. The effect of OSA therapy on glucose metabolism: it's all about CPAP adherence. J Clin Sleep Med 2017;13(3):365–7.

[106] Punjabi NM, Shahar E, Redline S, Gottlieb DJ, Givelber R, Resnick HE. Sleep-disordered breathing, glucose intolerance, and insulin resistance: the Sleep Heart Health Study. Am J Epidemiol 2004;160(6):521–30.

[107] Bakker JP, Weng J, Wang R, Redline S, Punjabi NM, Patel SR. Associations between obstructive sleep apnea, sleep duration, and abnormal fasting glucose. the multi-ethnic study of atherosclerosis. Am J Respir Crit Care Med 2015;192(6):745–53.

[108] Gu CJ, Li M, Li QY, Li N, Shi GC, Wan HY. Obstructive sleep apnea is associated with impaired glucose metabolism in Han Chinese subjects. Chin Med J 2013;126(1):5–10.

[109] Punjabi NM, Sorkin JD, Katzel LI, Goldberg AP, Schwartz AR, Smith PL. Sleep-disordered breathing and insulin resistance in middle-aged and overweight men. Am J Respir Crit Care Med 2002;165(5):677–82.

[110] Tahrani AA, Ali A, Raymond NT, Begum S, Dubb K, Altaf QA, et al. Obstructive sleep apnea and diabetic nephropathy: a cohort study. Diabetes Care 2013;36(11):3718–25.

[111] Aronsohn RS, Whitmore H, Van CE, Tasali E. Impact of untreated obstructive sleep apnea on glucose control in type 2 diabetes. Am J Respir Crit Care Med 2010;181(5):507–13.

[112] Tahrani AA, Ali A, Raymond NT, Begum S, Dubb K, Mughal S, et al. Obstructive sleep apnea and diabetic neuropathy: a novel association in patients with type 2 diabetes. Am J Respir Crit Care Med 2012;186(5):434–41.

[113] Ip MS, Lam B, Ng MM, Lam WK, Tsang KW, Lam KS. Obstructive sleep apnea is independently associated with insulin resistance. Am J Respir Crit Care Med 2002;165(5):670–6.

[114] Polotsky VY, Patil SP, Savransky V, Laffan A, Fonti S, Frame LA, et al. Obstructive sleep apnea, insulin resistance, and steatohepatitis in severe obesity. Am J Respir Crit Care Med 2009;179(3):228–34.

[115] Tanno S, Tanigawa T, Saito I, Nishida W, Maruyama K, Eguchi E, et al. Sleep-related intermittent hypoxemia and glucose intolerance: a community-based study. Sleep Med 2014;15(10):1212–8.

[116] Punjabi NM, Beamer BA. Alterations in Glucose Disposal in Sleep-disordered Breathing. Am J Respir Crit Care Med 2009;179(3):235–40.

[117] Pamidi S, Wroblewski K, Broussard J, Day A, Hanlon EC, Abraham V, et al. Obstructive sleep apnea in young lean men: impact on insulin sensitivity and secretion. Diabetes Care 2012;35(11):2384–9.

[118] Lin QC, Zhang XB, Chen GP, Huang DY, Din HB, Tang AZ. Obstructive sleep apnea syndrome is associated with some components of metabolic syndrome in nonobese adults. Sleep Breath 2011;.

[119] Foster GD, Borradaile KE, Sanders MH, Millman R, Zammit G, Newman AB, et al. A randomized study on the effect of weight loss on obstructive sleep apnea among obese patients with type 2 diabetes: the Sleep AHEAD study. Arch Intern Med 2009;169(17):1619–26.

[120] Appleton SL, Vakulin A, McEvoy RD, Wittert GA, Martin SA, Grant JF, et al. Nocturnal hypoxemia and severe obstructive sleep apnea are associated with incident type 2 diabetes in a population cohort of men. J Clin Sleep Med 2015;11(6):609–14.

[121] Kendzerska T, Gershon AS, Hawker G, Tomlinson G, Leung RS. Obstructive sleep apnea and incident diabetes. A historical cohort study. Am J Respir Crit Care Med 2014;190(2):218–25.

[122] Marshall NS, Wong KK, Phillips CL, Liu PY, Knuiman MW, Grunstein RR. Is sleep apnea an independent risk factor for prevalent and incident diabetes in the Busselton Health Study? J Clin Sleep Med 2009;5(1):15–20.

[123] Wang X, Bi Y, Zhang Q, Pan F. Obstructive sleep apnoea and the risk of type 2 diabetes: a meta-analysis of prospective cohort studies. Respirology (Carlton, VIC) 2013;18(1):140–6.

[124] Lindberg E, Theorell-Haglow J, Svensson M, Gislason T, Berne C, Janson C. Sleep apnea and glucose metabolism: a long-term follow-up in a community-based sample. Chest 2012;142(4):935–42.

[125] Hui P, Zhao L, Xie Y, Wei X, Ma W, Wang J, et al. Nocturnal hypoxemia causes hyperglycemia in patients with obstructive sleep apnea and type 2 diabetes mellitus. Am J Med Sci 2016;351(2):160–8.

[126] Elizur A, Maliar A, Shpirer I, Buchs AE, Shiloah E, Rapoport MJ. Decreased nocturnal glucose variability in non-diabetic patients with sleep apnea: a pilot study. Isr Med Assoc J 2013;15(9):465–9.

[127] Reutrakul S, Mokhlesi B. Obstructive sleep apnea and diabetes: a state of the art review. Chest 2017;152(5):1070–86.

[128] Guo LX, Zhao X, Pan Q, Sun X, Li H, Wang XX, et al. Effect of continuous positive airway pressure therapy on glycemic excursions and insulin sensitivity in patients with obstructive sleep apnea-hypopnea syndrome and type 2 diabetes. Chin Med J 2015;128(17):2301–6.

[129] Dawson A, Abel SL, Loving RT, Dailey G, Shadan FF, Cronin JW, et al. CPAP therapy of obstructive sleep apnea in type 2 diabetics improves glycemic control during sleep. J Clin Sleep Med 2008;4(6):538–42.

[130] Mokhlesi B, Grimaldi D, Beccuti G, Van Cauter E. Effect of one week of CPAP treatment of obstructive sleep apnoea on 24-hour profiles of glucose, insulin and counter-regulatory hormones in type 2 diabetes. Diabetes Obes Metab 2017;19(3):452–6.

[131] Pamidi S, Wroblewski K, Stepien M, Sharif-Sidi K, Kilkus J, Whitmore H, et al. Eight hours of nightly continuous positive airway pressure treatment of obstructive sleep apnea improves glucose metabolism in patients with prediabetes. a randomized controlled trial. Am J Respir Crit Care Med 2015;192(1):96–105.

[132] Ioachimescu OC, Anthony Jr. J, Constantin T, Ciavatta MM, McCarver K, Sweeney ME, VAMONOS (Veterans Affairs' Metabolism, Obstructed and Non-Obstructed Sleep Study). Effects of CPAP therapy on glucose metabolism in patients with obstructive sleep apnea. J Clin Sleep Med 2017;13(3):455–66.

[133] Coughlin SR, Mawdsley L, Mugarza JA, Wilding JP, Calverley PM. Cardiovascular and metabolic effects of CPAP in obese males with OSA. Eur Respir J 2007;29(4):720–7.

[134] Chopra S, Rathore A, Younas H, Pham LV, Gu C, Beselman A, et al. Obstructive sleep apnea dynamically increases nocturnal plasma free fatty acids, glucose, and cortisol during sleep. J Clin Endocrinol Metab 2017;.

[135] Salord N, Fortuna AM, Monasterio C, Gasa M, Perez A, Bonsignore MR, et al. A randomized controlled trial of continuous positive airway pressure on glucose tolerance in obese patients with obstructive sleep apnea. Sleep 2016;39(1):35–41.

[136] West SD, Nicoll DJ, Wallace TM, Matthews DR, Stradling JR. Effect of CPAP on insulin resistance and HbA1c in men with obstructive sleep apnoea and type 2 diabetes. Thorax 2007;62(11):969–74.

[137] Shaw JE, Punjabi NM, Naughton MT, Willes L, Bergenstal RM, Cistulli PA, et al. The effect of treatment of obstructive sleep apnea on glycemic control in type 2 diabetes. Am J Respir Crit Care Med 2016;194(4):486–92.

[138] Martinez-Ceron E, Barquiel B, Bezos AM, Casitas R, Galera R, Garcia-Benito C, et al. Effect of continuous positive airway pressure on glycemic control in patients with obstructive sleep apnea and type 2 diabetes. A randomized clinical trial. Am J Respir Crit Care Med 2016;194(4):476–85.

[139] Yang D, Liu Z, Yang H, Luo Q. Effects of continuous positive airway pressure on glycemic control and insulin resistance in patients with obstructive sleep apnea: a meta-analysis. Sleep Breath Schlaf Atmung 2013;17(1):33–8.

[140] Iftikhar IH, Hoyos CM, Phillips CL, Magalang UJ. Meta-analyses of the association of sleep apnea with insulin resistance, and the effects of CPAP on HOMA-IR, adiponectin, and visceral adipose fat. J Clin Sleep Med 2015;11(4):475–85.

[141] Theorell-Haglow J, Berne C, Janson C, Lindberg E. Obstructive sleep apnoea is associated with decreased insulin sensitivity in females. Eur Respir J 2008;31(5):1054–60.

[142] Chopra S, Polotsky VY, Jun JC. Sleep apnea research in animals: past, present, and future. Am J Respir Cell Mol Biol 2015;.

[143] Iiyori N, Alonso LC, Li J, Sanders MH, Garcia-Ocana A, O'Doherty RM, et al. Intermittent hypoxia causes insulin resistance in lean mice independent of autonomic activity. Am J Respir Crit Care Med 2007;175(8):851–7.

[144] Carreras A, Kayali F, Zhang J, Hirotsu C, Wang Y, Gozal D. Metabolic effects of intermittent hypoxia in mice: steady versus high-frequency applied hypoxia daily during the rest period. Am J Phys Regul Integr Comp Phys 2012;303(7):R700–9.

[145] Polak J, Shimoda LA, Drager LF, Undem C, McHugh H, Polotsky VY, et al. Intermittent hypoxia impairs glucose homeostasis in C57BL6/J mice: partial improvement with cessation of the exposure. Sleep 2013;36(10):1483–90. 90A-90B.

[146] Louis M, Punjabi NM. Effects of acute intermittent hypoxia on glucose metabolism in awake healthy volunteers. J Appl Physiol 2009;106(5):1538–44.

[147] Lambert GW, Straznicky NE, Lambert EA, Dixon JB, Schlaich MP. Sympathetic nervous activation in obesity and the metabolic syndrome—causes, consequences and therapeutic implications. Pharmacol Ther 2010;126(2):159–72.

[148] Faulenbach M, Uthoff H, Schwegler K, Spinas GA, Schmid C, Wiesli P. Effect of psychological stress on glucose control in patients with Type 2 diabetes. Diab Med 2012;29(1):128–31.

[149] Peltonen GL, Scalzo RL, Schweder MM, Larson DG, Luckasen GJ, Irwin D, et al. Sympathetic inhibition attenuates hypoxia induced insulin resistance in healthy adult humans. J Physiol 2012;590(Pt 11):2801–9.

[150] Shin MK, Yao Q, Jun JC, Bevans-Fonti S, Yoo DY, Han W, et al. Carotid body denervation prevents fasting hyperglycemia during chronic intermittent hypoxia. J Appl Physiol (Bethesda, MD: 1985) 2014;117(7):765–76.

[151] Shin MK, Han W, Bevans-Fonti S, Jun JC, Punjabi NM, Polotsky VY. The effect of adrenal medullectomy on metabolic responses to chronic intermittent hypoxia. Respir Physiol Neurobiol 2014;203C:60–7.

[152] Jun JC, Shin MK, Devera R, Yao Q, Mesarwi O, Bevans-Fonti S, et al. Intermittent hypoxia-induced glucose intolerance is abolished by alpha-adrenergic blockade or adrenal medullectomy. Am J Physiol Endocrinol Metab 2014;00373.

[153] Paquot N, Schneiter P, Jequier E, Tappy L. Effects of glucocorticoids and sympathomimetic agents on basal and insulin-stimulated glucose metabolism. Clin Physiol (Oxford, England) 1995;15(3):231–40.

[154] Kritikou I, Basta M, Vgontzas AN, Pejovic S, Fernandez-Mendoza J, Liao D, et al. Sleep apnoea and the hypothalamic-pituitary-adrenal axis in men and women: effects of continuous positive airway pressure. Eur Respir J 2015;.

[155] Karaca Z, Ismailogullari S, Korkmaz S, Cakir I, Aksu M, Baydemir R, et al. Obstructive sleep apnoea syndrome is associated with relative hypocortisolemia and decreased hypothalamo-pituitary-adrenal axis response to 1 and 250 mug ACTH and glucagon stimulation tests. Sleep Med 2013;14(2):160–4.

[156] Yokoe T, Alonso LC, Romano LC, Rosa TC, O'Doherty RM, Garcia-Ocana A, et al. Intermittent hypoxia reverses the diurnal glucose rhythm and causes pancreatic beta-cell replication in mice. J Physiol 2008;586(3):899–911.

[157] Rabe K, Lehrke M, Parhofer KG, Broedl UC. Adipokines and insulin resistance. Mol Med (Cambridge, MA) 2008;14(11–12):741–51.

[158] Lam JC, Xu A, Tam S, Khong PI, Yao TJ, Lam DC, et al. Hypoadiponectinemia is related to sympathetic activation and severity of obstructive sleep apnea. Sleep 2008;31(12):1721–7.

[159] Fu C, Jiang L, Zhu F, Liu Z, Li W, Jiang H, et al. Chronic intermittent hypoxia leads to insulin resistance and impaired glucose tolerance through dysregulation of adipokines in non-obese rats. Sleep Breath Schlaf Atmung 2015;19(4):1467–73.

[160] Magalang UJ, Cruff JP, Rajappan R, Hunter MG, Patel T, Marsh CB, et al. Intermittent hypoxia suppresses adiponectin secretion by adipocytes. Exp Clin Endocrinol Diab 2009;117(3):129–34.

[161] Dandona P, Aljada A, Bandyopadhyay A. Inflammation: the link between insulin resistance, obesity and diabetes. Trends Immunol 2004;25(1):4–7.

[162] Ryan S, Taylor CT, McNicholas WT. Selective activation of inflammatory pathways by intermittent hypoxia in obstructive sleep apnea syndrome. Circulation 2005;112(17):2660–7.

[163] Lavie L. Oxidative stress in obstructive sleep apnea and intermittent hypoxia—revisited—the bad ugly and good: implications to the heart and brain. Sleep Med Rev 2015;20:27–45.

[164] Nanduri J, Yuan G, Kumar GK, Semenza GL, Prabhakar NR. Transcriptional responses to intermittent hypoxia. Respir Physiol Neurobiol 2008;164(1-2):277–81.

[165] Greenberg H, Ye X, Wilson D, Htoo AK, Hendersen T, Liu SF. Chronic intermittent hypoxia activates nuclear factor-kappaB in cardiovascular tissues in vivo. Biochem Biophys Res Commun 2006;343(2):591–6.

[166] Ouchi N, Parker JL, Lugus JJ, Walsh K. Adipokines in inflammation and metabolic disease. Nat Rev Immunol 2011;11(2):85–97.

[167] Ryan S, Taylor CT, McNicholas WT. Predictors of elevated nuclear factor-kappaB-dependent genes in obstructive sleep apnea syndrome. Am J Respir Crit Care Med 2006;174(7):824–30.

[168] Murphy AM, Thomas A, Crinion SJ, Kent BD, Tambuwala MM, Fabre A, et al. Intermittent hypoxia in obstructive sleep apnoea mediates insulin resistance through adipose tissue inflammation. Eur Respir J 2017;49(4).

[169] Poulain L, Thomas A, Rieusset J, Casteilla L, Levy P, Arnaud C, et al. Visceral white fat remodelling contributes to intermittent hypoxia-induced atherogenesis. Eur Respir J 2014;43(2):513–22.

[170] Carreras A, Zhang SX, Almendros I, Wang Y, Peris E, Qiao Z, et al. Resveratrol attenuates intermittent hypoxia-induced macrophage migration to visceral white adipose tissue and insulin resistance in male mice. Endocrinology 2015;156(2):437–43.

[171] Taylor CT, Kent BD, Crinion SJ, McNicholas WT, Ryan S. Human adipocytes are highly sensitive to intermittent hypoxia induced NF-kappaB activity and subsequent inflammatory gene expression. Biochem Biophys Res Commun 2014;447(4):660–5.

[172] He Q, Yang QC, Zhou Q, Zhu H, Niu WY, Feng J, et al. Effects of varying degrees of intermittent hypoxia on proinflammatory cytokines and adipokines in rats and 3T3-L1 adipocytes. PLoS One 2014;9(1):e86326.

[173] Unnikrishnan D, Jun J, Polotsky V. Inflammation in sleep apnea: an update. Rev Endocr Metab Disord 2015;16(1):25–34.

[174] Xu X, Wang L, Zhang Y, Su T, Chen L, Zhang Y, et al. Effects of chronic sleep deprivation on glucose homeostasis in rats. Sleep Biol Rhythms 2016;14(4):321–8.

[175] Stamatakis KA, Punjabi NM. Effects of sleep fragmentation on glucose metabolism in normal subjects. Chest 2010;137(1):95–101.

[176] Nieto FJ, Young TB, Lind BK, Shahar E, Samet JM, Redline S, et al. Association of sleep-disordered breathing, sleep apnea, and hypertension in a large community-based study. Sleep Heart Health Study. JAMA 2000;283(14):1829–36.

[177] Tkacova R, McNicholas WT, Javorsky M, Fietze I, Sliwinski P, Parati G, et al. Nocturnal intermittent hypoxia predicts prevalent hypertension in the European Sleep Apnoea Database cohort study. Eur Respir J 2014;44(4):931–41.

[178] Jenner R, Fatureto-Borges F, Costa-Hong V, Lopes HF, Teixeira SH, Marum E, et al. Association of obstructive sleep apnea with arterial stiffness and nondipping blood pressure in patients with hypertension. J Clin Hypertens (Greenwich) 2017;.

[179] Walia HK, Li H, Rueschman M, Bhatt DL, Patel SR, Quan SF, et al. Association of severe obstructive sleep apnea and elevated blood pressure despite antihypertensive medication use. J Clin Sleep Med 2014;10(8):835–43.

[180] Peppard PE, Young T, Palta M, Skatrud J. Prospective study of the association between sleep-disordered breathing and hypertension. N Engl J Med 2000;342(19):1378–84.

[181] Mokhlesi B, Finn LA, Hagen EW, Young T, Hla KM, Van Cauter E, et al. Obstructive sleep apnea during REM sleep and hypertension. results of the Wisconsin Sleep Cohort. Am J Respir Crit Care Med 2014;190(10):1158–67.

[182] Fava C, Dorigoni S, Dalle Vedove F, Danese E, Montagnana M, Guidi GC, et al. Effect of CPAP on blood pressure in patients with OSA/hypopnea a systematic review and meta-analysis. Chest 2014;145(4):762–71.

[183] Hu X, Fan J, Chen S, Yin Y, Zrenner B. The role of continuous positive airway pressure in blood pressure control for patients with obstructive sleep apnea and hypertension: a meta-analysis of randomized controlled trials. J Clin Hypertens (Greenwich, CT) 2015;17(3):215–22.

[184] Liu L, Cao Q, Guo Z, Dai Q. Continuous positive airway pressure in patients with obstructive sleep apnea and resistant hypertension: a meta-analysis of randomized controlled trials. J Clin Hypertens (Greenwich, CT) 2016;18(2):153–8.

[185] Iftikhar IH, Valentine CW, Bittencourt LR, Cohen DL, Fedson AC, Gislason T, et al. Effects of continuous positive airway pressure on blood pressure in patients with resistant hypertension and obstructive sleep apnea: a meta-analysis. J Hypertens 2014;32(12):2341–50. discussion 50.

[186] Muxfeldt ES, Margallo V, Costa LM, Guimaraes G, Cavalcante AH, Azevedo JC, et al. Effects of continuous positive airway pressure treatment on clinic and ambulatory blood pressures in patients with obstructive sleep apnea and resistant hypertension: a randomized controlled trial. Hypertension (Dallas, TX: 1979) 2015;65(4):736–42.

[187] Barbe F, Duran-Cantolla J, Sanchez-de-la-Torre M, Martinez-Alonso M, Carmona C, Barcelo A, et al. Effect of continuous positive airway pressure on the incidence of hypertension and cardiovascular events in nonsleepy patients with obstructive sleep apnea: a randomized controlled trial. JAMA 2012;307(20):2161–8.

[188] Bratton DJ, Stradling JR, Barbe F, Kohler M. Effect of CPAP on blood pressure in patients with minimally symptomatic obstructive sleep apnoea: a meta-analysis using individual patient data from four randomised controlled trials. Thorax 2014;69(12):1128–35.

[189] Sanchez-de-la-Torre M, Khalyfa A, Sanchez-de-la-Torre A, Martinez-Alonso M, Martinez-Garcia MA, Barcelo A, et al. Precision medicine in patients with resistant hypertension and obstructive sleep apnea: blood pressure response to continuous positive airway pressure treatment. J Am Coll Cardiol 2015;66(9):1023–32.

[190] Fletcher EC, Lesske J, Qian W, Miller 3rd CC, Unger T. Repetitive, episodic hypoxia causes diurnal elevation of blood pressure in rats. Hypertension (Dallas, TX: 1979) 1992;19(6 Pt 1):555–61.

[191] Lesske J, Fletcher EC, Bao G, Unger T. Hypertension caused by chronic intermittent hypoxia--influence of chemoreceptors and sympathetic nervous system. J Hypertens 1997;15(12 Pt 2):1593–603.

[192] Brooks D, Horner RL, Kozar LF, Render-Teixeira CL, Phillipson EA. Obstructive sleep apnea as a cause of systemic hypertension. Evidence from a canine model. J Clin Investig 1997;99(1):106–9.

[193] Tamisier R, Pepin JL, Remy J, Baguet JP, Taylor JA, Weiss JW, et al. 14 nights of intermittent hypoxia elevate daytime blood pressure and sympathetic activity in healthy humans. Eur Respir J 2011;37(1):119–28.

[194] Hedner JA, Wilcox I, Laks L, Grunstein RR, Sullivan CE. A specific and potent pressor effect of hypoxia in patients with sleep apnea. Am Rev Respir Dis 1992;146(5 Pt 1):1240–5.

[195] Leuenberger U, Jacob E, Sweer L, Waravdekar N, Zwillich C, Sinoway L. Surges of muscle sympathetic nerve activity during obstructive apnea are linked to hypoxemia. J Appl Physiol (Bethesda, MD: 1985) 1995;79(2):581–8.

[196] Angheben JM, Schoorlemmer GH, Rossi MV, Silva TA, Cravo SL. Cardiovascular responses induced by obstructive apnea are enhanced in hypertensive rats due to enhanced chemoreceptor responsivity. PLoS One 2014;9(1):e86868.

[197] Moya EA, Arias P, Varela C, Oyarce MP, Del Rio R, Iturriaga R. Intermittent hypoxia-induced carotid body chemosensory potentiation and hypertension are critically dependent on peroxynitrite formation. Oxidative Med Cell Longev 2016;2016:9802136.

[198] Schulz R, Murzabekova G, Egemnazarov B, Kraut S, Eisele HJ, Dumitrascu R, et al. Arterial hypertension in a murine model of sleep apnea: role of NADPH oxidase 2. J Hypertens 2014;32(2):300–5.

[199] Del Rio R, Moya EA, Iturriaga R. Carotid body and cardiorespiratory alterations in intermittent hypoxia: the oxidative link. Eur Respir J 2010;36(1):143–50.

[200] Del Rio R, Moya EA, Parga MJ, Madrid C, Iturriaga R. Carotid body inflammation and cardiorespiratory alterations in intermittent hypoxia. Eur Respir J 2012;39(6):1492–500.

[201] Marcus NJ, Li YL, Bird CE, Schultz HD, Morgan BJ. Chronic intermittent hypoxia augments chemoreflex control of sympathetic activity: role of the angiotensin II type 1 receptor. Respir Physiol Neurobiol 2010;171(1):36–45.

[202] Lam SY, Liu Y, Ng KM, Liong EC, Tipoe GL, Leung PS, et al. Upregulation of a local renin-angiotensin system in the rat carotid body during chronic intermittent hypoxia. Exp Physiol 2014;99(1):220–31.

[203] Prabhakar NR. Carotid body chemoreflex: a driver of autonomic abnormalities in sleep apnoea. Exp Physiol 2016;101(8):975–85.

[204] Knight WD, Little JT, Carreno FR, Toney GM, Mifflin SW, Cunningham JT. Chronic intermittent hypoxia increases blood pressure and expression of FosB/DeltaFosB in central autonomic regions. Am J Physiol Regul Integr Comp Physiol 2011;301(1):R131–9.

[205] Cunningham JT, Knight WD, Mifflin SW, Nestler EJ. An essential role for DeltaFosB in the median preoptic nucleus in the sustained hypertensive effects of chronic intermittent hypoxia. Hypertension (Dallas, TX: 1979) 2012;60(1):179–87.

[206] Jameson H, Bateman R, Byrne P, Dyavanapalli J, Wang X, Jain V, et al. Oxytocin neuron activation prevents hypertension that occurs with chronic intermittent hypoxia/hypercapnia in rats. Am J Physiol Heart Circ Physiol 2016;310(11):H1549–57.

[207] Linz D, Mahfoud F, Linz B, Hohl M, Schirmer SH, Wirth KJ, et al. Effect of obstructive respiratory events on blood pressure and renal perfusion in a pig model for sleep apnea. Am J Hypertens 2014;27(10):1293–300.

[208] Zalucky AA, Nicholl DD, Hanly PJ, Poulin MJ, Turin TC, Walji S, et al. Nocturnal hypoxemia severity and renin-angiotensin system activity in obstructive sleep apnea. Am J Respir Crit Care Med 2015;192(7):873–80.

[209] Jin ZN, Wei YX. Meta-analysis of effects of obstructive sleep apnea on the renin-angiotensin-aldosterone system. J Geriatr Cardiol 2016;13(4):333–43.

[210] Nicholl DD, Hanly PJ, Poulin MJ, Handley GB, Hemmelgarn BR, Sola DY, et al. Evaluation of continuous positive airway pressure therapy on renin-angiotensin system activity in obstructive sleep apnea. Am J Respir Crit Care Med 2014;190(5):572–80.

[211] Fletcher EC, Bao G, Li R. Renin activity and blood pressure in response to chronic episodic hypoxia. Hypertension (Dallas, TX: 1979) 1999;34(2):309–14.

[212] Saxena A, Little JT, Nedungadi TP, Cunningham JT. Angiotensin II type 1a receptors in subfornical organ contribute towards chronic intermittent hypoxia-associated sustained increase in mean arterial pressure. Am J Physiol Heart Circ Physiol 2015;308(5):H435–46.

[213] Pepin JL, Tamisier R, Barone-Rochette G, Launois SH, Levy P, Baguet JP. Comparison of continuous positive airway pressure and valsartan in hypertensive patients with sleep apnea. Am J Respir Crit Care Med 2010;182(7):954–60.

[214] Thunstrom E, Manhem K, Rosengren A. Blood pressure response to losartan and continuous positive airway pressure in hypertension and obstructive sleep apnea. Am J Respir Crit Care Med 2016;193(3):310–20.

[215] Nadeem R, Singh M, Nida M, Waheed I, Khan A, Ahmed S, et al. Effect of obstructive sleep apnea hypopnea syndrome on lipid profile: a meta-regression analysis. J Clin Sleep Med 2014;10(5):475–89.

[216] Xu H, Guan J, Yi H, Zou J, Meng L, Tang X, et al. Elevated low-density lipoprotein cholesterol is independently associated with obstructive sleep apnea: evidence from a large-scale cross-sectional study. Sleep Breath Schlaf Atmung 2016;20(2):627–34.

[217] Liu A, Cardell J, Ariel D, Lamendola C, Abbasi F, Kim SH, et al. Abnormalities of lipoprotein concentrations in obstructive sleep apnea are related to insulin resistance. Sleep 2015;38(5):793–9.

[218] Nadeem R, Singh M, Nida M, Kwon S, Sajid H, Witkowski J, et al. Effect of CPAP treatment for obstructive sleep apnea hypopnea syndrome on lipid profile: a meta-regression analysis. J Clin Sleep Med 2014;10(12):1295–302.

[219] Phillips CL, Yee BJ, Marshall NS, Liu PY, Sullivan DR, Grunstein RR. Continuous positive airway pressure reduces postprandial lipidemia in obstructive sleep apnea: a randomized, placebo-controlled crossover trial. Am J Respir Crit Care Med 2011;184(3):355–61.

[220] Chirinos JA, Gurubhagavatula I, Teff K, Rader DJ, Wadden TA, Townsend R, et al. CPAP, weight loss, or both for obstructive sleep apnea. N Engl J Med 2014;370(24):2265–75.

[221] Li J, Thorne LN, Punjabi NM, Sun CK, Schwartz AR, Smith PL, et al. Intermittent hypoxia induces hyperlipidemia in lean mice. Circ Res 2005;97(7):698–706.

[222] Li J, Savransky V, Nanayakkara A, Smith PL, O'Donnell CP, Polotsky VY. Hyperlipidemia and lipid peroxidation are dependent on the severity of chronic intermittent hypoxia. J Appl Physiol 2007;102(2):557–63.

[223] Drager LF, Yao Q, Hernandez KL, Shin MK, Bevans-Fonti S, Gay J, et al. Chronic intermittent hypoxia induces atherosclerosis via activation of adipose angiopoietin-like 4. Am J Respir Crit Care Med 2013;188(2):240–8.

[224] Ginsberg HN, Zhang YL, Hernandez-Ono A. Regulation of plasma triglycerides in insulin resistance and diabetes. Arch Med Res 2005;36(3):232–40.

[225] Julius U. Influence of plasma free fatty acids on lipoprotein synthesis and diabetic dyslipidemia. Exp Clin Endocrinol Diab 2003;111(5):246–50.

[226] Barcelo A, Pierola J, de la Pena M, Esquinas C, Fuster A, Sanchez-de-la-Torre M, et al. Free fatty acids and the metabolic syndrome in patients with obstructive sleep apnoea. Eur Respir J 2011;37(6):1418–23.

[227] Jun JC, Drager LF, Najjar SS, Gottlieb SS, Brown CD, Smith PL, et al. Effects of sleep apnea on nocturnal free fatty acids in subjects with heart failure. Sleep 2011;34(9):1207–13.

[228] Jun J, Reinke C, Bedja D, Berkowitz D, Bevans-Fonti S, Li J, et al. Effect of intermittent hypoxia on atherosclerosis in apolipoprotein E-deficient mice. Atherosclerosis 2010;209(2):381–6.

[229] Weiszenstein M, Shimoda LA, Koc M, Seda O, Polak J. Inhibition of lipolysis ameliorates diabetic phenotype in a mouse model of obstructive sleep apnea. Am J Respir Cell Mol Biol 2016;55(2):299–307.

[230] Drager LF, Li J, Shin MK, Reinke C, Aggarwal NR, Jun JC, et al. Intermittent hypoxia inhibits clearance of triglyceride-rich lipoproteins and inactivates adipose lipoprotein lipase in a mouse model of sleep apnoea. Eur Heart J 2012;33(6):783–90.

[231] Jun JC, Shin MK, Yao Q, Bevans-Fonti S, Poole J, Drager LF, et al. Acute hypoxia induces hypertriglyceridemia by decreasing plasma triglyceride clearance in mice. In: Am J Physiol Endocrinol Metab; 2012.

[232] Jun JC, Shin MK, Yao Q, Devera R, Fonti-Bevans S, Polotsky VY. Thermoneutrality modifies the impact of hypoxia on lipid metabolism. Am J Physiol Endocrinol Metab 2013;304(4):E424–35.

[233] Li J, Nanayakkara A, Jun J, Savransky V, Polotsky VY. Effect of deficiency in SREBP cleavage-activating protein on lipid metabolism during intermittent hypoxia. Physiol Genomics 2007;31(2):273–80.

[234] Savransky V, Jun J, Li J, Nanayakkara A, Fonti S, Moser AB, et al. Dyslipidemia and atherosclerosis induced by chronic intermittent hypoxia are attenuated by deficiency of stearoyl coenzyme A desaturase. Circ Res 2008;103(10):1173–80.

[235] Harsch IA, Schahin SP, Radespiel-Troger M, Weintz O, Jahreiss H, Fuchs FS, et al. Continuous positive airway pressure treatment rapidly improves insulin sensitivity in patients with obstructive sleep apnea syndrome. Am J Respir Crit Care Med 2004;169(2):156–62.

Obesity and pulmonary hypertension

Maryellen C. Antkowiak*, Richard N. Channick[†]

Department of Medicine, University of Vermont Larner College of Medicine, Burlington, VT, United States* Department of Medicine, Massachusetts General Hospital, Harvard Medical School, Boston, MA, United States[†]

ABBREVIATIONS

AMPK	AMP-activated protein kinase
Apo-A1	apolipoprotein A-1
BMI	body mass index
BMPR-2	bone morphogenic protein receptor type 2
BNP	brain natriuretic peptide
CTEPH	chronic thromboembolic pulmonary hypertension
Cyclic AMP	cyclic adenosine monophosphate
Cyclic GMP	cyclic guanosine monophosphate
DPG	diastolic pulmonary gradient
GM-CSF	granulocyte-macrophage stimulating factor
HDL	high density lipoprotein
HFpEF	heart failure with preserved ejection fraction
HMG-CoA	3-hydroxy-3-methylglutaryl-coenzyme a
IL-6	interleukin-6
IPAH	idiopathic pulmonary arterial hypertension
KC	keratinocyte-derived cytokine
LVEDP	left ventricular end diastolic pressure
M2	macrophage 2 phenotype
miRNA	microribonucleic acid
mTOR	mammalian target of rapamycin
NF-κB	nuclear factor kappa B
NHANES	National Health and Nutrition Examination Survey
NIPPV	noninvasive positive pressure ventilation
NO	nitric oxide
NT-proBNP	N-terminal pro-brain natriuretic peptide
OSA	obstructive sleep apnea
OHS	obesity hypoventilation syndrome

Mechanisms and Manifestations of Obesity in Lung Disease. https://doi.org/10.1016/B978-0-12-813553-2.00009-9

PAH	pulmonary arterial hypertension
PAP	pulmonary artery pressure
PASMC	pulmonary artery smooth muscle cells
PASP	pulmonary artery systolic pressure
PAWP	pulmonary artery wedge pressure
PH	pulmonary hypertension
PPARγ	peroxisome proliferator-activated receptor-γ
PVH	pulmonary venous hypertension
PVR	pulmonary vascular resistance
REVEAL	Registry to Evaluate Early and Long-Term PAH Disease Management
ROS	reactive oxygen species
RVSP	right ventricular systolic pressure
sGC	soluable guanylate cyclase
SR-B1	scavenger receptor class B type 1
SRF-SRE	serum response factor-serum response element
TGF-β	transforming growth factor β
TNF-α	τumor necrosis factor alpha
TPG	transpulmonary gradient
Tregs	regulatory T-cell lymphocytes
VTE	venous thromboembolism
UCP2	uncoupling protein 2
WHO	World Health Organization

In this chapter, we will review the epidemiology of obesity and pulmonary hypertension, and explore potential metabolic and inflammatory mechanisms linking these two conditions.

INTRODUCTION

Unlike the systemic vasculature, the pulmonary circulation is a low pressure, low resistance circuit. A wide variety of disease processes can increase pulmonary arterial pressure, leading to the development of pulmonary hypertension (pH), defined as a mean pulmonary artery pressure (PAP) >25 mmHg. PH is broadly classified into five groups based on etiology. Patients with obesity may be at increased risk for the development of PH across several of the World Health Organization (WHO) groups. Greater understanding of the pathophysiology leading to pulmonary vascular remodeling and the development of PH has brought to light the significance of several signaling and metabolic pathways that may also be altered in the obese state. Obesity also presents unique challenges in the evaluation and management of patients presenting with PH (Figure 9.1).

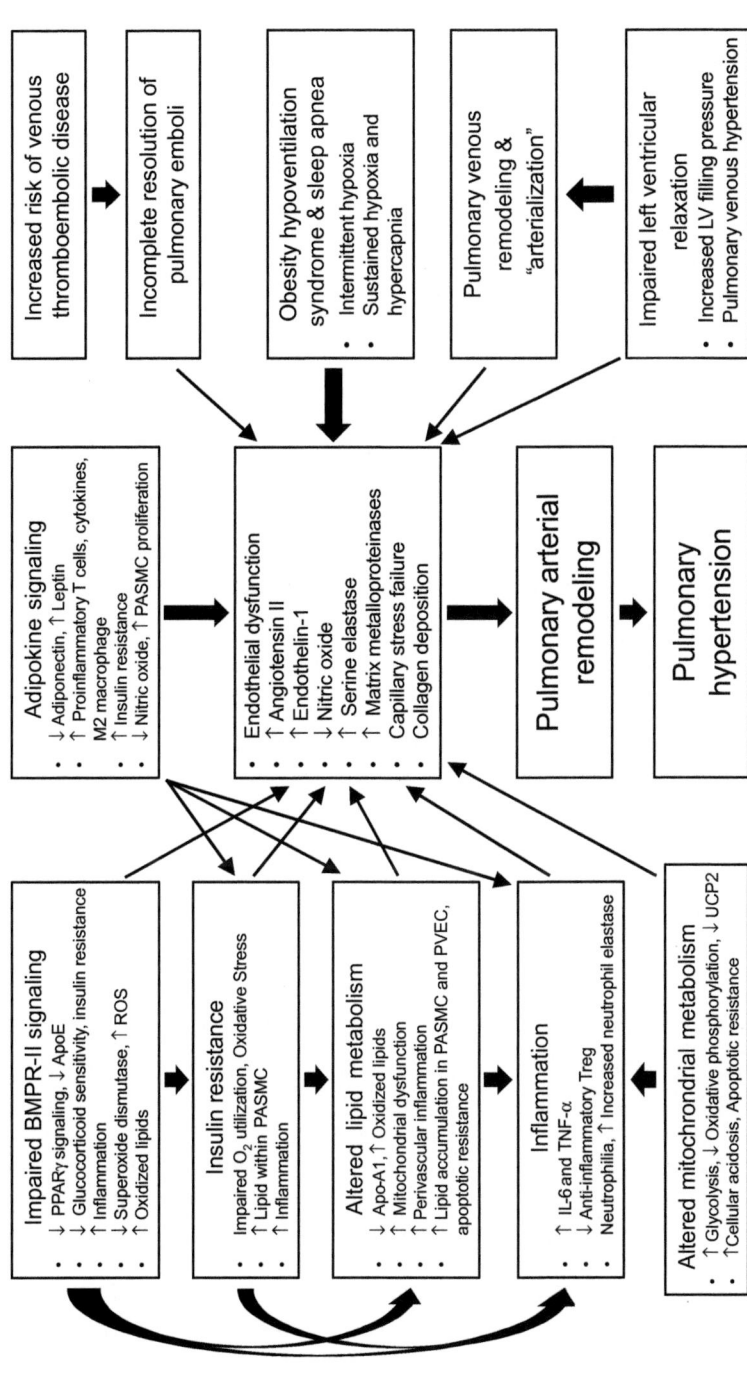

FIGURE 9.1

Risk factors and mechanisms contributing to pulmonary hypertension in obesity and metabolic syndrome. Patients with obesity are at risk for the development of pulmonary hypertension via several groups of the World Health Organization classification system. Additionally, components of the metabolic syndrome may contribute to pulmonary vascular remodeling through a variety of complex, intertwined mechanisms. ROS, reactive oxygen species; PASMC, pulmonary artery smooth muscle cells; PVEC, pulmonary vascular endothelial cells.

ASSOCIATION OF PULMONARY HYPERTENSION WITH OBESITY

The prevalence of PH in patients with obesity appears high, though the exact rates remain unclear. The limited number of reports assessing rates of PH in this population are largely single center, retrospective, or limited to a small number of subjects. The majority of these reports are based on echocardiographic estimates of PAP or right ventricular function rather than more accurate invasive hemodynamic assessment. Acknowledging these limitations, obesity has been reported to be present in 40%–50% of patients with severe PH [1]. One study of autopsy cases reported a higher prevalence of the histologic changes associated with PH in obese patients compared with matched lean controls [2], whereas echocardiographic studies controlling for other components of the metabolic syndrome show positive correlations between body mass index (BMI) and estimated pulmonary artery systolic pressure (PASP), as well as right ventricular size and worsening right ventricular function [3, 4]. Furthermore, in echocardiographic studies of otherwise healthy individuals, an elevated estimated PASP (defined as an estimated PASP >40 mmHg) has been found in up to 5% of subjects with a $BMI > 30 \, kg/m^2$. It should be noted that subjects in these series were not rigorously screened to rule out undiagnosed diastolic dysfunction, sleep disordered breathing, or other risk factors for the development of PH [3, 5].

CLASSIFICATION OF PULMONARY HYPERTENSION

Elevations in PAP can result from increased pulmonary vascular resistance (PVR), increased left atrial pressure leading to passive congestion, increased cardiac output and pulmonary blood flow, or any combination of these factors. PH has been classified into five groups by the WHO based on etiology, most recently updated at the Fifth World Symposium [6]. Patients with obesity may be at risk for PH in multiple WHO groups.

WHO GROUP I: PULMONARY ARTERIAL HYPERTENSION

This group includes a number of disorders that lead to remodeling of the pulmonary arterial vasculature in the absence of elevated left atrial filling pressure, parenchymal lung disease, or thromboembolic disease. This remodeling is characterized by intimal thickening, medial hypertrophy, and *in situ* thrombosis. Glomeruli-like lesions of dysregulated vascular proliferation surrounding a myofibroblastic core, termed "plexiform lesions," characterize advanced stages of the disease [7]. There are three distinct pathways implicated in the development of these pathologic changes and for which therapeutic targets are available [8]. Patients with PAH have reduced levels of prostacyclin and decreased signaling through the prostacyclin-cyclic AMP pathway, resulting in vasoconstriction and proliferation. PAH is also associated with decreased nitric oxide, resulting in impaired signaling through cyclic GMP, and loss of the vasoproliferative and vasodilatory effects normally induced by upregulation of cyclic GMP. Lastly, PAH is associated with increased levels of endothelin, which upon binding to endothelin receptors stimulates vasoconstriction and vasoproliferation. Derangements in these pathways are well described in obese and lean patients with pulmonary arterial hypertension (PAH). In patients whom a careful diagnostic evaluation is consistent with WHO group I PAH, therapy targeting one or more of these three pathways is indicated regardless of BMI [8].

Included in WHO group I are idiopathic and familial PH, as toxin-induced, congenital heart disease, and connective tissue disease-associated PAH, as well as infectious etiologies and portopulmonary hypertension [6]. The association between obesity and PAH remains a subject of debate. Toxin- or drug-induced PAH has historically been associated with increased BMI, owing to the high incidence of PAH observed with use of anorexigen agents prescribed for weight loss [9]. The risk of PAH was particularly high with fenfluramine derivatives prior to their removal from the market [9].

Although little evidence exists to support a causal role of obesity in the development of PAH, comparison of patients enrolled in the Registry to Evaluate Early and Long-Term PAH Disease Management (REVEAL) with age- and sex-matched controls from the National Health and Nutrition Examination Survey (NHANES) suggested a possible association between BMI and several subgroups of PAH. Notably, patients with idiopathic pulmonary arterial hypertension (IPAH) had a higher mean BMI than matched controls (29.1 vs. 28.1 kg/m^2, $P < 0.001$). A greater percentage of patients with IPAH had an obese BMI (>30 kg/m^2) than controls (40.1 vs. 32.5%, $P < 0.001$) [10]. Furthermore, these findings are intriguing given the increasing understanding of the role of the metabolic syndrome and disorders of metabolism and adipokine signaling in the pathogenesis of PAH. These will be discussed in detail later in this chapter.

WHO GROUP II: PULMONARY HYPERTENSION DUE TO LEFT HEART DISEASE

PH due to left heart disease, or pulmonary venous hypertension (PVH), is the most prevalent form of PH, accounting for an estimated 75% of cases [11]. Hemodynamically, group II PH is characterized by a mean PAP > 25 mmHg in the presence of a pulmonary artery occlusion pressure of >15 mmHg indicative of PVH due to long-standing elevations in left heart filling pressures. Group II PH can be the sequela of any cardiac pathology that results in chronic elevations of left heart filling pressures, including valvular disease or heart failure with reduced or preserved ejection fraction [6, 12–14]. Group II PH has also been termed "postcapillary pulmonary hypertension" characterized by lack of elevation in PVR, diastolic pulmonary gradient (DPG) or transpulmonary gradient (TPG). However, many patients with PH and elevated LV filling pressures will present with increases in PVR and elevated gradients, termed "combined pre- and postcapillary pulmonary hypertension," which are thought to evolve in the setting of chronic PVH [14, 15].

Obesity and other components of the metabolic syndrome are highly associated with PVH. In a 2009 study of patients with a hemodynamic profile consistent with PVH, almost 80% had a BMI > 30 kg/m^2 (compared with only 30% of patients suffering from PAH). They also exhibited significantly higher rates of diabetes mellitus, hypertension, hyperlipidemia, and coronary artery disease. Two or more components of the metabolic syndrome were observed in 94% of patients with PVH, but in only 34% of patients with PAH, and in 44% of the general population between the ages of 50 and 59 [16]. The prevalence of three or more components in the general population in this age group is 33%, whereas in patients of the same age with PVH, three or more components are present in 64% [16]. Heart failure with preserved ejection fraction (HFpEF) is increasingly recognized as a cause of PVH and group II PH. The prevalence of PH in patients with HFpEF may be higher than 80% [17–21]. Although obesity and the metabolic syndrome are highly associated with the development of HFpEF [22], obesity also appears to independently increase the risk of developing PH in association with HFpEF. In a 2009 study, a BMI > 40 kg/m^2 was associated with an odds ratio of 3.37 for the development of PH in a cohort of patients with HFpEF when controlling for the presence of metabolic syndrome [20]. Conversely, HFpEF patients with PH in this context also had higher mean BMI and higher rates of class I and II obesity as well [20]. Obesity and metabolic syndrome also may be associated with the development of combined pre- and postcapillary PH in patients with HFpEF, particularly in women [15, 22].

The mechanisms by which chronically elevated left-sided filling pressures lead to PH and increased PVR is a subject of much interest [13]. As increases in left atrial pressure are transmitted to the pulmonary venous system, these elevated pressures may result in endothelial dysfunction, leakage of signaling molecules into the venous wall, and ultimately stress failure of the pulmonary capillaries [23, 24]. Repeated episodes of stress failure may overcome normal repair mechanisms resulting in type IV collagen deposition, local activation of angiotensin II and endothelin-1, and loss of normal oscillations in endogenous nitric oxide, promoting further injury and remodeling [24–28]. Vascular wall serine elastases and matrix metalloproteinase are upregulated, resulting in medial hypertrophy, arterialization of the pulmonary venous system, and ultimately hypertrophy and remodeling of the pulmonary arterial system as well [23, 29]. Whether obesity or components of the metabolic syndrome contribute to or accelerate these processes leading to development or progression of combined pre- and postcapillary PH is not currently well understood.

WHO GROUP III: PULMONARY HYPERTENSION DUE TO LUNG DISEASE OR HYPOXIA

PH due to lung disease or hypoxia is the second-most prevalent form of PH, likely accounting for about a quarter of cases of PH [11, 30]. PH may be a complication of chronic obstructive pulmonary disease, interstitial lung disease, or sleep disordered breathing (such as obstructive sleep apnea [OSA]) [6]. Patients with obesity are at increased risk for the development of WHO group III PH largely due to an increased prevalence of sleep-disordered breathing in this population [30]. In fact, increased BMI is known to be the strongest risk factor for the development of OSA (see also Chapter 8) [31], and treatment of this disorder with noninvasive positive pressure ventilation correlates with a reduction in PAPs in cross-sectional studies, as well as decreased PAPs from baseline in small longitudinal studies [32, 33]. The prevalence of PH in patients with OSA may be increasing. Early studies reported prevalence of about 15%–20%, whereas more recently rates as high as 70% have been reported [32, 34–36]. This increase may be in part artefactual, due to differing definitions of PH and differing inclusion criteria among these various studies. More recent reports include a substantial number of patients with pulmonary artery occlusion pressures >15mmHg, suggesting a substantial role for concomitant PVH for which obese patients are also at increased risk, as discussed in the previous section [36]. Although severity of sleep apnea correlates poorly with the development of PH, BMI has consistently been a strong predictor of PH in patients with OSA, exhibiting a strong correlation with PAP [32, 34, 36, 37]. Observational studies of patients with OSA have also consistently demonstrated increased rates of PH in patients with concomitant daytime hypoxia or hypercapnia [34, 35]. In the absence of comorbid lung disease, daytime abnormalities of gas exchange in obese patients with sleep-disordered breathing may be more consistent with a diagnosis of obesity hypoventilation syndrome (OHS). OSA patients with OHS have been shown to have higher rates of PH than those without OSH, with rates reported between 45%–70% [33, 38, 39]. Additionally, the severity of PH in OSA/OHS is greater than that observed in patients with OSA alone [38]. Similar to patients with OSA, a strong positive correlation exists between BMI and PAP in patients with OSA [33], whereas weight loss surgery is associated with improved daytime hypoxia and hypercapnia, and decreases in PAP, left-sided filling pressures, and PVR [40, 41]. The development of PH in patients with sleep-disordered breathing is associated with decreased exercise capacity, lower functional scores, and increased mortality regardless of BMI [33, 36].

WHO GROUP IV: CHRONIC THROMBOEMBOLIC PULMONARY HYPERTENSION

Chronic thromboembolic pulmonary hypertension (CTEPH) develops as a result of incomplete resolution of pulmonary embolism. Persistent emboli become endothelialized and result in obstruction of the pulmonary vascular bed. Additionally, the unobstructed pulmonary arteries of patients with CTEPH exhibit remodeling similar to the histological changes described in patients with PAH [42]. These changes are proposed to be the results of increased flow, shear stress, upregulation of vasoconstrictor and proliferative signaling molecules by activated endothelial cells, and increased inflammatory cytokines [42]. Although the rates of CTEPH following acute pulmonary embolism remain unclear, multiple risk factors for the development the disease have been identified [43–47]. Obesity is an established risk factor for venous thromboembolic (VTE) disease [29]. Obesity correlates with a greater risk of VTE than other components of the metabolic syndrome [48], and the risk appears to increase with additional weight gain [49]. However, currently there are no data to suggest that obesity is a specific risk factor for progression to CTEPH following acute pulmonary embolism.

WHO GROUP V: PULMONARY HYPERTENSION DUE TO UNCLEAR OR MULTIFACTORIAL MECHANISMS

As the name implies, WHO group V PH includes a variety of conditions that lead to the development of PH as a result of unclear mechanisms or perhaps several mechanisms [6]. Group V represents the most poorly studied group of patients with PH, and there is little, if any, data to suggest a significant role of obesity in the development of PH in this group.

CHALLENGES IN THE DIAGNOSIS AND CLASSIFICATION OF PULMONARY HYPERTENSION IN THE OBESE PATIENT

There exist several important limitations of our ability to identify and characterize PH in obese subjects that should be noted. Although echocardiography is an essential tool in the initial evaluation of patients with suspected PH, obtaining diagnostically useful echocardiogram windows and images in patients with an obese body habitus can be challenging. Magnetic resonance imaging, increasingly recognized for its utility in the evaluation and prognostication of patients with PH, may not be feasible or may have impaired image quality in obese subjects [6]. Furthermore, invasive hemodynamic assessment via right heart catheterization can also be affected by obesity. Obese patients have increased respiratory variation in intrathoracic pressure, which is reflected in greater variations in assessment of pulmonary artery wedge pressure (PAWP) [50–52]. As this measurement is crucial in discriminating between pre- and postcapillary PH, errors in the assessment of this pressure can lead to misclassification of patients and even erroneous treatment. When guideline-based assessments of end-expiratory PAWP are compared with digital averaged mean PAWP reported by many modern catheterization lab software, end-expiratory measurements are consistently higher than means in obese patients [51, 52]. These differences increase with increasing BMI. Although end-expiratory PAWP correlates better with end-expiratory assessment of LVEDP, respiratory variations in LVEDP are not well studied in obese patients, and mean PAWP assessment may correlate better with clinical and echocardiographic PH phenotype [51, 52]. The appropriate method of assessment of PAWP in obese patients remains a subject in need of further investigation.

ASSOCIATION OF PULMONARY HYPERTENSION AND METABOLIC SYNDROME

Components of the metabolic syndrome are common in patients with PH. A 2009 study reported that in patients with WHO group I PH, 31% were obese [16]. Systemic hypertension was the most common component of the syndrome, observed in 54%. Diabetes mellitus, hyperlipidemia, and OSA were reported in 20%, 17%, and 14%, respectively [16]. As discussed previously, metabolic syndrome is much more common in patients with PVH (WHO group II). In the same study, 77% of patients with PVH were obese, and hypertension, diabetes mellitus, hyperlipidemia, and OSA were reported in 94%, 59%, 47%, and 29%, respectively [16]. Observational data suggesting significant comorbidity between PH, obesity, and the metabolic syndrome has garnered interest in investigating potential common pathophysiologic mechanisms.

THE OBESITY PARADOX IN PATIENTS WITH PULMONARY HYPERTENSION

The obesity paradox refers to the multiple studies that suggest that although obesity is associated with increased risk of chronic conditions, including cardiovascular disease and chronic heart failure, it may also confer a survival advantage in those patients who carry these diagnoses [53–55]. Studies that have investigated the paradox have been limited by retrospective nature, sample size, and confounding factors [56]. Acknowledging these limitations, several studies have demonstrated an "obesity paradox" in patients with PH. A retrospective study of 267 patients with WHO group I PAH from a national registry demonstrated that increasing BMI was the strongest predictor of survival, although nonobese but overweight (BMI 25–29.9) patients had the lowest overall mortality [57]. A prospective single center study of 105 patients with precapillary or combined pre- and postcapillary PH also reported increasing BMI as the strongest predictor of survival in multivariate analysis. Mortality rates were significantly lower in obese patients, both in precapillary and combined PH [58]. If, as these studies suggest, an obesity paradox does exist, the mechanisms by which obesity exerts this protective effect are yet to be elucidated. It may be that, although patients meet criteria to be classified as group 1 (PAH), they may, in reality, have group 2 (left heart disease) PH, a more common and often less severe form of PH.

BARIATRIC SURGERY IN PATIENTS WITH PULMONARY HYPERTENSION

The role of bariatric surgery in the management of obese patients with PH and other lung pathology is a subject of increasing interest, as are the mechanisms by which bariatric surgery exerts positive effects in a variety of lung diseases. In a retrospective case-control study of patients with PH who underwent weight loss surgery, average mean PAP and right ventricular systolic pressures (RVSPs) decreased by 18 mmHg, compared with no improvement in hemodynamics in controls with similar baseline hemodynamics and background medical therapy. Additionally, bariatric surgery was associated with a decreased need for pulmonary vasodilator therapy and diuretic dose whereas controls demonstrated no improvement in even increased dosing over the same period. There was a nonsignificant improvement in supplemental oxygen requirements in the surgical patients compared with medical therapy alone [41]. In patients with OHS, weight loss surgery is associated with improved daytime hypoxia and hypercapnia and decreases in PAP, left-sided filling pressures, and PVR [40]. These data are promising that weight loss surgery may be a safe and effective option in the management of patients with PH, but whether these improvements are a result of improvements in sleep disordered breathing, reduction of left ventricular filling pressures, or one or more of the additional direct metabolic or inflammatory mediated effects on the pulmonary vascular that will be described in the remained of this chapter is not yet known.

PATHOPHYSIOLOGIC MECHANISMS OF PULMONARY VASCULAR DISEASE IN OBESITY

Obesity and the metabolic syndrome have been associated with a number of derangements in metabolic pathways. Increasing evidence has suggested that these abnormalities may also play an important role in the development of pulmonary vascular disease and PH. Adipose tissue has been recognized for its role as an endocrine organ, and the signaling molecules it produces, known as adipokines, exhibit altered expression in obesity and the metabolic syndrome. Several of these adipokines have demonstrated effects on the pulmonary vasculature. Adipokines also significantly modulate inflammatory pathways and are important mediators of the chronic inflammatory state associated with obesity and the metabolic syndrome. As we have gained a greater understanding of the chronic inflammatory state of obesity, similar derangements of inflammatory pathways have been implicated in the pathophysiology of pulmonary vascular disease. Additionally, an increasing number of cytokines known to be elevated in obesity have been associated with pulmonary vascular remodeling.

ADIPOKINE SIGNALING IN PULMONARY VASCULAR DISEASE

It is now well recognized that adipose tissue's role includes more than simply energy storage. Adipose tissue also functions as an endocrine organ and secretes a number of cytokines and other signaling molecules with important effects on metabolic and inflammatory pathways. These cytokines are termed adipokines and, as adipose tissue accumulates, adipokine secretion and signaling patterns become altered. Several adipokines have been noted for their effects on the pulmonary vasculature. Two in particular, leptin and adiponectin, have well described roles in the development of pulmonary vascular disease [59–65].

Adiponectin, like all adipokines, is secreted from adipose tissue. However, levels of adiponectin are paradoxically higher in lean individuals and decrease with increasing adipose body mass. Levels also decrease in the presence of other components of the metabolic syndrome, including diabetes, and in patients with cardiovascular disease. Adiponectin has the capability to form a variety of oligomeric forms, particularly when circulating levels are significantly elevated. These forms differ in their interactions with cell surface receptors. As a result, adiponectin has pleiotropic effects not only on metabolism, but also on inflammatory pathways, pulmonary vascular tone and pulmonary artery smooth muscle proliferation. Adiponectin exerts several effects on pulmonary vascular endothelial cells, including increasing nitric oxide production and downregulation of adhesion factors [63, 64]. In animal models, adiponectin-deficient mice develop higher PAPs with aging than wild type controls. Additionally, these mice develop PH following hypoxic and allergic inflammatory insults while wild type mice do not exhibit this response [66–68].

The mechanisms by which the adiponectin deficiency observed in obesity and metabolic syndrome lead to the development of pulmonary vascular disease have not been well elucidated in human subjects. However, animal models have revealed a number of potentially beneficial effects of adiponectin in preventing the pathophysiologic changes that lead to PH. Adiponectin has been shown to have vasodilatory properties and endothelial cells of mice deficient in adiponectin exhibit reduced levels of nitric oxide. In addition to direct vasodilatory effects, adiponectin also exhibits antiproliferative effects on pulmonary artery smooth muscle cells. Adiponectin may inhibit this proliferation though several mechanisms. A number of growth factors stimulate smooth muscle proliferation via activation of the mammalian target of rapamycin (mTOR) pathway. Adiponectin inhibits activation of this pathway both by stimulating AMP-activated protein kinase (AMPK), a receptor with downstream mTOR inhibitory effects, and by binding to growth factors themselves, rendering them unavailable for receptor activation [69, 70]. Additionally, adiponectin has been shown to reduce activation of the serum response factor-serum response element (SRF-SRE) pathway [71]. This pathway plays a major role in promoting differentiation and growth of smooth muscle cells. By inhibiting this pathway, adiponectin prevents differentiation of progenitor smooth muscle cells into a proliferative phenotype. In the setting of adiponectin deficiency, these inhibitory effects on smooth muscle proliferation are lost, and adiponectin-deficient mice exhibit increased accumulation and proliferation of smooth muscle cells in response to vascular injury [64].

Adiponectin also appears to prevent pulmonary vascular remodeling via its antiinflammatory effects. In addition to developing increasing PAPs with aging, adiponectin-deficient mice demonstrate increased perivascular inflammation. This may be mediated by increased migration of leukocytes from the vasculature as a result of loss of adiponectin-mediated downregulation of endothelial leukocyte adhesion molecule expression [63]. Adiponectin receptors are widely expressed by leukocytes, and adiponectin appears to have particularly profound antiinflammatory effects on macrophages by promoting M2 polarization, inhibiting NF-κB driven transcription of proinflammatory cytokines, and upregulating the transcription of antiinflammatory cytokines [72]. The role of these inflammatory cytokines in the pathogenesis of pulmonary vascular disease will be further discussed in the next section.

Lastly, adiponectin has well-described roles in metabolic pathways [68]. Increased adiponectin levels have been associated with increased insulin sensitivity and decreased circulating free fatty acids. Of particular interest in the pathophysiology of pulmonary vascular disease, adiponectin is upregulated by activation of the peroxisome proliferator-activated receptor-γ (PPARγ) pathway, a pathway implicated in the effects of mutations in the bone morphogenic protein receptor (BMPR-II) on the development of PH. The role of these mutations will be discussed in subsequent sections of this chapter. Although the relationship between mutations in BMPR-II and adiponectin levels have not yet been described in models of pulmonary vascular remodeling, the common pathway of PPARγ signaling suggests this relationship warrants further investigation. Adiponectin levels can be increased by administration of the thiazolidinediones, which are PPARγ agonists [73]. Although it remains unclear exactly how adiponectin and insulin resistance affect the pulmonary vasculature, animal models have consistently demonstrated improvements in PH following thiazolidinedione administration, and these improvements correlate with increases in adiponectin and markers of insulin sensitivity [63].

Leptin is another adipokine for which there is a growing body of literature supporting a potential role in the pathophysiology of pulmonary vascular disease. Unlike adiponectin, leptin levels increase with increasing BMI. Investigations into the role of leptin in pulmonary vascular disease have yielded more conflicting data than those reported with adiponectin, and leptin's effects on the pulmonary vasculature may be even more complex than the multifactorial pathways described for adiponectin. Although patients with PAH exhibit higher plasma leptin levels than controls [61], low leptin levels have been associated with decreased survival in patients with PAH [65]. Such increased circulating levels of leptin may in fact be derived from the vascular endothelial cells of patients with PAH, as *in vitro* studies of these cells demonstrate higher levels of leptin production compared with those cultured from non-PAH controls. Additionally, circulating regulatory T cells (Tregs) isolated from these patients manifest higher levels of leptin receptor expression. Leptin may have an inhibitory effect on Treg function, as their function appears to be decreased despite higher numbers of surface receptors, with these decreases in function correlating with increasing leptin levels. Given the protective effects of Treg in autoimmune disease, decreased Treg function may be a mechanism by which leptin contributes to the development of pulmonary vascular disease [61]. Although leptin is associated with decreased antiinflammatory Treg function, it is also associated with proinflammatory effects including increased monocyte activation and increased pulmonary endothelial cell expression of leukocyte adhesion molecules [62]. Pulmonary artery smooth muscle cells (PASMCs) from patients with PAH also exhibit greater expression of leptin receptors and increased proliferation following leptin administration in cell culture. Conversely, the PASMCs exhibit decreased proliferation when cultured in the presence of a leptin neutralizer. Administration of leptin neutralizer is protective against hypoxia-induced vasoproliferation and PH in rodent models. [62]. Although these findings may help elucidate the role of leptin in the development of

PH, they do not shed light on the paradoxical finding that lower plasma leptin levels are associated with decreased survival.

Animal models of leptin deficiency have also led to somewhat paradoxical or conflicting results. Leptin receptor-deficient obese Zucker rats and leptin-deficient (ob/ob) mice manifest attenuated PH and lower PVR in response to chronic hypoxia than wild type littermates, with histologic studies showing less pulmonary artery and right ventricular hypertrophy in these leptin-deficient animals [60, 62]. In wild type mice and rats, administration of a leptin neutralizer or chronic dietary supplementation with a compound associated with decreased endothelial cell leptin production was also associated with a diminished pulmonary hypertensive response to chronic hypoxia [62]. Other investigators, however, have demonstrated that, in the absence of an additional insult or "second hit," leptin-deficient mice have increased pulmonary artery wall thickness along with smooth muscle and fibroblast proliferation, macrophage infiltration and hyaluronan deposition in pulmonary arterioles, and significant right ventricular hypertrophy [59]. One possible explanation for these discordant findings could be that leptin, like most adipokines, has pleiotropic effects on metabolic and inflammatory pathways. In the absence of pulmonary vascular insult, obese leptin-deficient animals may develop elevated PAP as a result of the hemodynamic and metabolic effects of chronic obesity and metabolic syndrome, yet at the same time, amelioration of leptin's proinflammatory affects may explain why leptin deficiency or signaling blockade is protective against the development of PH following injury.

INSULIN RESISTANCE IN PULMONARY VASCULAR DISEASE

Insulin resistance is an important component of the metabolic syndrome and frequently complicates obesity. Overt diabetes mellitus is common in patients with PVH, and is present in almost 60% of this most prevalent group in the WHO classification system [16]. Even in the absence of overt clinical diabetes, insulin resistance is associated with impaired left ventricular relaxation [74]. In contrast, diabetes is much less common in patients with PAH (group I), and is observed in about 20% of PAH [16]. Despite this lower incidence of clinical diabetes in patients, PAH is often associated with a state of insulin resistance [75, 76]. When compared with controls, patients with PAH demonstrate higher hemoglobin A1C and triglyceride levels and lower high-density lipoprotein levels independent of age and BMI [75, 76]. Rates of insulin resistance in these patients are more than twice those reported in matched controls (46% vs. 22%). Insulin resistant patients with PAH exhibit worse functional class and lower 6-month event-free survival compared with insulin-sensitive patients [76]. Despite these associations, insulin resistance has not been shown to be a strong predictor of hemodynamic parameters or right ventricular function, and studies have yielded somewhat conflicting data on the association of insulin resistance and exercise capacity in PAH [74, 75]. In animal models of PH, lipid accumulation and impaired oxygen utilization, both hallmarks of insulin resistance, have been observed not only in PASMCs but in skeletal muscle cells as well [77, 78]. These findings suggest the insulin-resistant state associated with PH is a systemic disorder, similar to obesity and the metabolic syndrome. It is not clear how this systemic state of impaired insulin sensitivity contributes to development of PAH, or even if insulin resistance has a true role in its pathogenesis. However, insulin resistance has well-established associations with increased inflammation and endothelial damage, both of which may contribute to the development of pulmonary vascular remodeling [77, 79]. Additionally, mutations in bone morphogenic protein receptor type 2 (BMPR-2), the most common mutation found in familial PH, have been shown in animal models to be associated with insulin resistance and pulmonary vascular remodeling [78]. These BMPR-2-induced changes appear to be mediated by signaling through PPARγ and apolipoprotein E [78, 79], and will be discussed in detail in the next section.

THE ROLE OF BMPR-2 IN PULMONARY HYPERTENSION

Although many genetic mutations have been associated with familial and sporadic cases of PH, the most common of these by far are mutations in BMPR-2. >300 different mutations in the gene encoding this receptor have been identified in patients with PAH. These mutations can be found in approximately 75% of familial cases and 25% of sporadic cases [80]. BMPR-2 is a member of the transforming growth factor β (TGF-β) family, which has well-established effects on glucose metabolism, mitochondrial function, and inflammatory response [80], potentially linking its function to several aspects of the metabolic syndrome. These metabolic derangements may be in part mediated by altered glucocorticoid signaling, as mutations in BMPR-2 are associated with higher constitutive levels of glucocorticoid receptor expression but diminished upregulation of these receptors following steroid administration, resulting in decreased sensitivity to glucocorticoids [24].

PASMCs isolated from patients with PAH with BMPR-2 mutations demonstrate a vasoproliferative phenotype. In contrast, in PASMCs isolated from healthy controls with normal BMPR-2 function, signaling via this receptor has antiproliferative effects. The effects of mutations in BMPR-2 on the pulmonary vasculature appear to be mediated via impaired downstream signaling though PPARγ and reduced expression of apolipoprotein E, and decreased PPARγ signaling and apolipoprotein E transcription have been observed in PASMCs isolated from patients with PAH [73, 81].

Mutations in BMPR-2 are more common in adult-onset cases of sporadic PH as compared with pediatric sporadic cases, prompting speculation that epigenetic factors may be required for the penetrance of the PH phenotype in patients with mutations in BMPR-2 [80]. These factors may interact with altered BMPR-2 to contribute to pulmonary vasoproliferation through a variety of mechanisms, mediated via PPARγ/apolipoprotein E and other pathways. In BMPR-2-deficient mouse models, exposure to an acute inflammatory insult results in increased serum concentrations of interleukin-6 (IL-6) and keratinocyte-derived cytokine (KC) compared with wild type littermates. These mice also exhibit decreased function of superoxide dismutase (leading to increased accumulation of reactive oxidative species and develop PH in the presence of chronic exposure to inflammatory stimuli). Administration of a superoxide dismutase mimetic results in attenuation of the inflammatory response and prevents the development of PH in BMPR-2-deficient mice [82]. Significant increases in lipid oxidation have been observed in the mitochondria of PASMCs, pulmonary vascular endothelial and skeletal muscle cells, and circulating in the serum of BMPR-2 mutated mice. These derangements of lipid metabolism are exacerbated by exposure to hyperoxia [83]. In murine models of BMPR-2 deficiency, chronic exposure to high levels of supplemental oxygen is associated with increased RVSP and decreased cardiac output, despite the pulmonary vasodilatory effects of oxygen. Wild type littermates exhibiting normal BMPR-2 function do not manifest these adverse hemodynamic responses to states of hyperoxia [83].

Lastly, BMPR-2 associated pulmonary vascular remodeling may be strongly influenced by diet. Mice expressing BMPR-2 mutations exhibit higher RVSP than wild type controls [78]. Exposure to a high-fat diet is associated with greater weight gain, insulin resistance, skeletal muscle lipid accumulation, and decreased cellular oxygen utilization in BMPR-2-deficient mice than wild type controls fed a similar diet. Additionally, when exposed to a high-fat diet, BMPR-2-deficient mice develop a higher RVSP than wild type controls or BMPR-2-deficient mice fed a diet of regular chow [78]. Whether dietary fat intake exacerbates BMPR-2-mediated pulmonary vascular remodeling in human subjects is currently unknown. Additionally, whether the chronic inflammatory state observed in obesity is associated with the development or potentiation of BMPR-2-mediated vascular remodeling remains a subject in need of further investigation.

The recognition of the significance of BMPR-2/PPARγ signaling in the pathogenesis of PH is of particular interest given the availability of therapeutic interventions targeting this pathway. Thiazolidinediones are a class of PPARγ approved for treatment of diabetes mellitus. Investigations into the potential benefit of these drugs in pulmonary vascular disease have demonstrated amelioration of PH induced by a high-fat diet in apolipoprotein *E*-deficient mice [81]. Additionally, *in vitro* exposure to thiazolidinediones has proapoptotic effects in human PASMCs [84]. Although these results are intriguing, they have not been recapitulated *in vivo* in human studies.

The therapeutic benefits of other modulators of lipid metabolism, insulin resistance, and inflammation on the pulmonary vasculature have also been studied in diet-induced animal models of obesity. Nitrooctadecanoic acid is a nitroalkene with antiinflammatory and metabolic modulatory effects produced via reactions between unsaturated fatty acids and nitric oxide. In murine models of diet-induced obesity associated with insulin resistance, dyslipidemia, pH, and elevated PVR, nitrooctadecanoic acid administration was associated with improvements in pulmonary hemodynamics and right ventricular hypertrophy, and reductions in levels of inflammatory cytokines and markers of oxidative stress [85]. These findings highlight the importance of the complex interactions between lipid and glucose metabolism, inflammatory pathways, diet composition, and pulmonary vascular remodeling in animal models of obesity but have not yet been investigated in human patients.

DYSLIPIDEMIA AND ALTERATIONS OF LIPID METABOLISM IN PULMONARY VASCULAR DISEASE

Dyslipidemia is another key component of the metabolic syndrome. Unlike insulin resistance, its links to pulmonary vascular disease are not well established. Indeed, there is a lower prevalence of dyslipidemia as compared with insulin resistance in patients with WHO group I PAH [16]. Despite promising data in animal models, randomized controlled trials of the use of lipid-lowering HMG CoA reductase inhibitors (statins) have not demonstrated benefit in patients with PAH [86]. However, as the pathways involved in the pathogenesis of pulmonary vascular disease become better understood, a multifactorial etiology with complex interactions between glucose and lipid metabolism, inflammation, and oxidant injury has become evident. Low levels of cardioprotective high density lipoprotein (HDL) are not only found in patients with obesity and the metabolic syndrome, but also in patients with PAH, and such decreased levels of HDL have been associated with worse functional class and decreased survival in PAH [87, 88]. These findings have led to increased interest in the role of lipid metabolism in pulmonary vascular inflammation and injury, and the development of PH.

One of the major constituents of HDL is apolipoprotein A-1 (apo A-1), also associated with cardiovascular protection. Apo-A1 levels are decreased in patients with PAH as compared with controls, and plasma levels correlate inversely with PAP [89]. Whether low-circulating apo A-1 is simply a marker of the metabolic derangement observed in patients with PAH or plays a pathophysiologic role in the development of PH is unknown. However, lower levels of apo A-1 are associated with decreased nitric oxide (NO) synthesis via decreased activation of scavenger receptor class B type 1 (SR-B1) and impaired vasodilatory response to acetylcholine [89, 90]. Furthermore, administration of apo A-1 mimetics has been shown to rescue monocrotaline- and hypoxia-induced PH in rodent models [91].

In addition to exhibiting lower levels of plasma HDL, patients with PAH have altered HDL composition. When compared with controls, HDL from patients with PAH contains higher levels of proinflammatory, oxidized lipids [92, 93]. This shift toward greater oxidation of circulating lipids is also seen in patients with metabolic syndrome and is associated with vascular injury [94]. Oxidized lipids and altered lipid metabolism have been associated with vasoconstriction, smooth muscle proliferation, and intimal fibrosis in patients with PAH and in animal models [93]. Oxidized lipids have also been associated with mitochondrial dysfunction, smooth muscle cell proliferation, endothelial dysfunction, apoptotic resistance, and perivascular inflammation, all of which contribute to the pathogenesis of pulmonary vascular disease [93]. These findings have led to interest in the use of protective or antiinflammatory lipid therapy in the treatment of PH, including apo A-1 mimetics, as well as microribonucleic acid (miRNA) therapy targeting the lipoxygenases, which has been shown to decrease lipid oxidation and ameliorate PH in animal models [93]. Whether apo A-1 mimetics or antioxidant miRNA therapy will prove to be of therapeutic benefit in patients with PAH, or in patients with PH associated with obesity and the metabolic syndrome, remains to be seen.

INFLAMMATORY PATHWAYS COMMON TO METABOLIC SYNDROME AND PULMONARY HYPERTENSION

Inflammation and immunity play important roles in the pathogenesis of PAH [95]. Perivascular inflammation and immune cell infiltration has been demonstrated in the lungs of patients with PAH. PAH has been associated with a chronic inflammatory state with increased circulating and infiltrating natural killer cells, macrophages, granulocytes, mast cells, and cytotoxic T cells, and downregulation of antiinflammatory Treg cells. Plasma samples from patients with PAH also demonstrate increased concentrations of a number of inflammatory cytokines [95]. Although many of these cytokines have been shown to be produced by pulmonary vascular endothelial cells, adipose tissue is also a source for production of many of the same inflammatory cytokines elevated in patients with PAH, including IL-6 and tumor necrosis factor alpha (TNF-α) [96]. Investigations into the chronic inflammatory states of both obesity and PAH have revealed overlaps between inflammatory and metabolic pathways. As discussed previously, the BMPR-2/ PPARγ/apolipoprotein E pathway illustrates this complex overlap. In addition to its well-established effects on insulin sensitivity and lipid metabolism, loss of normal BMPR-2 function has a number of proinflammatory effects, including increased production of IL-6 and TNF-α [82]. IL-6 is associated with the development of pulmonary vascular disease by inducing fibroblast growth factor-2 and other transcription factors associated with smooth muscle cell and vasoproliferation [97], whereas TNF-α is associated with endothelial apoptosis and vascular smooth muscle proliferation. BMPR-2 signaling through ERK/JNK pathways also plays a prominent role in promoting Treg differentiation, which, when functioning normally, is protective against the development of pulmonary vascular disease [95].

Loss of BMPR-2 signaling is also associated with increased sensitivity to cytokine signaling, such as increased secretion of granulocyte-macrophage colony-stimulating factor (GM-CSF) in response to TNF-α, which drives a state of peripheral neutrophilia [98], similar to that seen in. Although the role of neutrophils in the development of pulmonary vascular disease has historically been poorly understood, it has now been demonstrated that neutrophil elastase is present in higher concentrations in the PASMCs both of patients with PAH and mice exposed to chronic hypoxia. Administration of elafin, an elastase inhibitor, ameliorates PH in mice that overexpress elastase or in wild type mice exposed to chronic hypoxemic or monocrotaline models of PH [99].

Despite the purported mechanisms by which dysregulated inflammation contributes to pulmonary vascular disease, determining whether the inflammatory state observed in obesity contributes to the increased prevalence of PH in this population is challenging. As discussed, the pulmonary vascular bed of obese patients may be subject to increased flow and shear forces, long-standing PVH, chronic hypoventilation and hypoxia, and chronic insulin-resistant states and other metabolic derangements, in addition to increased inflammation. However, diet-induced and genetic models of obesity in rats have shown pulmonary vascular remodeling and resultant increased

PAP and right ventricular hypertrophy, which correlated with elevations in circulating inflammatory cytokines and a shift toward a proinflammatory lipid profile and oxidant injury, but not with hypoxia, hypercapnia, or insulin sensitivity [100].

Alterations of Mitochondrial Metabolism in Metabolic Syndrome and Pulmonary Hypertension.

Alterations of mitochondrial metabolism have been described in PASMCs in patients with PAH. There is data to suggest similar alterations can be found in skeletal muscle and other cells outside of the lung and vasculature in these patients. In the mitochondria of these cells, there is a shift from oxidative to glycolytic metabolism, resulting in a chronically acidotic intracellular environment and resistance to apoptosis. A similar mitochondrial phenomenon has been observed in the skeletal muscle cells and adipocytes of patients with obesity and the metabolic syndrome [77]. The search for pathophysiological similarities in the mitochondrial dysfunction observed in patients with PAH and those seen in metabolic syndrome has yielded some candidate mechanisms. Uncoupling protein 2 (UCP2) is one such candidate. UCP2 facilitates calcium entry into mitochondria, and PASMCs from mice deficient in UCP2 exhibit lower concentrations of mitochondrial calcium, and resultant decreases in the activity of pyruvate dehydrogenase and other mitochondrial enzymes critical to oxidative phosphorylation. The net effect of UCP2 deficiency is increased intracellular acidosis and resistance to apoptosis. UCP2-deficient mice have been shown to develop PH in the absence of hypoxia or other insult, whereas mutations in UCP2 have also been linked to increased risk of metabolic syndrome, cardiovascular disease, and several autoimmune diseases [77, 101, 102].

OXIDATIVE STRESS IN PULMONARY HYPERTENSION AND OBESITY

Human studies and animal models have established that PAH and metabolic syndrome are states of increased oxidative stress and oxidative injury. The precise role of oxidative injury in the pathophysiology of both PAH and in patients with metabolic syndrome remains a topic of ongoing investigation [75, 79, 103]. As discussed in the previous section, abnormalities of mitochondrial metabolism are present in both conditions, and such abnormalities are among the mechanisms that lead to development of increased levels of reactive oxygen species (ROS), in particular superoxide [77]. Mutations in BMPR2 have also been linked to increased levels of ROS, both at baseline and following exposure to states of hyperoxygenation [83]. The mitochondrial- and BMPR2-mediated states of oxidative stress seen in PAH are not limited to the pulmonary vasculature and PASMCs but are observed in multiple tissues, further suggesting that PAH represents a systemic disease [83]. This oxidative state results in increased oxidation of fatty acids and lipoproteins, which have an injurious role in both PH and metabolic syndrome [79, 103], and animal models have correlated obesity-associated elevations in PAP with oxidant injury [100]. Whether obesity and the metabolic syndrome exacerbate this state of oxidative injury and contributes to the development or progression of PH in human subjects is an area in need of further investigation.

CONCLUSIONS

PH is common in patients with obesity and is frequently seen as a consequence of increased left ventricular filling pressures, and the chronic hypoxia and hypercapnia associated with sleep disordered breathing and hypoventilation syndromes. Obesity is also associated with states of metabolic derangement, including insulin resistance, alterations in mitochondrial metabolism, shifts toward proinflammatory lipid profiles, and increased oxidative injury, the importance of which is becoming increasingly understood in the pathogenesis of pulmonary vascular disease. Additionally, both obesity and PH are associated with chronic inflammation. Adipokines and other cytokines present in increased circulating levels in obese patients have been implicated in pulmonary vascular injury and remodeling. Investigations into the benefit of therapies targeting many of these metabolic and inflammatory abnormalities remain in preliminary stages and may be of particular interest in patients who suffer from both obesity and PH. In addition to medical therapies, the benefits of preventative strategies, dietary modification, exercise therapy, and surgical weight loss require further study. Optimal diagnostic and monitoring strategies in obese patients with PH also remain unclear, and continued investigations into alternative imaging modalities, biomarkers, assessment of functional capacity, and even accurate invasive measurement of hemodynamics remain of the utmost importance.

REFERENCES

[1] Taraseviciute A, Voelkel NF. Severe pulmonary hypertension in postmenopausal obese women. Eur J Med Res 2006;11:198–202.

[2] Haque AK, Gadre S, Taylor J, Haque SA, Freeman D, Duarte A. Pulmonary and cardiovascular complications of obesity: an autopsy study of 76 obese subjects. Arch Pathol Lab Med 2008;132:1397–404.

[3] McQuillan BM, Picard MH, Leavitt M, Weyman AE. Clinical correlates and reference intervals for pulmonary artery systolic pressure among echocardiographically normal subjects. Circulation 2001;104:2797–802.

[4] Wong CY, O'Moore-Sullivan T, Leano R, Hukins C, Jenkins C, Marwick TH. Association of subclinical right ventricular dysfunction with obesity. J Am Coll Cardiol 2006;47:611–6.

[5] Dela Cruz CS, Matthay RA. Role of obesity in cardiomyopathy and pulmonary hypertension. Clin Chest Med 2009;30:509–23 [ix].

[6] Simonneau G, Gatzoulis MA, Adatia I, Celermajer D, Denton C, Ghofrani A, Gomez Sanchez MA, Krishna Kumar R, Landzberg M, Machado RF, Olschewski H, Robbins IM, Souza R. Updated clinical classification of pulmonary hypertension. J Am Coll Cardiol 2013;62:D34–41.

[7] Tuder RM, Marecki JC, Richter A, Fijalkowska I, Flores S. Pathology of pulmonary hypertension. Clin Chest Med 2007;28:23–42 [vii].

[8] Galie N, Corris PA, Frost A, Girgis RE, Granton J, Jing ZC, Klepetko W, McGoon MD, McLaughlin VV, Preston IR, Rubin LJ, Sandoval J, Seeger W, Keogh A. Updated treatment algorithm of pulmonary arterial hypertension. J Am Coll Cardiol 2013;62:D60–72.

[9] Abenhaim L, Moride Y, Brenot F, Rich S, Benichou J, Kurz X, Higenbottam T, Oakley C, Wouters E, Aubier M, Simonneau G, Begaud B. Appetite-suppressant drugs and the risk of primary pulmonary hypertension. International Primary Pulmonary Hypertension Study Group. N Engl J Med 1996;335:609–16.

[10] Burger CD, Foreman AJ, Miller DP, Safford RE, McGoon MD, Badesch DB. Comparison of body habitus in patients with pulmonary arterial hypertension enrolled in the Registry to Evaluate Early and Long-term PAH Disease Management with normative values from the National Health and Nutrition Examination Survey. Mayo Clin Proc 2011;86:105–12.

[11] Strange G, Playford D, Stewart S, Deague JA, Nelson H, Kent A, Gabbay E. Pulmonary hypertension: prevalence and mortality in the Armadale echocardiography cohort. Heart 2012;98:1805–11.

[12] Fang JC, DeMarco T, Givertz MM, Borlaug BA, Lewis GD, Rame JE, Gomberg-Maitland M, Murali S, Frantz RP, McGlothlin D, Horn EM, Benza RL. World Health Organization pulmonary hypertension group 2: pulmonary hypertension due to left heart disease in the adult—a summary statement from the Pulmonary Hypertension Council of the International Society for Heart and Lung Transplantation. J Heart Lung Transplant 2012;31:913–33.

[13] Guazzi M, Borlaug BA. Pulmonary hypertension due to left heart disease. Circulation 2012;126:975–90.

[14] Vachiery JL, Adir Y, Barbera JA, Champion H, Coghlan JG, Cottin V, De Marco T, Galie N, Ghio S, Gibbs JS, Martinez F, Semigran M, Simonneau G, Wells A, Seeger W. Pulmonary hypertension due to left heart diseases. J Am Coll Cardiol 2013;62:D100–8.

[15] Adir Y, Humbert M, Sitbon O, Wolf R, Lador F, Jais X, Simonneau G, Amir O. Out-of-proportion pulmonary hypertension and heart failure with preserved ejection fraction. Respiration 2013;85:471–7.

[16] Robbins IM, Newman JH, Johnson RF, Hemnes AR, Fremont RD, Piana RN, Zhao DX, Byrne DW. Association of the metabolic syndrome with pulmonary venous hypertension. Chest 2009;136:31–6.

[17] Klapholz M, Maurer M, Lowe AM, Messineo F, Meisner JS, Mitchell J, Kalman J, Phillips RA, Steingart R, Brown Jr. EJ, Berkowitz R, Moskowitz R, Soni A, Mancini D, Bijou R, Sehhat K, Varshneya N, Kukin M, Katz SD, Sleeper LA, Le Jemtel TH, New York Heart Failure C. Hospitalization for heart failure in the presence of a normal left ventricular ejection fraction: results of the New York Heart Failure Registry. J Am Coll Cardiol 2004;43:1432–8.

[18] Lam CS, Borlaug BA, Kane GC, Enders FT, Rodeheffer RJ, Redfield MM. Age-associated increases in pulmonary artery systolic pressure in the general population. Circulation 2009;119:2663–70.

[19] Lam CS, Roger VL, Rodeheffer RJ, Borlaug BA, Enders FT, Redfield MM. Pulmonary hypertension in heart failure with preserved ejection fraction: a community-based study. J Am Coll Cardiol 2009;53:1119–26.

[20] Leung CC, Moondra V, Catherwood E, Andrus BW. Prevalence and risk factors of pulmonary hypertension in patients with elevated pulmonary venous pressure and preserved ejection fraction. Am J Cardiol 2010;106:284–6.

[21] Mohammed SF, Hussain I, AbouEzzeddine OF, Takahama H, Kwon SH, Forfia P, Roger VL, Redfield MM. Right ventricular function in heart failure with preserved ejection fraction: a community-based study. Circulation 2014;130:2310–20.

[22] Thenappan T, Shah SJ, Gomberg-Maitland M, Collander B, Vallakati A, Shroff P, Rich S. Clinical characteristics of pulmonary hypertension in patients with heart failure and preserved ejection fraction. Circ Heart Fail 2011;4:257–65.

[23] Delgado JF, Conde E, Sanchez V, Lopez-Rios F, Gomez-Sanchez MA, Escribano P, Sotelo T, Gomez de la Camara A, Cortina J, de la Calzada CS. Pulmonary vascular remodeling in pulmonary hypertension due to chronic heart failure. Eur J Heart Fail 2005;7:1011–6.

[24] West JB, Mathieu-Costello O. Vulnerability of pulmonary capillaries in heart disease. Circulation 1995;92:622–31.

[25] Cody RJ, Haas GJ, Binkley PF, Capers Q, Kelley R. Plasma endothelin correlates with the extent of pulmonary hypertension in patients with chronic congestive heart failure. Circulation 1992;85:504–9.

[26] Guazzi M. Alveolar gas diffusion abnormalities in heart failure. J Card Fail 2008;14:695–702.

[27] Kerem A, Yin J, Kaestle SM, Hoffmann J, Schoene AM, Singh B, Kuppe H, Borst MM, Kuebler WM. Lung endothelial dysfunction in congestive heart failure: role of impaired Ca2+ signaling and cytoskeletal reorganization. Circ Res 2010;106:1103–16.

[28] Tsukimoto K, Mathieu-Costello O, Prediletto R, Elliott AR, West JB. Ultrastructural appearances of pulmonary capillaries at high transmural pressures. J Appl Physiol (1985) 1991;71:573–82.

[29] Hunt BJ. The effect of BMI on haemostasis: implications for thrombosis in women's health. Thromb Res 2017;151(1):S53–5. Suppl.

[30] Klinger JR. Group III pulmonary hypertension: pulmonary hypertension associated with lung disease: epidemiology, pathophysiology, and treatments. Cardiol Clin 2016;34:413–33.

[31] Vgontzas AN, Tan TL, Bixler EO, Martin LF, Shubert D, Kales A. Sleep apnea and sleep disruption in obese patients. Arch Intern Med 1994;154:1705–11.

[32] Ismail K, Roberts K, Manning P, Manley C, Hill NS. OSA and pulmonary hypertension: time for a new look. Chest 2015;147:847–61.

[33] Kauppert CA, Dvorak I, Kollert F, Heinemann F, Jorres RA, Pfeifer M, Budweiser S. Pulmonary hypertension in obesity-hypoventilation syndrome. Respir Med 2013;107:2061–70.

[34] Chaouat A, Weitzenblum E, Krieger J, Oswald M, Kessler R. Pulmonary hemodynamics in the obstructive sleep apnea syndrome. Results in 220 consecutive patients. Chest 1996;109:380–6.

[35] Kessler R, Chaouat A, Weitzenblum E, Oswald M, Ehrhart M, Apprill M, Krieger J. Pulmonary hypertension in the obstructive sleep apnoea syndrome: prevalence, causes and therapeutic consequences. Eur Respir J 1996;9:787–94.

[36] Minai OA, Ricaurte B, Kaw R, Hammel J, Mansour M, McCarthy K, Golish JA, Stoller JK. Frequency and impact of pulmonary hypertension in patients with obstructive sleep apnea syndrome. Am J Cardiol 2009;104:1300–6.

[37] Carratu P, Ventura VA, Maniscalco M, Dragonieri S, Berardi S, Ria R, Quaranta VN, Vacca A, Devito F, Ciccone MM, Phillips BA, Resta O. Echocardiographic findings and plasma endothelin-1 levels in obese patients with and without obstructive sleep apnea. Sleep Breath 2016;20:613–9.

[38] Almeneessier AS, Nashwan SZ, Al-Shamiri MQ, Pandi-Perumal SR, BaHammam AS. The prevalence of pulmonary hypertension in patients with obesity hypoventilation syndrome: a prospective observational study. J Thorac Dis 2017;9:779–88.

[39] Kessler R, Chaouat A, Schinkewitch P, Faller M, Casel S, Krieger J, Weitzenblum E. The obesity-hypoventilation syndrome revisited: a prospective study of 34 consecutive cases. Chest 2001;120:369–76.

[40] Sugerman HJ, Baron PL, Fairman RP, Evans CR, Vetrovec GW. Hemodynamic dysfunction in obesity hypoventilation syndrome and the effects of treatment with surgically induced weight loss. Ann Surg 1988;207:604–13.

[41] Sheu EG, Channick R, Gee DW. Improvement in severe pulmonary hypertension in obese patients after laparoscopic gastric bypass or sleeve gastrectomy. Surg Endosc 2016;30:633–7.

[42] Lang IM, Dorfmuller P, Vonk Noordegraaf A. The pathobiology of chronic thromboembolic pulmonary hypertension. Ann Am Thorac Soc 2016;13(Suppl 3):S215–21.

[43] Auger WR, Kim NH, Trow TK. Chronic thromboembolic pulmonary hypertension. Clin Chest Med 2010;31:741–58.

[44] Becattini C, Agnelli G, Pesavento R, Silingardi M, Poggio R, Taliani MR, Ageno W. Incidence of chronic thromboembolic pulmonary hypertension after a first episode of pulmonary embolism. Chest 2006;130:172–5.

[45] Bonderman D, Wilkens H, Wakounig S, Schafers HJ, Jansa P, Lindner J, Simkova I, Martischnig AM, Dudczak J, Sadushi R, Skoro-Sajer N, Klepetko W, Lang IM. Risk factors for chronic thromboembolic pulmonary hypertension. Eur Respir J 2009;33:325–31.

[46] Delcroix M, Kerr K, Fedullo P. Chronic thromboembolic pulmonary hypertension. Epidemiology and risk factors. Ann Am Thorac Soc 2016;13(Suppl 3):S201–6.

[47] Pengo V, Lensing AW, Prins MH, Marchiori A, Davidson BL, Tiozzo F, Albanese P, Biasiolo A, Pegoraro C, Iliceto S, Prandoni P, Thromboembolic Pulmonary Hypertension Study G. Incidence of chronic thromboembolic pulmonary hypertension after pulmonary embolism. N Engl J Med 2004;350:2257–64.

[48] Steffen LM, Cushman M, Peacock JM, Heckbert SR, Jacobs Jr. DR, Rosamond WD, Folsom AR. Metabolic syndrome and risk of venous thromboembolism: longitudinal investigation of thromboembolism etiology. J Thromb Haemost 2009;7:746–51.

[49] Horvei LD, Braekkan SK, Hansen JB. Weight change and risk of venous thromboembolism: the Tromso Study. PLoS One 2016;11:e0168878.

[50] Bitar A, Selej M, Bolad I, Lahm T. Poor agreement between pulmonary capillary wedge pressure and left ventricular end-diastolic pressure in a veteran population. PLoS One 2014;9:e87304.

[51] LeVarge BL, Pomerantsev E, Channick RN. Reliance on end-expiratory wedge pressure leads to misclassification of pulmonary hypertension. Eur Respir J 2014;44:425–34.

[52] Ryan JJ, Rich JD, Thiruvoipati T, Swamy R, Kim GH, Rich S. Current practice for determining pulmonary capillary wedge pressure predisposes to serious errors in the classification of patients with pulmonary hypertension. Am Heart J 2012;163:589–94.

[53] Curtis JP, Selter JG, Wang Y, Rathore SS, Jovin IS, Jadbabaie F, Kosiborod M, Portnay EL, Sokol SI, Bader F, Krumholz HM. The obesity paradox: body mass index and outcomes in patients with heart failure. Arch Intern Med 2005;165:55–61.

[54] Fonarow GC, Srikanthan P, Costanzo MR, Cintron GB, Lopatin M, Committee ASA, Investigators. An obesity paradox in acute heart failure: analysis of body mass index and inhospital mortality for 108,927 patients in the Acute Decompensated Heart Failure National Registry. Am Heart J 2007;153:74–81.

[55] Horwich TB, Fonarow GC, Hamilton MA, MacLellan WR, Woo MA, Tillisch JH. The relationship between obesity and mortality in patients with heart failure. J Am Coll Cardiol 2001;38:789–95.

[56] Ades PA, Savage PD. The obesity paradox: perception vs knowledge. Mayo Clin Proc 2010;85:112–4.

[57] Mazimba S, Holland E, Nagarajan V, Mihalek AD, Kennedy JL, Bilchick KC. Obesity paradox in group 1 pulmonary hypertension: analysis of the NIH-pulmonary hypertension registry. Int J Obes (Lond) 2017;.

[58] Zafrir B, Adir Y, Shehadeh W, Shteinberg M, Salman N, Amir O. The association between obesity, mortality and filling pressures in pulmonary hypertension patients; the "obesity paradox". Respir Med 2013;107:139–46.

[59] Aytekin M, Tonelli AR, Farver CF, Feldstein AE, Dweik RA. Leptin deficiency recapitulates the histological features of pulmonary arterial hypertension in mice. Int J Clin Exp Pathol 2014;7:1935–46.

[60] Chai S, Wang W, Liu J, Guo H, Zhang Z, Wang C, Wang J. Leptin knockout attenuates hypoxia-induced pulmonary arterial hypertension by inhibiting proliferation of pulmonary arterial smooth muscle cells. Transl Res 2015;166:772–82.

[61] Huertas A, Tu L, Gambaryan N, Girerd B, Perros F, Montani D, Fabre D, Fadel E, Eddahibi S, Cohen-Kaminsky S, Guignabert C, Humbert M. Leptin and regulatory T-lymphocytes in idiopathic pulmonary arterial hypertension. Eur Respir J 2012;40:895–904.

[62] Huertas A, Tu L, Thuillet R, Le Hiress M, Phan C, Ricard N, Nadaud S, Fadel E, Humbert M, Guignabert C. Leptin signalling system as a target for pulmonary arterial hypertension therapy. Eur Respir J 2015;45:1066–80.

[63] Medoff BD. Fat, fire and muscle—the role of adiponectin in pulmonary vascular inflammation and remodeling. Pulm Pharmacol Ther 2013;26:420–6.

[64] Summer R, Walsh K, Medoff BD. Obesity and pulmonary arterial hypertension: is adiponectin the molecular link between these conditions? Pulm Circ 2011;1:440–7.

[65] Tonelli AR, Aytekin M, Feldstein AE, Dweik RA. Leptin levels predict survival in pulmonary arterial hypertension. Pulm Circ 2012;2:214–9.

[66] Hansmann G, Rabinovitch M. The protective role of adiponectin in pulmonary vascular disease. Am J Physiol Lung Cell Mol Physiol 2010;298:L1–2.

[67] Medoff BD, Okamoto Y, Leyton P, Weng M, Sandall BP, Raher MJ, Kihara S, Bloch KD, Libby P, Luster AD. Adiponectin deficiency increases allergic airway inflammation and pulmonary vascular remodeling. Am J Respir Cell Mol Biol 2009;41:397–406.

[68] Summer R, Fiack CA, Ikeda Y, Sato K, Dwyer D, Ouchi N, Fine A, Farber HW, Walsh K. Adiponectin deficiency: a model of pulmonary hypertension associated with pulmonary vascular disease. Am J Physiol Lung Cell Mol Physiol 2009;297:L432–8.

[69] Arita Y, Kihara S, Ouchi N, Maeda K, Kuriyama H, Okamoto Y, Kumada M, Hotta K, Nishida M, Takahashi M, Nakamura T, Shimomura I, Muraguchi M, Ohmoto Y, Funahashi T, Matsuzawa Y. Adipocyte-derived plasma protein adiponectin acts as a platelet-derived growth factor-BB-binding protein and regulates growth factor-induced common postreceptor signal in vascular smooth muscle cell. Circulation 2002;105:2893–8.

[70] Wang Y, Lam KS, Xu JY, Lu G, Xu LY, Cooper GJ, Xu A. Adiponectin inhibits cell proliferation by interacting with several growth factors in an oligomerization-dependent manner. J Biol Chem 2005;280:18341–7.

[71] Weng M, Raher MJ, Leyton P, Combs TP, Scherer PE, Bloch KD, Medoff BD. Adiponectin decreases pulmonary arterial remodeling in murine models of pulmonary hypertension. Am J Respir Cell Mol Biol 2011;45:340–7.

[72] Yokota T, Oritani K, Takahashi I, Ishikawa J, Matsuyama A, Ouchi N, Kihara S, Funahashi T, Tenner AJ, Tomiyama Y, Matsuzawa Y. Adiponectin, a new member of the family of soluble defense collagens, negatively regulates the growth of myelomonocytic progenitors and the functions of macrophages. Blood 2000;96:1723–32.

[73] Hansmann G, de Jesus Perez VA, Alastalo TP, Alvira CM, Guignabert C, Bekker JM, Schellong S, Urashima T, Wang L, Morrell NW, Rabinovitch M. An antiproliferative BMP-2/PPARgamma/apoE axis in human and murine SMCs and its role in pulmonary hypertension. J Clin Invest 2008;118:1846–57.

[74] Brunner NW, Skhiri M, Fortenko O, Hsi A, Haddad F, Khazeni N, Zamanian RT. Impact of insulin resistance on ventricular function in pulmonary arterial hypertension. J Heart Lung Transplant 2014;33:721–6.

[75] Pugh ME, Robbins IM, Rice TW, West J, Newman JH, Hemnes AR. Unrecognized glucose intolerance is common in pulmonary arterial hypertension. J Heart Lung Transplant 2011;30:904–11.

[76] Zamanian RT, Hansmann G, Snook S, Lilienfeld D, Rappaport KM, Reaven GM, Rabinovitch M, Doyle RL. Insulin resistance in pulmonary arterial hypertension. Eur Respir J 2009;33:318–24.

[77] Paulin R, Michelakis ED. The metabolic theory of pulmonary arterial hypertension. Circ Res 2014;115:148–64.

[78] West J, Niswender KD, Johnson JA, Pugh ME, Gleaves L, Fessel JP, Hemnes AR. A potential role for insulin resistance in experimental pulmonary hypertension. Eur Respir J 2013;41:861–71.

[79] Pugh ME, Hemnes AR. Metabolic and hormonal derangements in pulmonary hypertension: from mouse to man. Int J Clin Pract Suppl 2010;5–13.

[80] Soubrier F, Chung WK, Machado R, Grunig E, Aldred M, Geraci M, Loyd JE, Elliott CG, Trembath RC, Newman JH, Humbert M. Genetics and genomics of pulmonary arterial hypertension. J Am Coll Cardiol 2013;62:D13–21.

[81] Hansmann G, Wagner RA, Schellong S, Perez VA, Urashima T, Wang L, Sheikh AY, Suen RS, Stewart DJ, Rabinovitch M. Pulmonary arterial hypertension is linked to insulin resistance and reversed by peroxisome proliferator-activated receptor-gamma activation. Circulation 2007;115:1275–84.

[82] Soon E, Crosby A, Southwood M, Yang P, Tajsic T, Toshner M, Appleby S, Shanahan CM, Bloch KD, Pepke-Zaba J, Upton P, Morrell NW. Bone morphogenetic protein receptor type II deficiency and increased inflammatory cytokine production. A gateway to pulmonary arterial hypertension. Am J Respir Crit Care Med 2015;192:859–72.

[83] Fessel JP, Flynn CR, Robinson LJ, Penner NL, Gladson S, Kang CJ, Wasserman DH, Hemnes AR, West JD. Hyperoxia synergizes with mutant bone morphogenic protein receptor 2 to cause metabolic stress, oxidant injury, and pulmonary hypertension. Am J Respir Cell Mol Biol 2013;49:778–87.

[84] Green DE, Murphy TC, Kang BY, Bedi B, Yuan Z, Sadikot RT, Hart CM. Peroxisome proliferator-activated receptor-gamma enhances human pulmonary artery smooth muscle cell apoptosis through microRNA-21 and programmed cell death 4. Am J Physiol Lung Cell Mol Physiol 2017;313:L371–83.

[85] Kelley EE, Baust J, Bonacci G, Golin-Bisello F, Devlin JE, St Croix CM, Watkins SC, Gor S, Cantu-Medellin N, Weidert ER, Frisbee JC, Gladwin MT, Champion HC, Freeman BA, Khoo NK. Fatty acid nitroalkenes ameliorate glucose intolerance and pulmonary hypertension in high-fat diet-induced obesity. Cardiovasc Res 2014;101:352–63.

[86] Kawut SM, Bagiella E, Lederer DJ, Shimbo D, Horn EM, Roberts KE, Hill NS, Barr RG, Rosenzweig EB, Post W, Tracy RP, Palevsky HI, Hassoun PM, Girgis RE, Group A-SS. Randomized clinical trial of aspirin and simvastatin for pulmonary arterial hypertension: ASA-STAT. Circulation 2011;123:2985–93.

[87] Al-Naamani N, Palevsky HI, Lederer DJ, Horn EM, Mathai SC, Roberts KE, Tracy RP, Hassoun PM, Girgis RE, Shimbo D, Post WS, Kawut SM, Group A-SS. Prognostic significance of biomarkers in pulmonary arterial hypertension. Ann Am Thorac Soc 2016;13:25–30.

[88] Heresi GA, Aytekin M, Newman J, DiDonato J, Dweik RA. Plasma levels of high-density lipoprotein cholesterol and outcomes in pulmonary arterial hypertension. Am J Respir Crit Care Med 2010;182:661–8.

[89] Anjum F, Lazar J, Soh J, Albitar M, Gowda S, Hussain MM, Wadgaonkar R. Dysregulation of ubiquitin-proteasome pathway and apolipoprotein A metabolism in sickle cell disease-related pulmonary arterial hypertension. Pulm Circ 2013;3:851–5.

[90] Yuditskaya S, Tumblin A, Hoehn GT, Wang G, Drake SK, Xu X, Ying S, Chi AH, Remaley AT, Shen RF, Munson PJ, Suffredini AF, Kato GJ. Proteomic identification of altered apolipoprotein patterns in pulmonary hypertension and vasculopathy of sickle cell disease. Blood 2009;113:1122–8.

[91] Sharma S, Umar S, Potus F, Iorga A, Wong G, Meriwether D, Breuils-Bonnet S, Mai D, Navab K, Ross D, Navab M, Provencher S, Fogelman AM, Bonnet S, Reddy ST, Eghbali M. Apolipoprotein A-I mimetic peptide 4F rescues pulmonary hypertension by inducing microRNA-193-3p. Circulation 2014;130:776–85.

[92] Ross DJ, Hough G, Hama S, Aboulhosn J, Belperio JA, Saggar R, Van Lenten BJ, Ardehali A, Eghbali M, Reddy S, Fogelman AM, Navab M. Proinflammatory high-density lipoprotein results from oxidized lipid mediators in the pathogenesis of both idiopathic and associated types of pulmonary arterial hypertension. Pulm Circ 2015;5:640–8.

[93] Sharma S, Ruffenach G, Umar S, Motayagheni N, Reddy ST, Eghbali M. Role of oxidized lipids in pulmonary arterial hypertension. Pulm Circ 2016;6:261–73.

[94] Hurtado-Roca Y, Bueno H, Fernandez-Ortiz A, Ordovas JM, Ibanez B, Fuster V, Rodriguez-Artalejo F, Laclaustra M. Oxidized LDL is associated with metabolic syndrome traits independently of central obesity and insulin resistance. Diabetes 2017;66:474–82.

[95] Rabinovitch M, Guignabert C, Humbert M, Nicolls MR. Inflammation and immunity in the pathogenesis of pulmonary arterial hypertension. Circ Res 2014;115:165–75.

[96] Majka SM, Barak Y, Klemm DJ. Concise review: adipocyte origins: weighing the possibilities. Stem Cells 2011;29:1034–40.

[97] Hagen M, Fagan K, Steudel W, Carr M, Lane K, Rodman DM, West J. Interaction of interleukin-6 and the BMP pathway in pulmonary smooth muscle. Am J Physiol Lung Cell Mol Physiol 2007;292:L1473–9.

[98] Sawada H, Saito T, Nickel NP, Alastalo TP, Glotzbach JP, Chan R, Haghighat L, Fuchs G, Januszyk M, Cao A, Lai YJ, Perez Vde J, Kim YM, Wang L, Chen PI, Spiekerkoetter E, Mitani Y, Gurtner GC, Sarnow P, Rabinovitch M. Reduced BMPR2 expression induces GM-CSF translation and macrophage recruitment in humans and mice to exacerbate pulmonary hypertension. J Exp Med 2014;211:263–80.

[99] Kim YM, Haghighat L, Spiekerkoetter E, Sawada H, Alvira CM, Wang L, Acharya S, Rodriguez-Colon G, Orton A, Zhao M, Rabinovitch M. Neutrophil elastase is produced by pulmonary artery smooth muscle cells and is linked to neointimal lesions. Am J Pathol 2011;179:1560–72.

[100] Irwin DC, Garat CV, Crossno Jr. JT, MacLean PS, Sullivan TM, Erickson PF, Jackman MR, Harral JW, Reusch JE, Klemm DJ. Obesity-related pulmonary arterial hypertension in rats correlates with increased circulating inflammatory cytokines and lipids and with oxidant damage in the arterial wall but not with hypoxia. Pulm Circ 2014;4:638–53.

[101] Dromparis P, Paulin R, Sutendra G, Qi AC, Bonnet S, Michelakis ED. Uncoupling protein 2 deficiency mimics the effects of hypoxia and endoplasmic reticulum stress on mitochondria and triggers pseudohypoxic pulmonary vascular remodeling and pulmonary hypertension. Circ Res 2013;113:126–36.

[102] Fisler JS, Warden CH. Uncoupling proteins, dietary fat and the metabolic syndrome. Nutr Metab (Lond) 2006;3:38.

[103] Fessel JP, West JD. Redox biology in pulmonary arterial hypertension (2013 Grover Conference Series). Pulm Circ 2015;5:599–609.

Influence of obesity on the response to influenza infection and vaccination

10

Erik A. Karlsson*, J. Justin Milner[†], William D. Green[‡], Jennifer Rebeles[§], Stacey Schultz-Cherry[¶], Melinda A. Beck[‡]

Virology Unit, Institut Pasteur du Cambodge, Phnom Penh, Cambodia* Division of Biological Sciences, University of California, San Diego, CA, United States[†] Department of Nutrition, Gilling's School of Global Public Health, University of North Carolina, Chapel Hill, NC, United States[‡] National Research Council Postdoctoral Research Associate, US Army Institute of Surgical Research, San Antonio, TX, United States[§] Department of Infectious Diseases, St. Jude Children's Research Hospital, Memphis, TN, United States[¶]

ABBREVIATIONS

ALI	acute lung injury
APC	antigen presenting cell
ARDS	acute respiratory distress syndrome
BMI	body mass index
COX	cyclooxygenase
DC	dendritic cell
DIO	diet-induced obese
GLUT	glucose transporter
HA	hemagglutinin
HAI	hemagglutination inhibition
ICU	intensive care unit
IFN	interferon
LepR	leptin receptor
M	matrix protein
MHC	major histocompatibility complex
Mφ	macrophage
NA	neuraminidase
NAI	neuraminidase inhibitor
NK	natural killer cell
PA	polymerase acidic protein
PB1	polymerase basic protein 1
PB2	polymerase basic protein 2
pdmH1N1	2009 H1N1 pandemic strain of influenza
PR8	A/Puerto Rico/8/1934
PUFA	polyunsaturated fatty acid

Mechanisms and Manifestations of Obesity in Lung Disease. https://doi.org/10.1016/B978-0-12-813553-2.00010-5

T2D	type 2 diabetes
TIV	trivalent, inactivated influenza vaccine
TLR	Toll-like receptor
TNF	tumor necrosis factor
Treg	regulatory T cell

OBESITY AT THE HOST-PATHOGEN INTERFACE

Malnutrition and infection are the major causes of preventable deaths and disabilities worldwide, especially in children. Nutrition has long been known to have both synergistic and antagonistic effects on infectious disease, and nutritional status is considered to be a key factor in host-pathogen interactions [1,2]. Overall, malnutrition is the principal source of immunodeficiency worldwide [3]. Historically, times of famine have frequently been associated with outbreaks of pestilence and plague; however, detailed scientific descriptions of the interaction between infection and nutrition have only appeared in the middle to late part of the last century [4–6]. Nutrients have been shown to be crucial cofactors in the development, maintenance, and expression of the immune response. Also of importance is the ability of the infection to influence host nutritional status. Infection can lead to dramatic changes in nutritional status [1,2]. Indeed, higher frequency of exposure to infectious diseases has been shown to increase the risk of poor nutrition [7].

Although numerous studies have focused on the effect of nutritional deficiencies on immune status, overnutrition has also begun to be appreciated for its effects on the immune system. Overweight (Body Mass Index; $BMI \geq 25\,kg/m^2$) and obesity ($BMI \geq 30\,kg/m^2$) have reached epidemic proportions worldwide. Globally, as of 2014, it is estimated that 1.9 billion adults (>18 years of age) were overweight or obese. In children (<5 years of age), ~42 million were overweight or obese [8,9]. These numbers continue to expand on a yearly basis. In the United States alone, 67% of the adult population is overweight/obese. Interestingly, obesity is not limited to resource-rich countries. In lower- to middle-income countries, rate of increase of childhood overweight and obesity is >30% higher versus more developed countries, and >65% of the global population lives in countries where overweight and obesity kill more people than underweight [8].

Obesity has been shown to have an enormous influence on the immune response as well as susceptibility and severity of infections. Hospitalized obese patients have increased susceptibility to nosocomial infections, and even so-called "metabolically healthy" obese individuals appear to be at increased risk for community-acquired infections. Obese mouse models have also demonstrated increased susceptibility and poorer outcomes following pathogen challenge. Although a link between obesity, immune response, and infection susceptibility and severity are relatively well established, the exact mechanisms for each pathogen are unknown [10–13]. Therefore there is a pressing need for more research on the mechanisms behind obesity-associated infection morbidity and mortality as well as potential therapeutic and prophylactic effectiveness in this expanding population.

WHAT IS INFLUENZA?

One particular disease that seems to be greatly impacted by nutritional status, especially obesity, is influenza. The influenza virus is a segmented, negative-strand RNA virus of the family *Orthomyxoviridae*. Of the four currently known types of influenza, influenza A and B viruses are typically responsible for seasonal human infections. Influenza C strains are rare and typically associated with mild illness, and influenza D has only been associated with infections in cattle [14]. Spread through airborne transmission, influenza viruses are highly contagious and are responsible for a great deal of morbidity and mortality worldwide [15]. In any given year, 5%–15% of the world population is infected with influenza virus, resulting in 3–5 million cases of severe illness and 500,000 deaths from influenza and influenza-related complications [16]. Young children, the elderly, and people with chronic diseases are particularly susceptible to influenza-related complications including viral or bacterial pneumonia, dehydration, and death [17]. Due to their highly mutable nature, influenza viruses undergo continual antigenic drift resulting in flu outbreaks that typically occur during the winter months. Occasionally these viruses can also go through a major antigenic shift resulting in worldwide pandemic outbreaks as most recently evidenced by the 2009 H1N1 pandemic [18,19].

Influenza strains A and B contain 8 RNA segments, and influenza C contains 7 segments encoding 11 and 9 proteins, respectively [20]. The outer structure of the virus consists of a lipid bilayer obtained from the host during virus budding from an infected host cell. The lipid membrane contains viral-encoded glycoproteins hemagglutinin (HA) and neuraminidase (NA), crucial for viral entry and exit from the host cell, respectively [14]. HA is the most abundant protein in the lipid bilayer and binds to sialic acid receptors on host cells, facilitating viral entry. Viral HA has been the target of influenza vaccination strategies since their inception [21–23]. NA proteins function as enzymes that cleave sialic acids to allow budding and release of the virion to infect neighboring cells [22]. The NA protein is the target of the majority of antiviral medications used to treat influenza infection [20]. Other important proteins include the ion channel matrix (M) protein [24] and the RNA-dependent RNA polymerase complex made up of three proteins (PA, PB1, and PB2) that produce the positive strand mRNA that serves as a template for the transcription of viral proteins in the cytoplasm [22]. The matrix protein is the target of the adamantine family of antiviral drugs; however, the majority of currently circulating influenza strains are resistant to these antivirals [24].

THE IMMUNE RESPONSE IS CRITICAL FOR RESOLVING AN INFLUENZA VIRUS INFECTION

The immune response is crucial for sensing and resolution of an influenza virus infection in the respiratory tract; however, it is also associated with lung damage and other factors contributing to severity and mortality from influenza infection. Overall, the immune system is composed of two arms, innate and adaptive, that form a complex network of cells, tissues, and organs that respond to invading pathogens or injury. Upon infection, the innate immune system is immediately activated to provide a first line of nonspecific host defense against an invading pathogen. The innate immune system consists of both physical and chemical barriers such as: mucosal epithelium; innate immune cells such as macrophages (Mφ), neutrophils, dendritic cells (DC), and natural killer (NK) cells; and soluble proteins and cytokines that coordinate interactions between immune cells and the host environment [25]. The coordinated actions of the innate immune system also provide support for the activation of the adaptive immune system. The presentation of viral peptides by antigen presenting cells (APCs) of the innate immune system (i.e., DC) to T cells in secondary lymphoid organs initiates B- and T-cell activation and coordinated intracellular signaling cascades. The adaptive immune system consists of the humoral and cellular immune response that correlates with B- and T-cell-driven immunity, respectively. Unlike its innate counterpart, adaptive immunity is specific for an invading pathogen. Both arms of the immune system play critical roles in the response to influenza infection. Recently, the adaptive immune system has received tremendous attention in modern medicine and public health intervention for its specificity and long-term memory.

During an influenza infection, innate immune cells coordinate efforts to minimize damage to the host and work with the adaptive immune system to initiate the most effective and efficient response to resolve the infection. Influenza virus infection begins with the virus infecting airway and alveolar epithelial cells. Viral replication in epithelial cells triggers innate sensing pathways that result in the release of antiviral compounds, and proinflammatory cytokines and chemokines that promote viral clearance, such as interferon type I and III (IFN-α/β and λ) and tumor necrosis factor α (TNF-α). In addition, viral proteins also trigger intrinsic apoptotic pathways to induce clearance of infected cells [26,27]. Concurrently, tissue-resident alveolar Mφ residing in the alveolar lumen phagocytose infected cells [28], or secrete nitric oxide synthase 2 and TNF-α [29,30] to limit the spread of infection. Release of proinflammatory cytokines induces the recruitment and differentiation of neutrophils and APCs, Mφ, and DC, which endocytose viral components and migrate to the draining lymph nodes to present viral peptides to T cells through specialized cell surface receptors called major histocompatibility complexes (MHC). It is these interactions that bridge the coordination of the innate and adaptive immune arms to enhance a more specific and targeted immune response to influenza infection.

There are two classes of MHCs that present antigens to different subsets of T cells. The uptake of an extracellular pathogen or protein in lysosome/endosomal vesicles are degraded by proteases, bound to MHC class II receptors, transported to the cell surface, and presented to CD4$^+$ T cells to proliferate and initiate helper T-cell functions [25]. Cytosolic viral proteins are degraded by proteasomes and transported to the endoplasmic reticulum where they bind to MHC class I receptors to be exported to the cell surface and presented to CD8$^+$ T cells, inducing CD8$^+$ T cell activation, clonal expansion, and differentiation into cytotoxic effector cells. CD8$^+$ T cells are then responsible for inducing lysis of influenza-infected cells and continual viral clearance [25]. The presentation of viral peptides by APCs to T cells in secondary lymphoid organs also initiates B-cell-signaling cascades responsible for generation or stimulation of the humoral response. When there has been no previous exposure to the virus, a primary immune response is initiated that relies heavily on the innate immune response before the adaptive immune response can be activated. However, immunological memory to the infection is generated following a primary infection, allowing for a more rapid and efficient response following reexposure. Overall, a functional and robust immune response is critical for resolution of influenza virus infection. Figure 10.1 provides an outline of the immune response to influenza infection and indicates where obesity affects this response (discussed in detail in section "The Weight of Obesity on Influenza Infection").

WHY DO PEOPLE DIE FROM INFLUENZA INFECTION?

Generally, mortality from influenza virus infection itself is due to development of acute respiratory distress syndrome (ARDS), which is associated with lung damage and injury to the alveolar capillary membrane [26,27]. Indeed, findings from individuals who have succumbed to influenza infection reveal extensive lung damage, but viral RNA is only present in a small subset, suggesting that severe lung injury from the immune response to infection drives subsequent mortality more than viral infection itself does [31]. Lung damage resulting from antiviral and apoptotic pathways in the lung as well as cytotoxic T-cell response results in acute lung injury that can lead to alveolar permeability, edema, impaired oxygen exchange, and in severe cases, ultimately death [32]. In mice, research has shown that influenza infection is survivable until the loss of approximately 10% of type I epithelial cells [33]. These losses are generally attributed to overly exuberant immune responses or impaired viral clearance resulting in excessive activation of inflammation [26,27]. These inflammatory responses have been shown to persist up to 30 days postinfluenza virus clearance [33]. In addition to viral complications alone, influenza virus infection results in extensive areas of denuded basement membranes in the airways. Exposure of basement membrane and an already over-worked immune response provides opportunity for secondary bacterial infections, which are a major cause of influenza-associated morbidity and mortality during seasonal epidemics as well as pandemics [34,35].

Although the mechanisms behind complete repair of damaged airway epithelium are not fully understood, it is clear that resolution of lung injury and regeneration of epithelium requires coordination by the immune system. Antiinflammatory cytokines (e.g., IL-10) released from $CD8^+$ T cells attenuate and resolve inflammation and suppress Mφ scavenging [36,37], whereas Mφ-associated factors promote the expansion of regulatory T cells required to suppress neutrophil-driven cytokines [38]. Following suppression of the immune response, rapid regeneration of epithelium is necessary to restore gas exchange and prevent secondary microbial infection, and is achieved through proliferation of airway progenitor cells present in the lung

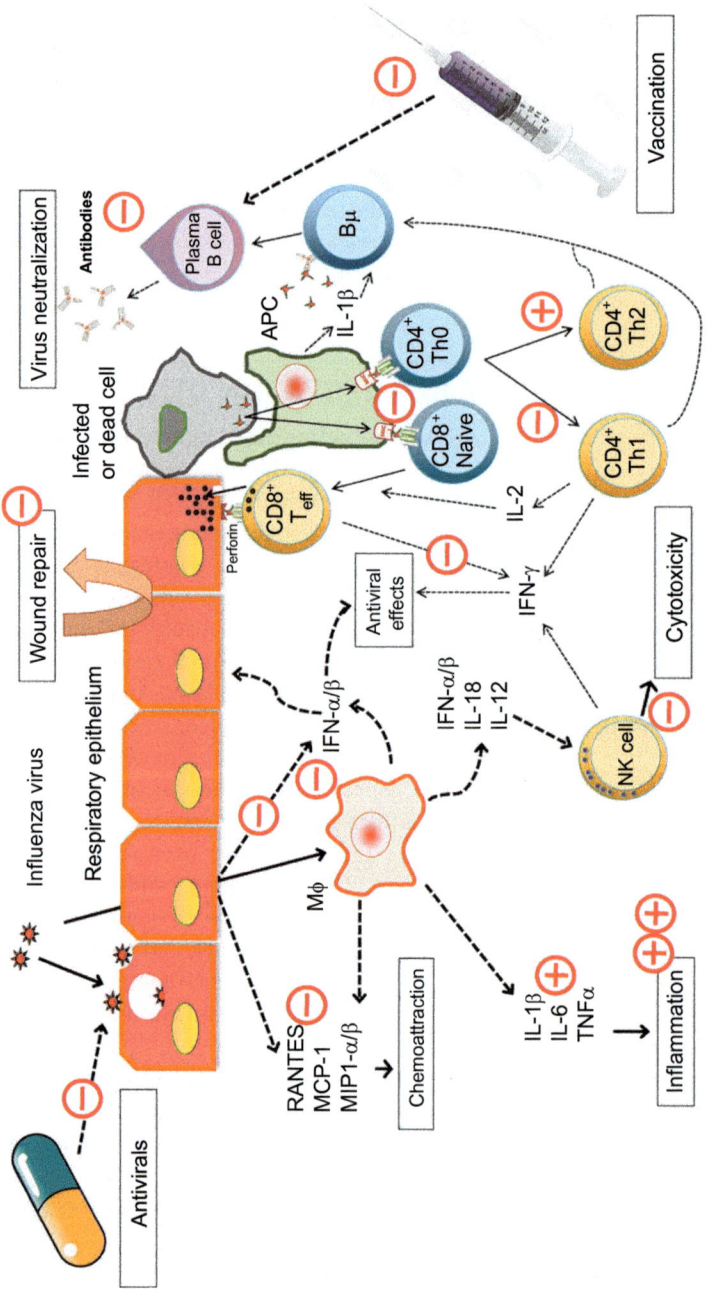

FIGURE 10.1

Overview of the immune response to infection with influenza virus. Obesity's effect on the response, either increasing (+) or decreasing (−) specific aspects of the response is shown in red on the diagram.

Modified from Karlsson EA, Beck MA. Exp Biol Med 2010;235:1412–24.

in response to environmental cues [39]. From the initial response to influenza virus infection to the replacement of airway epithelium, all systems must work together to balance resolution and repair, and to prevent significant morbidity and mortality in the host. Any perturbation of this system could result in ARDS and/or secondary infection, and greatly increase the chance of mortality from influenza virus infection.

PROPHYLACTIC AND THERAPEUTIC STRATEGIES AGAINST INFLUENZA

Effective prophylactic and therapeutic strategies are critical for controlling infection and transmission of influenza virus. Currently, vaccination remains the primary method of combating influenza epidemics and pandemics. Since the inception of vaccination as a modern public health practice, developing an effective influenza vaccine has been a priority for its potential to reduce medical and economic burden through disease prevention. Following the 1918 Spanish Influenza pandemic and its widespread devastation, influenza vaccines have undergone dramatic changes in their preparation and composition from a monovalent, live-attenuated virus in the 1940s to today's quadrivalent, inactivated, split vaccine composed of two influenza A and two influenza B strains [40,41]. Influenza vaccination campaigns target seasonal peaks in virus transmission, occurring predominately in colder months in temperate zones when dry air and communal living promote the spread of the virus in contrast to year-round circulation in the tropics [42]. Thus incident cases of seasonal influenza vary by location, with flu season in northern hemisphere countries occurring between October and May, and in the southern hemisphere, June through September [43]. Despite this seasonal shift in influenza transmission, infection can occur at any point in the year, as the virus is easily spread via water droplets from coughing and sneezing, or contaminating respiratory mucus present on environmental surfaces (e.g., countertops and door handles) [42].

Influenza virus is constantly undergoing structural changes in viral HA and NA peptides due to evolutionary pressures from host immune systems, resulting in antigenic drift, and thus driving the need for yearly reevaluation of vaccine effectiveness [44,45]. In general, vaccines provide protective immunity against pathogens by stimulating the adaptive immune system to establish immunological memory specific to the pathogen without actual infection. Thus following vaccination and upon subsequent exposure to peptides of a specific influenza virus included in the vaccine, the immune system mounts a swift and robust response, clearing the virus with attenuated or absent symptoms. Generally, vaccination strategies rely on the humoral immune response to generate production of neutralizing antibodies from antibody-secreting plasma B cells. These antibodies, mainly directed against the HA, bind to viral surface proteins to prevent or attenuate infection [25,46]. Hemagglutination inhibition (HAI) is currently the only accepted assay of predicting protection against influenza virus; however, this measurement may not be accurate for high-risk populations or for preventing severe disease. Generally, an HAI titer of ≥40 is considered to be seroprotective, and increased HAI positively correlates with increased rates of protection. Seroconversion is also a measure of vaccine responsiveness, indicated as at least fourfold increase of the HAI titer postvaccination [47]. As opposed to seasonal drift, pandemic influenza occurs when dramatic changes occur through mutation or reassortment resulting in antigenic shift, rendering current vaccines completely ineffective [48]. Vaccines that offer cross-protection against multiple strains of influenza virus may reduce the likelihood of influenza pandemics and could eliminate the need for annual reformulation of influenza vaccines. Although this next generation of influenza vaccines is still under development, strides are being made at generating cross-protective immunity toward conserved regions of the influenza virus including the HA stalk, the matrix, and the nucleoproteins [49,50].

Although vaccines remain the most effective prevention against influenza infection, vaccine mismatch or poor response in certain groups may result in vaccine failure. In contrast, influenza antiviral drugs remain efficacious despite antigenic drift or shift. Currently, two types of antivirals are licensed for use against influenza virus, the adamantanes and neuraminidase inhibitors (NAIs) [51]. Used for >30 years, adamantane compounds, amantadine and rimantadine, target the viral M protein. However, these compounds are no longer recommended for treatment of seasonal human influenza A viruses due to widespread antiviral resistance (>99.99% of viruses) and extremely fast generation of resistant mutants in clinical usage (2–3 days) in >30% of patients [52]. NAIs (oseltamivir, zanamivir, laninamivir, peramivir) impair viral budding from infected host cells by targeting the viral NA. Frequency of NAI resistance in currently circulating human seasonal strains is low (<1%); however, oseltamivir-resistance has quickly developed in previously circulating strains, and localized clusters of oseltamivir-resistant H1N1pdm09 viruses have been detected [53]. In addition, NAIs are only effective when delivered within 48 h after the onset of symptoms. Aside from these current drugs, a number of other antivirals targeting the viral polymerase, among other factors, are in preclinical and clinical testing. Overall, further research is warranted for antiviral drugs for influenza that maximize viral clearance while balancing ease of delivery with low selection for resistance mutations [51].

THE WEIGHT OF OBESITY ON INFLUENZA INFECTION

Obesity has long been associated with a higher risk of chronic diseases, such as type II diabetes (T2D), heart disease, high blood pressure, and certain types of cancer [54]. However, obesity has recently received tremendous attention for its association with higher risks of infectious disease [10,11,55]. The majority of people who become infected with influenza virus only experience mild illness and recover in 2 weeks or less; however, certain groups of people, such as children under 5, pregnant women, and the elderly, are more likely to have complications such as pneumonia, bronchitis, sinusitis, and otitis from influenza infection [56]. All of these groups are tied together by their altered immune state. Because obesity can also be considered an immunocompromised state [10,11], it is therefore hypothesized that obese individuals may also be at risk for increased severity of influenza infection and possibly a higher risk for mortality.

EPIDEMIOLOGICAL STUDIES ON INFLUENZA INFECTION IN OBESE HUMANS

Sometime in 2008 or 2009, a previously unknown H1N1 influenza A virus went through zoonotic transmission from swine to humans, quickly established human-to-human transmission, and led to a global pandemic. Since then, this pandemic H1N1 virus (pdmH1N1) has overtaken the previously circulating human H1N1 viruses and is currently considered to be the predominant seasonal human strain [57]. Interestingly, epidemiological studies indicated that a significant number of cases across the globe occurred in obese adults [58–61]. Using public health surveillance data from hospitals and clinics in California following the outbreak of pdmH1N1 "swine flu", Louie *et al.* found that, in a cohort of 534 influenza cases, of the 91 influenza-related deaths, 56 were in obese adults (BMI $\geq 30\,kg/m^2$) [62]. This retrospective study introduced obesity as a novel risk factor for severe influenza infection and identified obesity as an independent risk factor for influenza morbidity and mortality [62]. Subsequently, further prospective and retrospective epidemiological studies have found similar results in cohorts from the 2009 pandemic and with the pdmH1N1 strain in general. Overall, obese individuals had significantly increased hospitalizations, morbidity, and/or mortality [63–65]. Taken together, these findings demonstrated a novel relationship between obesity and pandemic influenza infection that had previously only been suggested in mouse models [66]. Aside from pandemic influenza, seasonal influenza has also been investigated for increased risk of morbidity and mortality in obese populations. One recent study found the odds of influenza-related hospitalization were 1.45 times higher for obese adults (BMI $\geq 30\,kg/m^2$) compared with lean adults (BMI = 18.5 to $24.9\,kg/m^2$) over the study period. Class II and/or class III obesity (BMI $\geq 35.0\,kg/m^2$) increased the odds of hospitalization to 2.2 times higher than that of lean adults. Controlling for other known risk factors for hospitalization from seasonal influenza, there remains a strong association between increased risk of influenza-related hospitalization and obesity in adults [67].

However, not all studies have found an association between influenza virus morbidity/mortality and obesity. In 2011, Diaz *et al.* investigated the association of obesity as a risk factor for mortality in a prospective, observational, cohort in Spain. Using real-time polymerase chain reaction to confirm H1N1 infection, the authors reported 36.1% of intensive care unit (ICU) cases occurred in the obese [68]. However, despite finding a positive association with greater incidence of mechanical ventilation, longer need for ventilation, and longer mean length of ICU stay, there was no significant difference between obese and nonobese influenza case mortality rates in their cohort [68]. Similarly for seasonal influenza infection prior to the 2009 pandemic, no association was found between BMI and influenza hospitalization [69]. Despite some nonassociations, overall, epidemiological data suggests that obesity is associated with influenza-related hospitalization and poor health outcomes in adults for both seasonal as well as pandemic influenza infections [70]. Currently, the CDC considers people with a BMI above 40 to be at high risk for developing influenza-related complications.

EPIDEMIOLOGICAL STUDIES ON INFLUENZA VACCINATION AND ANTIVIRALS IN OBESE HUMANS

Given that obesity is associated with increased severity and mortality from influenza infection, it is also critical to understand how obese individuals respond to prophylactic and therapeutic strategies against influenza infection. As stated previously, adequate response to influenza vaccination is crucial for protection against seasonal and pandemic influenza. Obesity has previously been associated with decreased vaccine responses to hepatitis B and tetanus vaccines [10,71]. In addition, due to study design, epidemiological data could be skewed due to differences in influenza vaccine coverage in obese hosts. Indeed, changes in influenza vaccine effectiveness in the obese versus lean populations could account for differences in influenza incidence and mortality rates in these populations.

Several studies have attempted to address the influenza vaccine response in obese populations. In one study, although there were no differences in antibody titers due to BMI at 30 days postvaccination, increasing BMI was associated with a greater decline in influenza specific antibody titers at 1-year postvaccination [72]. Thus although obesity did not impair the initial antibody response to vaccination, it was associated with a steeper decline in titer over time [72]. In addition, Paich *et al.* demonstrated that obesity alters the cellular response to *ex vivo* influenza challenge in previously influenza-vaccinated adults. In this study, peripheral blood mononuclear cells, isolated from influenza-vaccinated adults, were stimulated in culture with pd-mH1N1 influenza virus and the cellular response to virus challenge was assessed using flow cytometry. $CD4^+$ and $CD8^+$ T cells from overweight and obese adults expressed less activation markers and functional markers [73]. Interestingly, DC activation and function was unaltered between lean, overweight, and obese subjects [73]. These data suggest alterations in $CD4^+$ and $CD8^+$ T-cell activation and function in response to influenza virus challenge in the obese may account for the higher degree of influenza morbidity and mortality. However, this study lacks mechanistic evidence as to how obesity may alter the T-cell response to influenza infection.

In terms of vaccine effectiveness, data appear to be conflicted. One study by Talbot *et al.* found no correlation between obesity, vaccine response, and susceptibility to infection in elderly individuals [74]. In contrast, Neidich *et al.* found vaccinated obese adults were twice as likely to develop influenza or influenza-like illness despite a robust serological response to influenza vaccination [75]. This finding challenges the current standard of protection for antibody titers in obese adults. Furthermore, it supports previously established epidemiological data identifying a higher risk of influenza infection in obese adults, suggesting impairments in the cellular but not humoral immune response may account for higher risk of influenza infection in obese adults. To better understand the mechanisms by which obesity alters the immune response to influenza infection and vaccination, targeted studies using animal models of infection to investigate the adaptive immune response provide an opportunity to delineate this complex physiological condition.

In addition to understanding the influence of obesity on vaccination, an understanding of obesity's effects on antiviral effectiveness, pharmacokinetic processes, and toxicity is also needed. Diet and nutrition are extremely important in the pharmatoxicological properties of chemicals, and obesity has been shown to alter therapeutic properties due to changes in drug absorption and systemic exposure [76–78]. In the case of influenza antivirals, because obese individuals were at increased risk for pdmH1N1 infection, recommendations have been considered for increased dosages in this expanding high-risk population [79]. Several studies have investigated systemic oseltamivir levels in uninfected obese versus lean individuals and found no differences, suggesting increased dosage may be unnecessary [80–84]. However, human studies of influenza antiviral effectiveness during influenza infection in obese individuals are scarce, and further work is needed to determine how systemic levels correlate to exposure in the lung epithelium.

STUDIES ON INFLUENZA SEVERITY IN OBESE ANIMAL MODELS

Studies in obese mice infected with influenza virus demonstrate impaired innate and adaptive immune defenses resulting in increased morbidity and mortality. Starting in 2007, Smith *et al.* first demonstrated impairments in the primary immune response to influenza virus infection in diet-induced obese (DIO) mice using a mouse-adapted strain of influenza virus, A/Puerto Rico/8/34 (PR8). Despite showing no differences in viral titer, DIO mice experienced 6.6-fold greater mortality compared with lean mice and were found to have reduced NK cytotoxicity and reduced expression of antiviral cytokines [85]. Further studies showed DIO mice had delayed mononuclear infiltration in the lung and decreased DC numbers as well as impaired DC APC function [86]. Following the 2009 pandemic and the finding of an epidemiological association between obesity and influenza infection severity and mortality, several groups also investigated infection of obese mice with pandemic influenza [87,88]. Similar to findings with mouse-adapted influenza infection (PR8), both DIO and genetically obese (*ob/ob*; B6·Cg-Lep^{ob}/J) mice were found to be more susceptible to pandemic strains of influenza than their lean counterparts with altered immune responses despite showing no differences in virus titers. In addition, O'Brien *et al.* were able to show that obese mice had significant decreases in lung epithelial regeneration, which may contribute to the increased mortality seen in the obese host [87]. Further work has gone on to show that this increased exposure of basement membrane may contribute to increased susceptibility to secondary bacterial infection in the obese host regardless of influenza or bacterial strain [89].

STUDIES ON INFLUENZA VACCINATION AND ANTIVIRAL EFFECTIVENESS IN OBESE ANIMAL MODELS

Several studies have examined vaccination responses and effectiveness in obese mouse models, and overall, unlike the equivalent antibody titers observed in obese and lean humans, traditional vaccination strategies against seasonal, pandemic, and avian influenza viruses appear to result in decreased antibody responses in obese mice compared with lean controls. As expected, following viral challenge, vaccinated obese mice show no protection against influenza virus and quickly succumb to infection with increased viral titers, lung injury, and inflammation compared with lean controls [90–92]. Further, in an attempt to boost antibody production following vaccination, one study tested how the addition of adjuvant (alum or a squalene-based adjuvant) to an influenza vaccine would improve neutralizing antibody responses as well as protection from influenza challenge. Although addition of an adjuvant successfully boosted antibody response to both pandemic and avian influenza vaccines over the theoretical level of protection, antibody responses were still significantly decreased in obese mice versus lean controls. In addition, obese mice were not protected from influenza challenge despite serological responses. Increasing the vaccine dose fourfold or decreasing viral challenge by 100-fold did not lead to full protection in obese mice, indicating hypersusceptibility in obesity outweighs vaccine response [92]. This study correlates nicely with the findings in humans by Neidich *et al.* showing increased risk of influenza infection in obese humans despite similar antibody vaccine responses to lean controls [75]. To date, only one study has looked at the use of antivirals in the obese mouse during influenza infection. O'Brien *et al.* found that standard dosages of the NAI oseltamivir were protective against influenza challenge. However, 100% survival was not achieved unless obese mice were administered a fivefold higher dosage, indicating that standard dosages may be effective but not fully protective in obese individuals [87].

STUDIES ON IMMUNOLOGICAL MEMORY IN OBESE ANIMAL MODELS

Aside from vaccination responses, prior exposure to influenza virus can generate cross-protective immunity in which T cells recognize the internal conserved regions of the influenza A virus and can thereby recognize and protect against influenza A strains with differing external cell surface proteins [93–97]. DIO mice have impaired generation and maintenance of effector and central effector T-cell memory populations [90,98,99]. In addition, obese mice show impaired memory T-cell responses to heterologous secondary influenza infection. Lean and DIO mice were infected with X-31, a mouse-adapted H3N2 influenza strain and 4 weeks later infected with PR8. Although lean mice were fully protected from the secondary infection, obese mice had a 25% mortality rate, dysregulated lung cytokine and chemokine expression, and reduced influenza-specific CD8+ T cells expressing IFNγ. Memory T cells in obese mice also had decreased IFNγ response to antigen presentation by DCs [99]. Interestingly, although mortality was not found to differ, studies on cross-protective responses (similar to vaccination studies discussed earlier) did show decreased neutralizing antibodies, increased viral titer, and increased lung inflammation and injury. In addition, following secondary challenge, lungs of obese mice also had increased immune cell infiltration and increased levels of albumin from bronchoalveolar lavage fluid, indicating damage to the alveolar capillary membrane. Evaluation of regulatory T-cell (Treg) populations (CD25 + Foxp3+) demonstrated that obese mice had increased Tregs in the lung, but these populations were less suppressive, indicating impaired Treg functionality [100].

IS THE DECREASED IMMUNE RESPONSE IN ANIMAL MODELS DUE TO DIETARY EFFECTS OR TO OBESE STATE ITSELF?

Overall, it appears to be well established that obesity impairs the humoral and cell-mediated immune response to influenza infection. However, in mice, it is unclear whether obesity itself was driving altered immune responses or whether it was diet-related factors associated with high-fat diet feeding experiments. High-fat diets can be proinflammatory, and therefore it could be that dietary components, and not obesity, are driving the altered immune responses to influenza infection. To address this question, a genetic mouse model of hyperphagia-induced obesity in which the leptin receptor (*Lepr*) was deleted only in the hypothalamic neurons ($Lepr^{H-/-}$) was examined [101]. Compared with lean, wild type and heterozygous mice, all of which consumed identical, low-fat chow diets, obese $LepR^{H-/-}$ mice had significantly increased mortality to influenza virus infection [102]. Thus these results suggest that the obese state itself, independent of diet, is sufficient to impair the immune response to influenza infection.

WHY DOES OBESITY WEIGH DOWN HOST IMMUNE DEFENSE? MACRONUTRIENT METABOLISM AND INFLUENZA IMMUNITY

Underlying conditions associated with metabolic perturbations such as obesity, diabetes, pregnancy, hypertension, and chronic kidney, heart, and liver disease can increase the risk for severe influenza infection [103]. Given that dynamic changes in cellular metabolism is linked to effective immune responses, systemic metabolic perturbations likely affect immune cell function in metabolic diseases. Therefore obesity-induced altered systemic metabolism might be expected to impair immune cell function and understanding how alterations in metabolism, both cellular and systemic, can directly affect immune cell function may reveal explanatory mechanisms for greater influenza severity in the obese host.

Macronutrients in the human diet consist of carbohydrates, lipids, and protein. These molecules are defined as such because they are consumed in a greater quantity in comparison to vitamins, minerals, and other nutrients. The overall balance of energy intake through macronutrient consumption can profoundly affect the immune response to influenza infection. Carbohydrates (primarily glucose), amino acids, and fatty acids can directly affect immune cell function by fueling energetic demands, serving as biosynthetic precursors, and directly regulating gene expression through modulation of a variety of metabolic signaling pathways [11,104,105]. During infections or inflammatory conditions, immune cells undergo a variety of bioactive processes including cell growth, functional activation, migration, and proliferation [105–107]. An abundance of evidence has emerged that suggests cellular energy metabolism, in part regulated by macronutrient availability, modulates these essential immune processes. In recent years, the effect of macronutrient metabolism and metabolic fate on function, differentiation, and inflammatory capacity has primarily been focused on Mφ and T cells [105].

Manipulation of cellular metabolism has been shown to enhance or diminish the inflammatory and functional capacity of T cells and Mφ [105,108]. Both T cells and Mφ can influence influenza infection outcome, and therefore understanding how macronutrient metabolism directly regulates their function could provide insight into how obesity may increase risk for severe influenza infection [109,110]. In CD4$^+$ and CD8$^+$ T cells, macronutrient metabolism can directly impact T-cell differentiation and function [104,105,111]. To fuel the energetic and molecular synthesis demands of cell growth and clonal expansion, activated effector T cells engage in Warburg metabolism, undergoing glycolysis and producing lactic acid despite the availability of oxygen. This metabolic reprogramming allows effector T cells to rapidly acquire biosynthetic precursors required to support growth and clonal expansion [107,112]. Further, memory T cells and CD4$^+$ T-cell subsets (Th17, Tregs, Th1, and Th2) have differing metabolic profiles, and metabolic manipulation or altering substrate availability can affect the differentiation and function of these T cells [104,107].

In addition to T cells, macronutrient metabolism can alter the polarization of Mφ, a spectrum ranging from antiinflammatory and wound-healing "M2" cells to proinflammatory "M1" cells. M1 Mφ display a preference for glycolytic metabolism in contrast to M2 cells, which prefer β-oxidation of fatty acids for generation of ATP [108]. Further, genetic or chemical manipulation of these intrinsic metabolic pathways can enhance or attenuate inflammatory responses, affecting immunity [113,114]. Taken together, macronutrient metabolism is a critical regulator of Mφ and T-cell function. However, it remains unclear how dietary consumption of macronutrients and macronutrient concentrations in the immune cell microenvironment regulate immunity. Further understanding of these complex processes may generate explanations for the increased influenza severity observed in populations with preexisting metabolic diseases, such as obesity.

GLUCOSE/DIABETES

The World Health Organization (WHO) reports that 347 million individuals are living with diabetes worldwide, a chronic disease characterized by hyperglycemia due to an inability of the pancreas to secrete insulin (diabetes mellitus type I) or an inability of cells to properly respond to insulin (diabetes mellitus type II, T2D) [115]. As obesity is a risk factor for developing T2D, it is no surprise that the rates of T2D are increasing along with increasing levels of obesity worldwide. Diabetes is a well-established risk factor for increased influenza infection severity, even when well managed [16,59,116]. Further, it has been reported that hyperglycemia may increase risk for severe infection to avian influenza viruses [117]. However, the exact mechanisms by which diabetes and elevated blood glucose may modulate immune cell function during influenza infection are unclear. It is known that illness can raise blood sugar levels, and thus influenza infection may complicate management of blood glucose levels in diabetics [116]. Further, several investigations have reported diabetes or hyperglycemia can impair immune cell function [118–122]. The immunomodulatory effect of diabetes may be derived from the toxic effects of hyperglycemia itself, and/or secondary consequences from hyperglycemia such as oxidants and glycation products. Additionally, alterations in signaling pathways due to hyperglycemia may directly affect immune cell metabolism and function [123].

In vitro studies have revealed that low glucose concentrations impair T-cell proliferation and enhance T-cell apoptosis [124]. Conversely, excessive glucose uptake in T cells, through transgenic overexpression of the primary glucose transporter 1, GLUT1, increased T-cell size and cytokine production [125]. In 2013, Chang *et al.* showed that aerobic glycolysis is required for effector function, but not proliferation or survival, of activated effector T cells [126], demonstrating through stimulation and blocking of aerobic glycolysis that IFN-γ translation and production is dependent on glycolytic metabolism [126]. Other studies have confirmed these findings through knocking out GLUT1, thereby abolishing effector T-cell activation [127]. Additionally, exposure of T cells to elevated glucose concentrations in culture enhances oxidative stress [128].

Despite emerging evidence for the regulatory effects of glucose on immune cell function, *in vivo* models investigating the effect of blood glucose levels on immune cell function during influenza infection are lacking. However, blood glucose levels have been shown to correlate with influenza virus titer in a mouse model of diabetes and influenza infection [129]. In addition, patients with T2D have been found to respond favorably to influenza vaccination, and antidiabetic drugs have been shown to be protective against lethal influenza infection in mice [130,131].

DIETARY LIPIDS

Lipid molecules have been shown to directly regulate influenza virus infectivity and modulate host immunity and influenza infection outcome [132,133]. In addition to classical roles such as providing an abundant source of ATP or being used for biosynthesis in immune cells, fatty acids can modulate immune responses through a number of other mechanisms. Saturated fatty acids tend to be proinflammatory, in part, by activating Toll-like receptors 2 and 4 [134]; however, there is limited evidence of how altering the level of saturated fatty acids in the diet, independent of obesity, increases risk for severe influenza infection. DIO mice fed a diet high in saturated fat are more susceptible to influenza infection, although this is likely primarily due to the pathophysiological complications of obesity; therefore it is difficult to tease out the immunological effect of greater saturated fat consumption [66,99,100,135].

Prostaglandins are fatty acid-derived mediators of inflammation, and manipulation of prostaglandin production can affect influenza infection outcome. The cyclooxygenase (COX) enzymes catalyze the initial step in the synthesis of prostaglandins, and mouse models using pharmacological inhibitors or genetic depletion have demonstrated that reduced COX-1 activity enhances influenza morbidity and mortality. However, COX-2 deficient mice have improved infection outcomes compared with control mice [136,137]. Further, synthesis of inflammatory lipid molecules can be modulated through dietary means, as consumption of omega-3 poly unsaturated fatty acids (PUFAs) can induce antiinflammatory responses through COX pathways and via other mechanisms, whereas omega-6 fatty acids can mediate proinflammatory responses [138–140].

Whether reducing the level of inflammatory lipid molecules is beneficial during influenza infection is unclear. Omega-3 fatty acids have consistently been a part of the human diet and the ratio of omega-6 to omega-3 fatty acids in the diet of early humans is estimated to be relatively equal [141]. This ratio in the United States rose to approximately 10:1 due to reduced omega-3 fatty acid intake and the widespread use of vegetable oils [142]. Due to the recent explosion of research associating PUFA intake with several health benefits [143], fish oil supplements and fish consumption (a natural source of omega-3 fatty acids), as well as the use of omega-3-enriched products, has significantly increased [144]. In some studies, mice fed a fish oil diet exhibit altered immune defenses [145,146], delayed viral clearance, and enhanced morbidity and mortality upon influenza infection [147]. In contrast, a recent investigation demonstrated the omega-3 PUFA-derived protectin D1 attenuates influenza virus replication and improves infection survival in mice [132]. Although the results of these studies are unclear and more research is necessary into the effects of dietary lipids on influenza infection, the regulatory effect of lipid metabolites on influenza infection has recently been reflected in a recent report by Tam *et al.*, which identified lipid metabolites as mediators of mouse and human influenza infection outcome [133].

Fatty acids can also directly influence cell-mediated immunity through modulation of T-cell function and survival. CD4$^+$ and CD8$^+$ T cells utilize amino acids and fatty acids to fuel oxidative phosphorylation, generating ATP during cell surveillance [148]. This efficient generation of energy intermediates allows for the maintenance of naïve and memory T-cell populations. Upon antigen exposure, these naïve and memory populations transition to effector T cells, where glycolysis again fuels ATP and other intermediates for daughter-cell generation [149]. However, inhibiting fatty acid oxidation impairs memory T-cell development, as they cannot revert back to a quiescent anabolic state [150,151]. Furthermore, enhancing fatty acid oxidation improves memory formation of T cells [150], thereby sustaining these essential immune cell populations should exposure to a pathogen such as influenza occur.

Immune cells, like many other cell types, can generate *de novo* fatty acids for storage as long-chain fatty acid species or use them for oxidative ATP generation much like exogenous fatty acids. Work by O'Sullivan *et al.* suggests memory CD8$^+$ T cells utilize *de novo* fatty acid synthesis followed by cell intrinsic lipolysis to fuel fatty acid oxidation of memory T cells [152]. This counterintuitive futile cycle utilizes glucose to generate pyruvate and thus synthesize fatty acids through the TCA cycle only to immediately hydrolyze these fatty acids for mitochondrial oxidative phosphorylation. The authors argue this process maintains mitochondrial health and glycolytic as well as lipogenic pathways through constant cycling of fatty acid synthesis followed by fatty acid oxidation.

Other studies have shown the importance of exogenous fatty acids in the maintenance of tissue-resident memory T cells. This critical population of memory T cells serves as the primary memory defense system for pathogens by residing in high-exposure epithelial tissues, such as the lung, so as to immediately and robustly respond upon exposure. Pan *et al.* demonstrated how tissue-resident memory T cells, unlike their central memory T counterparts, which use the futile cycle of cell-intrinsic lipolysis to fuel fatty acid oxidation, utilize exogenous fatty acids to fuel beta-oxidation through actions of *Fabp4/Fabp5* uptake [153]. Furthermore, one study showed that the exogenous fatty acid palmitate can directly influence CD4+ T-cell differentiation into proinflammatory effector memory T cells through a PI3K-Akt-dependent activation pathway [154].

This difference in substrate usage pertains to the varying milieu of macronutrient-rich lymphoid and circulatory systems compared with that of lipid-rich but macronutrient-poor epithelial cells. How these processes are influenced by conditions of obesity (i.e., excess exogenous fatty acid, cholesterol, glucose, etc.) remains a critical question to understanding how obesity influences the generation and maintenance of varying memory T-cell populations.

In addition to the fatty acids, cholesterol is another lipid molecule that may affect influenza virus infection outcomes. High cholesterol remains a significant public health problem in the United States with more than one-quarter of individuals aged 40–74 at risk [155]. Cholesterol is an essential constituent in cell membranes and, given that influenza virus is coated in a lipid envelope derived from host cells, the influenza virus is also coated with host-derived cholesterol. Cholesterol can regulate influenza virus entry and budding, and influenza virus cholesterol depletion decreases influenza infectivity [156]. Although dietary consumption has relatively little effect on circulating levels of cholesterol in most individuals [157,158], the percentage of adults taking cholesterol-lowering medication (statins) has skyrocketed with an increase of 4% to approximately 23% of adults in the past 3 decades in the United States alone [155]. Several studies have suggested that statins may reduce influenza mortality [159,160]; however, more research is necessary before any definite conclusions are drawn on the pleiotropic effect of statins.

Macronutrient consumption is a major public health concern, given that excessive calorie intake can increase risk for infectious disease. As discussed previously, immune cell function can be altered by distinct changes in cellular metabolism, which can be influenced by host nutrition status. Therefore it is not surprising that obesity increases susceptibility to severe influenza infection [161]. Figure 10.2 provides a summary of T-cell metabolism and the potential effects of obesity on altering metabolic cell-signaling pathways necessary for T-cell function.

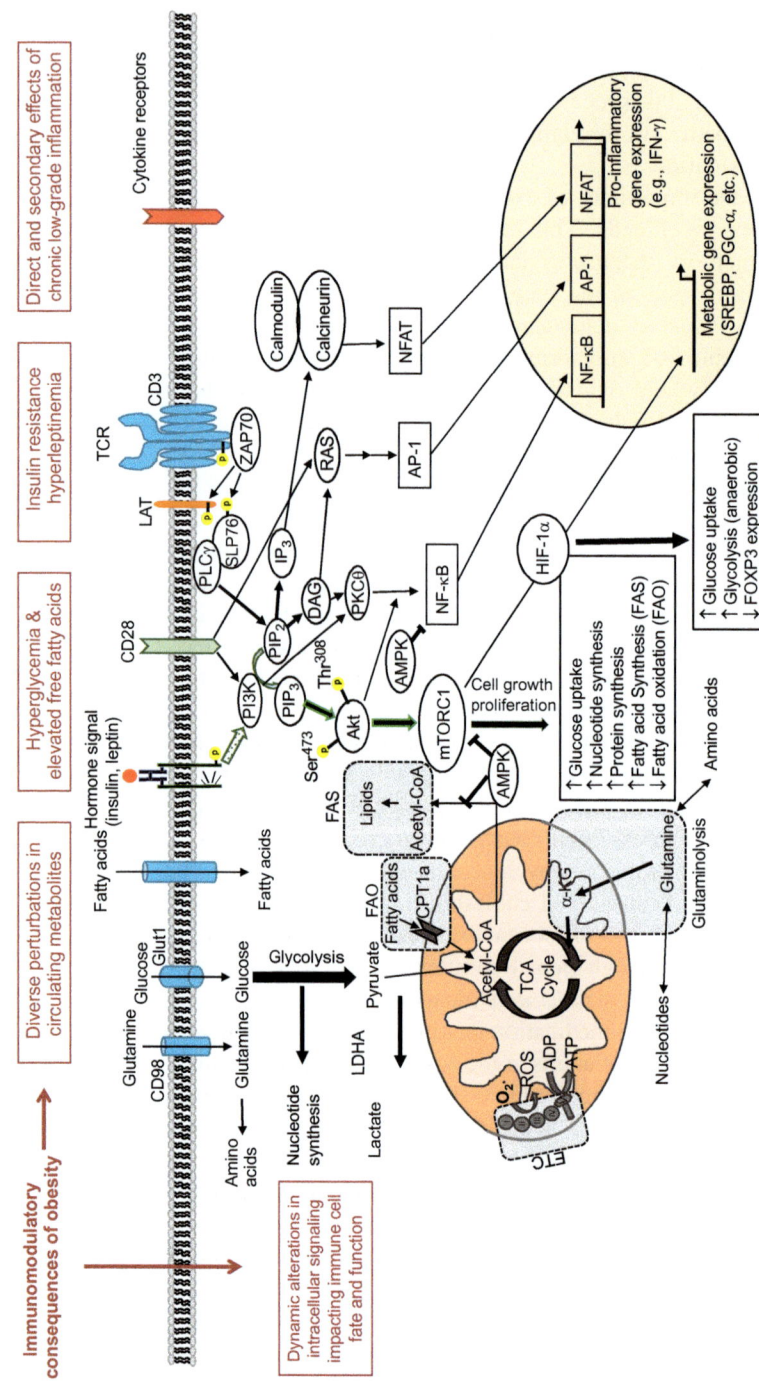

FIGURE 10.2

Overview of T-cell metabolism. While quiescent, a metabolically healthy T cell primarily utilizes lipids for fatty acid oxidation. Following activation of the T-cell receptor (TCR) with its coreceptor CD28, a signaling cascade occurs resulting in increased aerobic glycolysis through increased glucose uptake, increased nucleotide, protein and fatty acid synthesis and decreased fatty acid oxidation. The increase in glycolysis requires increased uptake of glucose, accomplished through insulin-stimulated expression of Glut1 on the plasma membrane. Obesity may impact this response as noted on the diagram. Abbreviations: TCR, T cell receptor; LAT, linker for activation; Glut, glucose transporter; PI3K, phophoinositide 3-kinase; IP3, inositol-1,4,5-triphosphate, PKCθ, protein kinase Cθ, PLCγ, phospholipase Cγ; SLP76, SH2-domain-containing leukocyte protein of 65 kDa; ZAP70, chain associated protein kinase of 70 kDaL; PIP$_2$, phosphatidylinositol biphosphate; HIF-1α, hypoxia-inducible factor 1α; NFAT, nuclear factor of activated T cells; NF-κB, nuclear factor-κB; mTORC1, mammalian target of rapamycin 1; AP-1, activator protein 1; DAG, diacylglycerol; ROS, reactive oxygen species; TCA, tricarboxylic citrate acid cycle; CPT1α, carnitine palmitoyl-transferase 1α; FAO, fatty acid oxidation; FAS, fatty acid synthesis; AMPK, AMP-activated protein kinase.

WHY DOES OBESITY WEIGH DOWN HOST IMMUNE DEFENSE? OBESITY-RELATED HORMONE CHANGES, CHRONIC INFLAMMATION, AND MICROBIOME

Obesity is a complex condition. In addition to alterations in metabolism, other obesity-related conditions may also affect the response to influenza. These include alterations in hormone levels such as insulin, leptin, and adiponectin; chronic inflammation due to increased visceral adipose tissue mass; and alterations in the microbiome. Each of these obesity-related alterations are discussed in the following text.

OBESITY-RELATED HORMONE CHANGES

Long-term excess caloric intake characteristic of obesity influences the regulation of several key hormones responsible for controlling nutrient uptake and storage, and hunger/satiety. These hormones include insulin and the adipokines leptin and adiponectin. In addition to their traditional role in regulating short-term and long-term nutrient control, these hormones have recently been shown to directly influence the adaptive immune response to infection.

Typically, insulin secretion from pancreatic beta cells facilitates peripheral glucose uptake through insulin-dependent Glut4 transporters in tissues such as muscle and adipose tissue [162]. Immune cells primarily utilize other glucose transporters such as Glut1 and Glut3 for glucose uptake in an insulin-independent manner [163]. However, insulin signaling in immune cells can augment glucose uptake, supporting the glycolytic functions of activated immune cells. Fischer *et al.* demonstrated that knocking down insulin receptor in CD4$^+$ T cells limited these cells' ability to uptake glucose for glycolysis, thereby impairing cytokine production, proliferation, migration, and effector cell survival [164]. The authors conclude that upregulation of insulin receptor on T cells during activation provides a bioenergetic advantage to support effector T-cell glycolysis and thus T-cell-mediated immunity. However, in the case of obesity, in which prolonged caloric intake often leads to elevated serum insulin and insulin resistance, whether or not defects in insulin signaling account for obesity-impaired immune response of T cells to influenza remains unknown.

Obesity has long been associated with higher circulating leptin levels as adipocyte mass increases [165]. This hormone is secreted by adipose tissue and binds hypothalamic leptin receptors to regulate satiety, indicating long-term nutritional status of the host [166]. Recently, similar to insulin, leptin has also been shown to support the glycolytic activity of effector T cells through increasing glucose uptake and supporting cytokine production [167]. Furthermore, leptin signaling has been shown to be required for Th17 differentiation [168] but not naïve or regulatory T-cell function or maintenance. In a study investigating the impact of leptin on the immune response to influenza infection in DIO mice, Zhang *et al.* found high serum leptin levels increase lung injury following pdmH1N1 influenza virus infection [169]. Additionally, the authors demonstrated that blocking leptin signaling with a leptin-specific antibody improved the survival of obese mice and reduced proinflammatory cytokines IL-6 and IL-1β [169].

Taken together, these data suggest that systemic metabolism can alter the metabolic profile of T cells through excess insulin and/or leptin signaling. Alterations in these hormone signals could possibly account for impairments in cytokine production through changes in the balance of transcription factors (NFAT, AP-1, and/or NF-κB) via alterations in upstream activation signals. Evidence for this hypothesis has been demonstrated through leptin- and insulin-mediated activation of the PI3K-Akt-mTOR pathway in T cells [170,171]. Furthermore, the counterregulatory hormone adiponectin, which opposes leptin's action, has also been suggested as a possible factor reducing severity to influenza through suppressing antigen-activated effector T-cell action and reducing proinflammatory cytokine production [172,173]. Given that adiponectin levels are reduced in the obese, inverse to levels of its counterregulatory hormone partner leptin, perturbations to these hormones, along with insulin, may account for higher obesity-associated morbidity and mortality due to influenza infection.

OBESITY AS A STATE OF CHRONIC INFLAMMATION

Chronic inflammation is another major contributing factor often proposed as a possible mechanism for impairments in the obese immune responses to influenza infection. Low-grade inflammation can be characterized by increased circulating proinflammatory cytokines TNF-α, IL-6, and IL-1β, as well as acute phase reactants such as C-reactive protein and fibrinogen [174]. Activated immune cells, namely Mφ, T cells, and DC, produce these proinflammatory cytokines upon infiltration into adipocytes [154]. Recently, large adipocytes have been shown to activate CD4+ T cells via MHCII recognition [175,176]. These activated T cells recruit other pro-inflammatory immune cells, leading to the establishment of an inflammatory microenvironment. Proinflammatory cytokines can also activate immune cells through similar signaling pathways (Erk/MAPK/Akt), thereby skewing the immune population to an inflammatory state [154]. This pro-inflammatory skewing of immune cell populations in obesity could contribute to the increased lung injury and complications associated with severe influenza infection. Furthermore, less effective immune resolution might also result due to an imbalance in immune cytokine signals characteristic of low-grade chronic inflammation.

THE OBESE MICROBIOME

Another possible contributor to the differences observed in influenza infection severity between lean and obese subjects could be the intestinal microbiome. Commensal bacteria are essential for shaping the immune response in both health and disease, especially in the intestinal microenvironment, and gut-commensal bacteria support intestinal immune homeostasis by regulating Tregs and Th17 cells [177–179]. However, it is less clear how changes in intestinal flora can influenza immune function at a nonintestinal site such as the lung (see also Chapter 12). Several recent studies have shown that changes in the microbiome can cause major differences in response to respiratory virus challenge, specifically influenza A virus, by modulating helper T-cell responses and inflammation [180,181]. In addition, a number of studies have shown that oral treatment of mice and humans with microbiome-altering probiotics can significantly decrease incidence and severity of respiratory virus infection, mainly through modulation of the innate immune response and modulation of lung damage [182–188]. Overall, the microbiome is made up mostly (~92.6%) of two divisions of bacteria, Bacteriodetes and Firmicutes, in a delicate balance. Obesity has been shown to significantly alter this balance of gut-microbial ecology and, in both mice and humans, obese individuals have increased Firmicute numbers versus Bacteriodetes [189,190]. Although the exact etiology of the differences between the lean and obese microbiome has yet to be determined, the increased ratio of Firmicutes to Bacteroidetes may help promote adiposity or could represent a host-mediated response to limit energy uptake and storage [189] Given that Firmicutes appear to affect inflammatory responses, it could be surmised that they contribute to some of the low-grade, chronic inflammation in the obese microenvironment. In addition to the intestinal microenvironment, the respiratory microbiome has been conjectured to be involved in response to influenza infection [180,191,192]. However, very little is known about the contribution of bacterial populations to immune defense at this site. To date, no studies have addressed how changes in the gut or respiratory microbiome of the obese host could affect severity of respiratory infection.

Additionally, studies have also shown the importance of the microbiome to influenza vaccination. In a study examining the effect of trivalent inactivated influenza vaccine (TIV), Oh *et al.* found Toll-like receptor-5 (TLR5) mediated flagellin sensing by gut microbiota was associated with plasma cell generation of influenza-specific antibody titers in humans. Additionally, they reported that TLR5 signaling supports short-lived plasma cell generation, memory B-cell activation and function, and importantly, that certain classes of gut microbiota are required for the antibody response to TIV vaccination [193]. Given how obesity has been shown to influence the composition of gut microbiota [189], these findings suggest variations in gut microbiota communities impair the immune response to vaccination and therefore weaken the immune system's ability to respond to influenza infection.

SUMMARY AND PUBLIC HEALTH IMPLICATIONS

The global burden of obesity is expanding. In the United States alone, severe obesity is projected to increase by 130% in the next 20 years, and >70% of the European population will be overweight by 2030 [194,195]. Undeniably, obesity has been determined to be a risk factor for increased severity and mortality for influenza virus infection. Obesity increases susceptibility to and severity of influenza virus infection, dampens primary and secondary immune responses, and decreases wound healing in the lungs. Concerningly, standard vaccination strategies appear to be inefficient in this high-risk group making it extremely difficult to protect against infection. Given the yearly seasonal outbreaks of influenza coupled with the threat of another pandemic and the continued evolution, globalization, and zoonotic transmission of avian influenza, it is imperative to continue research on obesity and influenza virus infection. Indeed, the vast majority of obese children live in low- to middle-income developing countries [8] where the threat from seasonal and zoonotic influenza is greatest. Therefore future studies should focus on the development of novel strategies to treat and prevent influenza virus infection in this weighty portion of the population.

REFERENCES

[1] Thurnham DI, Northrop-Clewes CA. Effects of infection on nutritional and immune status. In: Hughes DA, Darlington LG, Bendich A, editors. Diet and human immune function. Nutrition and health. Totowa, NJ: Humana Press; 2004. p. 35–66.

[2] Keusch GT. The history of nutrition: malnutrition, infection and immunity. J Nutr 2003;133(1):336S–40S.

[3] Katona P, Katona-Apte J. The interaction between nutrition and infection. Clin Infect Dis 2008;46(10):1582–8.

[4] Clausen SW. Nutrition and infection. JAMA 1935;104:793–8.

[5] Scrimshaw NS, Taylor CE, Gordon JE. Interactions of nutrition and infecton. Am J Med Sci 1959;237:367–403.

[6] Mata LJ, Kromal RA, Urrutia JJ, Garcia B. Effect of infection on food intake and the nutritional state: perspectives as viewed from the village. Am J Clin Nutr 1977;30:1215–27.

[7] Calder PC, Jackson AA. Undernutrition, infection and immune function. Nutr Res Rev 2002;13:3–29.

[8] World Health Organization. Fact sheet 311: obesity [cited 2017 July 31]; 2016. Available from: http://www.who.int/mediacentre/factsheets/fs311/en/.

[9] World Health Organization. Fact sheet: malnutrition [cited 2017 July 31]; 2017. Available from: http://www.who.int/mediacentre/factsheets/malnutrition/en/.

[10] Karlsson EA, Beck MA. The burden of obesity on infectious disease. Exp Biol Med (Maywood) 2010;235(12):1412–24.

[11] Milner JJ, Beck MA. The impact of obesity on the immune response to infection. Proc Nutr Soc 2012;71(2):298–306.

[12] Mancuso P. Obesity and respiratory infections: does excess adiposity weigh down host defense? Pulm Pharmacol Ther 2013;26(4):412–9.

[13] Falagas ME, Kompoti M. Obesity and infection. Lancet Infect Dis 2006;6(7):438–46.

[14] Center for Disease Control. Types of influenza viruses [updated September 15, 2016; cited 2017 May 10]; 2017. Available from: https://www.cdc.gov/flu/about/viruses/types.htm.

[15] Murphy B, Webster R. In: Fields B, editor. Orthomyxoviruses. 3rd ed. Philadelphia: Lippincott-Raven; 1997.

[16] World Health Organization. WHO factsheet 211: influenza (seasonal) [cited 2017 July 31]. Geneva: World Health Organization; 2009. Available from: http://www.who.int/mediacentre/factsheets/fs211/en/index.htm.

[17] Advisory Committee on Immunization Practices Prevention and Control of Influenza. Recommendations of the Advisory Committee on Immunization Practices (ACIP). MMWR Recomm Rep 2006;55(RR-10):1–42.

[18] Michaelis M, Doerr H, Cinatl J. An influenza A H1N1 virus revival—pandemic H1N1/09 virus. Infection 2009;37(5):381–9.

[19] Cohen J, Enserink M. After delays, WHO agrees: the 2009 pandemic has begun. Science 2009;324(5934):1496–7.

[20] Kamps BS, Hoffmann C, Preiser W, Behrens G. Influenza report 2006: ©2006 [cited 2017 May 10]. Paris, France: Flying Publisher; 2006. Available from: http://www.influenzareport.com.

[21] Webster RG, editor. Textbook of influenza [electronic resource]. Chichester, West Sussex, UK: Wiley Blackwell; 2013.

[22] Shi Y, Wu Y, Zhang W, Qi J, Gao GF. Enabling the 'host jump': structural determinants of receptor-binding specificity in influenza A viruses. Nat Rev Microbiol 2014;12(12):822–31.

[23] Ito T, Couceiro JN, Kelm S, Baum LG, Krauss S, Castrucci MR, et al. Molecular basis for the generation in pigs of influenza A viruses with pandemic potential. J Virol 1998;72(9):7367–73.

[24] Pielak RM, Schnell JR, Chou JJ. Mechanism of drug inhibition and drug resistance of influenza A M2 channel. Proc Natl Acad Sci U S A 2009;106(18):7379–84.

[25] Abbas AK. In: Lichtman AH, Pillai S, editors. Cellular and molecular immunology. Philadelphia: Elsevier/Saunders; 2012.

[26] Herold S, Becker C, Ridge KM, Budinger GR. Influenza virus-induced lung injury: pathogenesis and implications for treatment. Eur Respir J 2015;45(5):1463–78.

[27] Gregory DJ, Kobzik L. Influenza lung injury: mechanisms and therapeutic opportunities. Am J Physiol Lung Cell Mol Physiol 2015;309(10):L1041–6.

[28] Kim HM, Lee YW, Lee KJ, Kim HS, Cho SW, van Rooijen N, et al. Alveolar macrophages are indispensable for controlling influenza viruses in lungs of pigs. J Virol 2008;82(9):4265–74.

[29] Jayasekera JP, Vinuesa CG, Karupiah G, King NJ. Enhanced antiviral antibody secretion and attenuated immunopathology during influenza virus infection in nitric oxide synthase-2-deficient mice. J Gen Virol 2006;87:3361–71. Pt 11.

[30] Peper RL, Van Campen H. Tumor necrosis factor as a mediator of inflammation in influenza A viral pneumonia. Microb Pathog 1995;19(3):175–83.

[31] Writing Committee of the WHOCoCAoPI, Bautista E, Chotpitayasunondh T, Gao Z, Harper SA, Shaw M, et al. Clinical aspects of pandemic 2009 influenza A (H1N1) virus infection. N Engl J Med 2010;362(18):1708–19.

[32] Meliopoulos VA, Van de Velde LA, Van de Velde NC, Karlsson EA, Neale G, Vogel P, et al. An epithelial integrin regulates the amplitude of protective lung interferon responses against multiple respiratory pathogens. PLoS Pathog 2016;12(8):e1005804.

[33] Sanders CJ, Vogel P, McClaren JL, Bajracharya R, Doherty PC, Thomas PG. Compromised respiratory function in lethal influenza infection is characterized by the depletion of type I alveolar epithelial cells beyond threshold levels. Am J Physiol Lung Cell Mol Physiol 2013;304(7):L481–8.

[34] Morris DE, Cleary DW, Clarke SC. Secondary bacterial infections associated with influenza pandemics. Front Microbiol 2017;8:1041.

[35] Bellinghausen C, Rohde GGU, Savelkoul PHM, Wouters EFM, Stassen FRM. Viral-bacterial interactions in the respiratory tract. J Gen Virol 2016;97(12):3089–102.

[36] Sun J, Madan R, Karp CL, Braciale TJ. Effector T cells control lung inflammation during acute influenza virus infection by producing IL-10. Nat Med 2009;15(3):277–84.

[37] Sun K, Metzger DW. Inhibition of pulmonary antibacterial defense by interferon-gamma during recovery from influenza infection. Nat Med 2008;14(5):558–64.

[38] Moser EK, Hufford MM, Braciale TJ. Late engagement of CD86 after influenza virus clearance promotes recovery in a FoxP3+ regulatory T cell dependent manner. PLoS Pathog 2014;10(8):e1004315.

[39] Hogan BL, Barkauskas CE, Chapman HA, Epstein JA, Jain R, Hsia CC, et al. Repair and regeneration of the respiratory system: complexity, plasticity, and mechanisms of lung stem cell function. Cell Stem Cell 2014;15(2):123–38.

[40] Hannoun C. The evolving history of influenza viruses and influenza vaccines. Expert Rev Vaccines 2013;12(9):1085–94.

[41] LC L, AS F. Influenza vaccines for the future. N Engl J Med 2010;363(21):2036–44.

[42] Lowen AC, Mubareka S, Steel J, Palese P. Influenza virus transmission is dependent on relative humidity and temperature. PLoS Pathog 2007;3(10):e151.

[43] World Health Organization. Influenza (seasonal) [cited 2017 November 15]. Available from: http://www.who.int/mediacentre/factsheets/fs211/en/index.html; 2014.

[44] Oldstone MBA. Influenza pathogenesis and control—volume II, 386.

[45] Whitley Richard J, Monto Arnold S. Prevention and treatment of influenza in high-risk groups: children, pregnant women, immunocompromised hosts, and nursing home residents. J Infect Dis 2006;194(Supplement_2):S133–8.

[46] Chiu C, Openshaw PJ. Antiviral B cell and T cell immunity in the lungs. Nat Immunol 2015;16(1):18–26.

[47] Reber A, Katz J. Immunological assessment of influenza vaccines and immune correlates of protection. Expert Rev Vaccines 2013;12(5):519–36.

[48] Trombetta C, Piccirella S, Perini D, Kistner O, Montomoli E. Emerging influenza strains in the last two decades: a threat of a new pandemic? Vaccine 2015;3(1):172–85.

[49] Zhang H, Wang L, Compans RW, Wang BZ. Universal influenza vaccines, a dream to be realized soon. Viruses 2014;6(5):1974–91.

[50] Nachbagauer R, Krammer F. Universal influenza virus vaccines and therapeutic antibodies. Clin Microbiol Infect 2017;23(4):222–8.

[51] Koszalka P, Tilmanis D, Hurt AC. Influenza antivirals currently in late-phase clinical trial. Influenza Other Respi Viruses 2017;11(3):240–6.

[52] Dong G, Peng C, Luo J, Wang C, Han L, Wu B, et al. Adamantane-resistant influenza A viruses in the world (1902–2013): frequency and distribution of M2 gene mutations. PloS One 2015;10(3):e0119115.

[53] Spanakis N, Pitiriga V, Gennimata V, Tsakris A. A review of neuraminidase inhibitor susceptibility in influenza strains. Expert Rev Anti Infect Ther 2014;12(11):1325–36.

[54] Center for Disease Control. Adult obesity facts [updated September 1, 2016; cited 2017 May 3]. Division of Nutrition, Physical Activity, and Obesity, National Center for Chronic Disease Prevention and Health Promotion; 2017. Available from: https://www.cdc.gov/obesity/data/adult.html.

[55] Huttunen R, Syrjanen J. Obesity and the risk and outcome of infection. Int J Obes (Lond) 2013;37(3):333–40.

[56] Centers for Disease Control and Prevention. Influenza (Flu) [cited 2017 Aug 30th]. Available from: https://www.cdc.gov/flu/about/disease/high_risk.htm; 2016.

[57] York I, Donis RO. The 2009 pandemic influenza virus: where did it come from, where is it now, and where is it going? In: Richt JA, Webby RJ, editors. Swine influenza. Berlin, Heidelberg: Springer Berlin Heidelberg; 2013. p. 241–57.

[58] Jain S, Kamimoto L, Bramley AM, Schmitz AM, Benoit SR, Louie J, et al. Hospitalized patients with 2009 H1N1 influenza in the United States, April–June 2009. N Engl J Med 2009;361(20):1935–44.

[59] Louie JK, Acosta M, Winter K, Jean C, Gavali S, Schechter R, et al. Factors associated with death or hospitalization due to pandemic 2009 influenza A (H1N1) infection in California. JAMA 2009;302(17):1896–902.

[60] Kumar A, Zarychanski R, Pinto R, Cook DJ, Marshall J, Lacroix J, et al. Critically ill patients with 2009 influenza A (H1N1) infection in Canada. JAMA 2009;302(17):1872–9.

[61] Webb SA, Pettila V, Seppelt I, Bellomo R, Bailey M, Cooper DJ, et al. Critical care services and 2009 H1N1 influenza in Australia and New Zealand. N Engl J Med 2009;361(20):1925–34.

[62] Louie JK, Acosta M, Samuel MC, Schechter R, Vugia DJ, Harriman K, et al. A novel risk factor for a novel virus: obesity and 2009 pandemic influenza A (H1N1). Clin Infect Dis 2011;52(3):301–12.

[63] Santa-Olalla Peralta P, Cortes-Garcia M, Vicente-Herrero M, Castrillo-Villamandos C, Arias-Bohigas P, Panchon-del Amo I, et al. Risk factors for disease severity among hospitalised patients with 2009 pandemic influenza A (H1N1) in Spain, April–December 2009. Euro Surveill 2010;15(38).

[64] Morgan OW, Bramley A, Fowlkes A, Freedman DS, Taylor TH, Gargiullo P, et al. Morbid obesity as a risk factor for hospitalization and death due to 2009 pandemic influenza A (H1N1) disease. PLoS One 2010;5(3):e9694.

[65] Fezeu L, Julia C, Henegar A, Bitu J, Hu FB, Grobbee DE, et al. Obesity is associated with higher risk of intensive care unit admission and death in influenza A (H1N1) patients: a systematic review and meta-analysis. Obes Rev 2011;12(8):653–9.

[66] Smith AG, Sheridan PA, Harp JB, Beck MA. Diet-induced obese mice have increased mortality and altered immune responses when infected with influenza virus. J Nutr 2007;137(5):1236–43.

[67] Kwong JC, Campitelli MA, Rosella LC. Obesity and respiratory hospitalizations during influenza seasons in Ontario, Canada: a cohort study. Clin Infect Dis 2011;53(5):413–21.

[68] Diaz E, Rodriguez A, Martin-Loeches I, Lorente L, Del Mar Martin M, Pozo JC, et al. Impact of obesity in patients infected with 2009 influenza A (H1N1). Chest 2011;139(2):382–6.

[69] Coleman LA, Waring SC, Irving SA, Vandermause M, Shay DK, Belongia EA. Evaluation of obesity as an independent risk factor for medically attended laboratory-confirmed influenza. Influenza Other Respir Viruses 2013;7(2):160–7.

[70] Phung DT, Wang Z, Rutherford S, Huang C, Chu C. Body mass index and risk of pneumonia: a systematic review and meta-analysis. Obes Rev 2013;14(10):839–57.

[71] Painter SD, Ovsyannikova IG, Poland GA. The weight of obesity on the human immune response to vaccination. Vaccine 2015;33(36):4422–9.

[72] Sheridan PA, Paich HA, Handy J, Karlsson EA, Hudgens MG, Sammon AB, et al. Obesity is associated with impaired immune response to influenza vaccination in humans. Int J Obes (Lond) 2012;36(8):1072–7.

[73] Paich HA, Sheridan PA, Handy J, Karlsson EA, Schultz-Cherry S, Hudgens MG, et al. Overweight and obese adult humans have a defective cellular immune response to pandemic H1N1 influenza A virus. Obesity (Silver Spring) 2013;21(11):2377–86.

[74] Talbot HK, Coleman LA, Crimin K, Zhu Y, Rock MT, Meece J, et al. Association between obesity and vulnerability and serologic response to influenza vaccination in older adults. Vaccine 2012;30(26):3937–43.

[75] Neidich SD, Green WD, Rebeles J, Karlsson EA, Schultz-Cherry S, Noah TL, et al. Increased risk of influenza among vaccinated adults who are obese. Int J Obes (Lond) 2017;.

[76] Krishnaswamy K. Drug metabolism and pharmacokinetics in malnourished children. Clin Pharmacokinet 1989;17(Suppl 1):68–88.

[77] Bartelink IH, Savic RM, Dorsey G, Ruel T, Gingrich D, Scherpbier HJ, et al. The effect of malnutrition on the pharmacokinetics and virologic outcomes of Lopinavir, Efavirenz and Nevirapine in food insecure HIV-infected children in Tororo, Uganda. Pediatr Infect Dis J 2015;34(3):e63–70.

[78] Sampson MR, Cohen-Wolkowiez M, Benjamin DK, Capparelli EV, Watt KM. Pharmacokinetics of antimicrobials in obese children. GaBi J 2013;2(2):76–81.

[79] World Health Organization. WHO guidelines for pharmacological management of pandemic (H1N1) 2009 influenza and other influenza viruses [cited 2017 Aug 30th]. Available from: http://www.who.int/csr/resources/publications/swineflu/h1n1_use_antivirals_20090820/en/; 2010.

[80] Jittamala P, Pukrittayakamee S, Tarning J, Lindegardh N, Hanpithakpong W, Taylor WRJ, et al. Pharmacokinetics of orally administered oseltamivir in healthy obese and nonobese Thai subjects. Antimicrob Agents Chemother 2014;58(3):1615–21.

[81] Pai MP, Lodise TP. Oseltamivir and oseltamivir carboxylate pharmacokinetics in obese adults: dose modification for weight is not necessary. Antimicrob Agents Chemother 2011;55(12):5640–5.

[82] Thorne-Humphrey LM, Goralski KB, Slayter KL, Hatchette TF, Johnston BL, McNeil SA. Oseltamivir pharmacokinetics in morbid obesity (OPTIMO trial). J Antimicrob Chemother 2011;66(9):2083–91.

[83] Castañeda-Hernández G, Martínez-Talavera A, Barranco-Garduño LM, Cervantes-Nevárez A, León-Molina H, Carrasco-Portugal MC, et al. Oseltamivir pharmacokinetics in Mexican obese and non-obese healthy subjects and patients. Evidence for an absence of interethnic variability. Br J Clin Pharmacol 2016;82(3):890–1.

[84] Chairat K, Jittamala P, Hanpithakpong W, Day NPJ, White NJ, Pukrittayakamee S, et al. Population pharmacokinetics of oseltamivir and oseltamivir carboxylate in obese and non-obese volunteers. Br J Clin Pharmacol 2016;81(6):1103–12.

[85] Smith A. Diet-induced obese mice have increased mortality and altered immune responses when infected with influenza virus. J Nutr Nutr Immunol 2007;.

[86] Smith AG, Sheridan PA, Tseng RJ, Sheridan JF, Beck MA. Selective impairment in dendritic cell function and altered antigen-specific CD8+ T-cell responses in diet-induced obese mice infected with influenza virus. Immunology 2009;126(2):268–79.

[87] O'Brien KB, Vogel P, Duan S, Govorkova EA, Webby RJ, McCullers JA, et al. Impaired wound healing predisposes obese mice to severe influenza virus infection. J Infect Dis 2012;205(2):252–61.

[88] Easterbrook JD, Dunfee RL, Schwartzman LM, Jagger BW, Sandouk A, Kash JC, et al. Obese mice have increased morbidity and mortality compared to non-obese mice during infection with the 2009 pandemic H1N1 influenza virus. Influenza Other Respi Viruses 2011;5(6):418–25.

[89] Karlsson EA, Meliopoulos V, NCvd V, L-Avd V, Mann B, Gao G, et al. A perfect storm: increased colonization and failure of vaccination leads to severe secondary bacterial infection in influenza infected obese mice. mBio 2017;8(5). e00889-17.

[90] Park H-L, Shim S-H, Lee E-Y, Cho W, Park S, Jeon H-J, et al. Obesity-induced chronic inflammation is associated with the reduced efficacy of influenza vaccine. Hum Vaccin Immunother 2014;10(5):1181–6.

[91] Kim Y-H, Kim J-K, Kim D-J, Nam J-H, Shim S-M, Choi Y-K, et al. Diet-induced obesity dramatically reduces the efficacy of a 2009 pandemic H1N1 vaccine in a mouse model. J Infect Dis 2012;205(2):244–51.

[92] Karlsson EA, Hertz T, Johnson C, Mehle A, Krammer F, Schultz-Cherry S. Obesity outweighs protection conferred by adjuvanted influenza vaccination. mBio 2016;7(4).

[93] Sridhar S, Begom S, Bermingham A, Hoschler K, Adamson W, Carman W, et al. Cellular immune correlates of protection against symptomatic pandemic influenza. Nat Med 2013;19(10):1305–12.

[94] McMichael AJ, Gotch FM, Noble GR, Beare PA. Cytotoxic T-cell immunity to influenza. N Engl J Med 1983;309(1):13–7.

[95] Wang Z, Wan Y, Qiu C, Quinones-Parra S, Zhu Z, Loh L, et al. Recovery from severe H7N9 disease is associated with diverse response mechanisms dominated by CD8(+) T cells. Nat Commun 2015;6:6833.

[96] Liang S, Mozdzanowska K, Palladino G, Gerhard W. Heterosubtypic immunity to influenza type A virus in mice. Effector mechanisms and their longevity. J Immunol 1994;152(4):1653–61.

[97] Hillaire ML, Vogelzang-van Trierum SE, Kreijtz JH, de Mutsert G, Fouchier RA, Osterhaus AD, et al. Human T-cells directed to seasonal influenza A virus cross-react with 2009 pandemic influenza A (H1N1) and swine-origin triple-reassortant H3N2 influenza viruses. J Gen Virol 2013;94(Pt 3):583–92.

[98] Karlsson EA, Sheridan PA, Beck MA. Diet-induced obesity in mice reduces the maintenance of influenza-specific CD8+ memory T cells. J Nutr 2010;140(9):1691–7.

[99] Karlsson EA, Sheridan PA, Beck MA. Diet-induced obesity impairs the T cell memory response to influenza virus infection. J Immunol 2010;184(6):3127–33.

[100] Milner JJ, Sheridan PA, Karlsson EA, Schultz-Cherry S, Shi Q, Beck MA. Diet-induced obese mice exhibit altered heterologous immunity during a secondary 2009 pandemic H1N1 infection. J Immunol 2013;191(5):2474–85.

[101] Ring LE, Zeltser LM. Disruption of hypothalamic leptin signaling in mice leads to early-onset obesity, but physiological adaptations in mature animals stabilize adiposity levels. J Clin Invest 2010;120(8):2931–41.

[102] Milner JJ, Rebeles J, Dhungana S, Stewart DA, Sumner SC, Meyers MH, et al. Obesity increases mortality and modulates the lung metabolome during pandemic H1N1 influenza virus infection in mice. J Immunol 2015;194(10):4846–59.

[103] Van Kerkhove MD, Vandemaele KA, Shinde V, Jaramillo-Gutierrez G, Koukounari A, Donnelly CA, et al. Risk factors for severe outcomes following 2009 influenza A (H1N1) infection: a global pooled analysis. PLoS Med 2011;8(7):e1001053.

[104] Michalek RD, Gerriets VA, Jacobs SR, Macintyre AN, MacIver NJ, Mason EF, et al. Cutting edge: distinct glycolytic and lipid oxidative metabolic programs are essential for effector and regulatory CD4+ T cell subsets. J Immunol 2011;186(6):3299–303.

[105] Pearce EL, Pearce EJ. Metabolic pathways in immune cell activation and quiescence. Immunity 2013;38(4):633–43.

[106] Finlay D, Cantrell DA. Metabolism, migration and memory in cytotoxic T cells. Nat Rev Immunol 2011;11(2):109–17.

[107] Gerriets VA, Rathmell JC. Metabolic pathways in T cell fate and function. Trends Immunol 2012;33(4):168–73.

[108] Johnson AR, Milner JJ, Makowski L. The inflammation highway: metabolism accelerates inflammatory traffic in obesity. Immunol Rev 2012;249(1):218–38.

[109] Ghosh S, Gregory D, Smith A, Kobzik L. MARCO regulates early inflammatory responses against influenza: a useful macrophage function with adverse outcome. Am J Respir Cell Mol Biol 2011;45(5):1036–44.

[110] Sun J, Braciale TJ. Role of T cell immunity in recovery from influenza virus infection. Curr Opin Virol 2013;3(4):425–9.

[111] Sukumar M, Liu J, Ji Y, Subramanian M, Crompton JG, Yu Z, et al. Inhibiting glycolytic metabolism enhances CD8+ T cell memory and antitumor function. J Clin Invest 2013;123(10):4479–88.

[112] Vander Heiden MG, Cantley LC, Thompson CB. Understanding the Warburg effect: the metabolic requirements of cell proliferation. Science 2009;324(5930):1029–33.

[113] Vats D, Mukundan L, Odegaard JI, Zhang L, Smith KL, Morel CR, et al. Oxidative metabolism and PGC-1beta attenuate macrophage-mediated inflammation. Cell Metab 2006;4(1):13–24.

[114] Odegaard JI, Chawla A. Alternative macrophage activation and metabolism. Annu Rev Pathol 2011;6:275–97.

[115] World Health Organization. Diabetes Programme Geneva: World Health Organization [cited 2017 October 22]. Available from: http://www.who.int/diabetes/en/; 2013.

[116] Centers for Disease Control. Flu and People with Diabetes Atlanta: Centers for Disease Control [cited 2017 November 6]. Available from: http://www.cdc.gov/flu/diabetes/; 2013.

[117] Wiwanitkit V. Hyperglycemia in the recent reported cases of bird flu infection in Thailand and Vietnam. J Diabetes Complicat 2008;22(1):76.

[118] Stegenga ME, van der Crabben SN, Blumer RM, Levi M, Meijers JC, Serlie MJ, et al. Hyperglycemia enhances coagulation and reduces neutrophil degranulation, whereas hyperinsulinemia inhibits fibrinolysis during human endotoxemia. Blood 2008;112(1):82–9.

[119] Black CT, Hennessey PJ, Andrassy RJ. Short-term hyperglycemia depresses immunity through nonenzymatic glycosylation of circulating immunoglobulin. J Trauma 1990;30(7):830–2. [discussion 2-3].

[120] Kwoun MO, Ling PR, Lydon E, Imrich A, Qu Z, Palombo J, et al. Immunologic effects of acute hyperglycemia in nondiabetic rats. JPEN J Parenter Enteral Nutr 1997;21(2):91–5.

[121] Bagdade JD, Root RK, Bulger RJ. Impaired leukocyte function in patients with poorly controlled diabetes. Diabetes 1974;23(1):9–15.

[122] Jakelic J, Kokic S, Hozo I, Maras J, Fabijanic D. Nonspecific immunity in diabetes: hyperglycemia decreases phagocytic activity of leukocytes in diabetic patients. Med Arh 1995;49(1–2):9–12.

[123] Sheetz MJ, King GL. Molecular understanding of hyperglycemia's adverse effects for diabetic complications. JAMA 2002;288(20):2579–88.

[124] Maciver NJ, Jacobs SR, Wieman HL, Wofford JA, Coloff JL, Rathmell JC. Glucose metabolism in lymphocytes is a regulated process with significant effects on immune cell function and survival. J Leukoc Biol 2008;84(4):949–57.

[125] Jacobs SR, Herman CE, Maciver NJ, Wofford JA, Wieman HL, Hammen JJ, et al. Glucose uptake is limiting in T cell activation and requires CD28-mediated Akt-dependent and independent pathways. J Immunol 2008;180(7):4476–86.

[126] Chang CH, Curtis JD, Maggi Jr. LB, Faubert B, Villarino AV, O'Sullivan D, et al. Posttranscriptional control of T cell effector function by aerobic glycolysis. Cell 2013;153(6):1239–51.

[127] Macintyre AN, Gerriets VA, Nichols AG, Michalek RD, Rudolph MC, Deoliveira D, et al. The glucose transporter Glut1 is selectively essential for CD4 T cell activation and effector function. Cell Metab 2014;20(1):61–72.

[128] Stentz FB, Kitabchi AE. Hyperglycemia-induced activation of human T-lymphocytes with de novo emergence of insulin receptors and generation of reactive oxygen species. Biochem Biophys Res Commun 2005;335(2):491–5.

[129] Reading PC, Allison J, Crouch EC, Anders EM. Increased susceptibility of diabetic mice to influenza virus infection: compromise of collectin-mediated host defense of the lung by glucose? J Virol 1998;72(8):6884–7.

[130] Frasca D, Diaz A, Romero M, Mendez NV, Landin AM, Ryan JG, et al. Young and elderly patients with type 2 diabetes have optimal B cell responses to the seasonal influenza vaccine. Vaccine 2013;31(35):3603–10.

[131] Moseley CE, Webster RG, Aldridge JR. Peroxisome proliferator-activated receptor and AMP-activated protein kinase agonists protect against lethal influenza virus challenge in mice. Influenza Other Respi Viruses 2010;4(5):307–11.

[132] Morita M, Kuba K, Ichikawa A, Nakayama M, Katahira J, Iwamoto R, et al. The lipid mediator protectin D1 inhibits influenza virus replication and improves severe influenza. Cell 2013;153(1):112–25.

[133] Tam VC, Quehenberger O, Oshansky CM, Suen R, Armando AM, Treuting PM, et al. Lipidomic profiling of influenza infection identifies mediators that induce and resolve inflammation. Cell 2013;154(1):213–27.

[134] Huang S, Rutkowsky JM, Snodgrass RG, Ono-Moore KD, Schneider DA, Newman JW, et al. Saturated fatty acids activate TLR-mediated proinflammatory signaling pathways. J Lipid Res 2012;53(9):2002–13.

[135] O'Brien KB, Vogel P, Duan S, Govorkova EA, Webby RJ, McCullers JA, et al. Impaired wound healing predisposes obese mice to severe influenza virus infection. J Infect Dis 2012;205(2):252–61.

[136] Carey MA, Bradbury JA, Rebolloso YD, Graves JP, Zeldin DC, Germolec DR. Pharmacologic inhibition of COX-1 and COX-2 in influenza A viral infection in mice. PloS One 2010;5(7):e11610.

[137] Carey MA, Bradbury JA, Seubert JM, Langenbach R, Zeldin DC, Germolec DR. Contrasting effects of cyclooxygenase-1 (COX-1) and COX-2 deficiency on the host response to influenza A viral infection. J Immunol 2005;175(10):6878–84.

[138] Calder PC. N-3 polyunsaturated fatty acids, inflammation, and inflammatory diseases. Am J Clin Nutr 2006;83(6):1505S–19S. Suppl.

[139] Calder PC. Polyunsaturated fatty acids and inflammation. Prostaglandins Leukot Essent Fatty Acids 2006;75(3):197–202.

[140] Norris PC, Dennis EA. Omega-3 fatty acids cause dramatic changes in TLR4 and purinergic eicosanoid signaling. Proc Natl Acad Sci U S A 2012;109(22):8517–22.

[141] Simopoulos AP. Evolutionary aspects of omega-3 fatty acids in the food supply. Prostaglandins Leukot Essent Fatty Acids 1999;60(5–6):421–9.

[142] Kris-Etherton PM, Taylor DS, Yu-Poth S, Huth P, Moriarty K, Fishell V, et al. Polyunsaturated fatty acids in the food chain in the United States. Am J Clin Nutr 2000;71(1 Suppl):179S–88S.

[143] Oehlenschlager J. Seafood: nutritional benefits and risk aspects. Int J Vitam Nutr Res 2012;82(3):168–76.

[144] Simopoulos AP. An increase in the Omega-6/Omega-3 fatty acid ratio increases the risk for obesity. Nutrients 2016;8(3):128.

[145] Byleveld PM, Pang GT, Clancy RL, Roberts DC. Fish oil feeding delays influenza virus clearance and impairs production of interferon-gamma and virus-specific immunoglobulin A in the lungs of mice. J Nutr 1999;129(2):328–35.

[146] Byleveld M, Pang GT, Clancy RL, Roberts DC. Fish oil feeding enhances lymphocyte proliferation but impairs virus-specific T lymphocyte cytotoxicity in mice following challenge with influenza virus. Clin Exp Immunol 2000;119(2):287–92.

[147] Schwerbrock NM, Karlsson EA, Shi Q, Sheridan PA, Beck MA. Fish oil-fed mice have impaired resistance to influenza infection. J Nutr 2009;139(8):1588–94.

[148] Buck MD, O'Sullivan D, Pearce EL. T cell metabolism drives immunity. J Exp Med 2015;212(9):1345–60.

[149] MacIver NJ, Michalek RD, Rathmell JC. Metabolic regulation of T lymphocytes. Annu Rev Immunol 2013;31:259–83.

[150] Pearce EL, Walsh MC, Cejas PJ, Harms GM, Shen H, Wang LS, et al. Enhancing CD8 T-cell memory by modulating fatty acid metabolism. Nature 2009;460(7251):103–7.

[151] Lee J, Walsh MC, Hoehn KL, James DE, Wherry EJ, Choi Y. Regulator of fatty acid metabolism, acetyl coenzyme a carboxylase 1, controls T cell immunity. J Immunol 2014;192(7):3190–9.

[152] O'Sullivan D, van der Windt GJ, Huang SC, Curtis JD, Chang CH, Buck MD, et al. Memory CD8(+) T cells use cell-intrinsic lipolysis to support the metabolic programming necessary for development. Immunity 2014;41(1):75–88.

[153] Pan Y, Tian T, Park CO, Lofftus SY, Mei S, Liu X, et al. Survival of tissue-resident memory T cells requires exogenous lipid uptake and metabolism. Nature 2017;543(7644):252–6.

[154] Mauro C, Smith J, Cucchi D, Coe D, Fu H, Bonacina F, et al. Obesity-induced metabolic stress leads to biased effector memory CD4+ T cell differentiation via PI3K p110delta-Akt-mediated signals. Cell Metab 2017;25(3):593–609.

[155] Kuklina EV, Carroll MD, Shaw KM, Hirsch R. Trends in high LDL cholesterol, cholesterol-lowering medication use, and dietary saturated-fat intake: United States, 1976–2010. In: Statistics NCfH, editor. Hyattsville, MD2013.

[156] Sun X, Whittaker GR. Role for influenza virus envelope cholesterol in virus entry and infection. J Virol 2003;77(23):12543–51.

[157] Fernandez ML, Calle M. Revisiting dietary cholesterol recommendations: does the evidence support a limit of 300 mg/d? Curr Atheroscler Rep 2010;12(6):377–83.

[158] Lecerf JM, de Lorgeril M. Dietary cholesterol: from physiology to cardiovascular risk. Br J Nutr 2011;106(1):6–14.

[159] Fedson DS. Pandemic influenza: a potential role for statins in treatment and prophylaxis. Clin Infect Dis 2006;43(2):199–205.

[160] Fedson DS. Treating influenza with statins and other immunomodulatory agents. Antiviral Res 2013;99(3):417–35.

[161] Gardner EM, Beli E, Clinthorne JF, Duriancik DM. Energy intake and response to infection with influenza. Annu Rev Nutr 2011;31:353–67.

[162] Chang L, Chiang SH, Saltiel AR. Insulin signaling and the regulation of glucose transport. Mol Med 2004;10(7–12):65–71.

[163] Maratou E, Dimitriadis G, Kollias A, Boutati E, Lambadiari V, Mitrou P, et al. Glucose transporter expression on the plasma membrane of resting and activated white blood cells. Eur J Clin Invest 2007;37(4):282–90.

[164] Fischer HJ, Sie C, Schumann E, Witte AK, Dressel R, van den Brandt J, et al. The insulin receptor plays a critical role in T cell function and adaptive immunity. J Immunol 2017;198(5):1910–20.

[165] Hukshorn CJ, Lindeman JH, Toet KH, Saris WH, Eilers PH, Westerterp-Plantenga MS, et al. Leptin and the proinflammatory state associated with human obesity. J Clin Endocrinol Metab 2004;89(4):1773–8.

[166] Trayhurn P, Beattie JH. Physiological role of adipose tissue: white adipose tissue as an endocrine and secretory organ. Proc Nutr Soc 2001;60(3):329–39.

[167] Saucillo DC, Gerriets VA, Sheng J, Rathmell JC, Maciver NJ. Leptin metabolically licenses T cells for activation to link nutrition and immunity. J Immunol 2014;192(1):136–44.

[168] Reis BS, Lee K, Fanok MH, Mascaraque C, Amoury M, Cohn LB, et al. Leptin receptor signaling in T cells is required for Th17 differentiation. J Immunol 2015;194(11):5253–60.

[169] Zhang AJ, To KK, Li C, Lau CC, Poon VK, Chan CC, et al. Leptin mediates the pathogenesis of severe 2009 pandemic influenza A (H1N1) infection associated with cytokine dysregulation in mice with diet-induced obesity. J Infect Dis 2013;207(8):1270–80.

[170] Procaccini C, De Rosa V, Galgani M, Carbone F, Cassano S, Greco D, et al. Leptin-induced mTOR activation defines a specific molecular and transcriptional signature controlling CD4+ effector T cell responses. J Immunol 2012;189(6):2941–53.

[171] Han JM, Patterson SJ, Speck M, Ehses JA, Levings MK. Insulin inhibits IL-10-mediated regulatory T cell function: implications for obesity. J Immunol 2014;192(2):623–9.

[172] Wilk S, Scheibenbogen C, Bauer S, Jenke A, Rother M, Guerreiro M, et al. Adiponectin is a negative regulator of antigen-activated T cells. Eur J Immunol 2011;41(8):2323–32.

[173] Tsatsanis C, Margioris AN, Kontoyiannis DP. Association between H1N1 infection severity and obesity-adiponectin as a potential etiologic factor. J Infect Dis 2010;202(3):459–60.

[174] Minihane AM, Vinoy S, Russell WR, Baka A, Roche HM, Tuohy KM, et al. Low-grade inflammation, diet composition and health: current research evidence and its translation. Br J Nutr 2015;114(7):999–1012.

[175] Xiao L, Yang X, Lin Y, Li S, Jiang J, Qian S, et al. Large adipocytes function as antigen-presenting cells to activate CD4(+) T cells via upregulating MHCII in obesity. Int J Obes (Lond) 2016;40(1):112–20.

[176] Deng T, Lyon CJ, Minze LJ, Lin J, Zou J, Liu JZ, et al. Class II major histocompatibility complex plays an essential role in obesity-induced adipose inflammation. Cell Metab 2013;17(3):411–22.

[177] Round JL, Mazmanian SK. Inducible Foxp3+ regulatory T-cell development by a commensal bacterium of the intestinal microbiota. Proc Natl Acad Sci U S A 2010;107(27):12204–9.

[178] Gaboriau-Routhiau V, Rakotobe S, Lecuyer E, Mulder I, Lan A, Bridonneau C, et al. The key role of segmented filamentous bacteria in the coordinated maturation of gut helper T cell responses. Immunity 2009;31(4):677–89.

[179] Niess JH, Leithauser F, Adler G, Reimann J. Commensal gut flora drives the expansion of proinflammatory CD4 T cells in the colonic lamina propria under normal and inflammatory conditions. J Immunol 2008;180(1):559–68.

[180] Ichinohe T, Pang IK, Kumamoto Y, Peaper DR, Ho JH, Murray TS, et al. Microbiota regulates immune defense against respiratory tract influenza A virus infection. Proc Natl Acad Sci U S A 2011;108(13):5354–9.

[181] Yu B, Dai CQ, Chen J, Deng L, Wu XL, Wu S, et al. Dysbiosis of gut microbiota induced the disorder of helper T cells in influenza virus-infected mice. Hum Vaccin Immunother 2015;11(5):1140–6.

[182] Zelaya H, Tsukida K, Chiba E, Marranzino G, Alvarez S, Kitazawa H, et al. Immunobiotic lactobacilli reduce viral-associated pulmonary damage through the modulation of inflammation-coagulation interactions. Int Immunopharmacol 2014;19(1):161–73.

[183] Park MK, Ngo V, Kwon YM, Lee YT, Yoo S, Cho YH, et al. Lactobacillus plantarum DK119 as a probiotic confers protection against influenza virus by modulating innate immunity. PloS One 2013;8(10):e75368.

[184] Goto H, Sagitani A, Ashida N, Kato S, Hirota T, Shinoda T, et al. Anti-influenza virus effects of both live and non-live lactobacillus acidophilus L-92 accompanied by the activation of innate immunity. Br J Nutr 2013;110(10):1810–8.

[185] Leyer GJ, Li S, Mubasher ME, Reifer C, Ouwehand AC. Probiotic effects on cold and influenza-like symptom incidence and duration in children. Pediatrics 2009;124(2):e172–9.

[186] Waki N, Matsumoto M, Fukui Y, Suganuma H. Effects of probiotic lactobacillus brevis KB290 on incidence of influenza infection among schoolchildren: an open-label pilot study. Lett Appl Microbiol 2014;59(6):565–71.

[187] Nakayama Y, Moriya T, Sakai F, Ikeda N, Shiozaki T, Hosoya T, et al. Oral administration of *Lactobacillus gasseri* SBT2055 is effective for preventing influenza in mice. Sci Rep 2014;4:4638.

[188] Vouloumanou EK, Makris GC, Karageorgopoulos DE, Falagas ME. Probiotics for the prevention of respiratory tract infections: a systematic review. Int J Antimicrob Agents 2009;34(3):197 e1–10.

[189] Ley RE, Backhed F, Turnbaugh P, Lozupone CA, Knight RD, Gordon JI. Obesity alters gut microbial ecology. Proc Natl Acad Sci U S A 2005;102(31):11070–5.

[190] Kallus SJ, Brandt LJ. The intestinal microbiota and obesity. J Clin Gastroenterol 2012;46(1):16–24.

[191] Chaban B, Albert A, Links MG, Gardy J, Tang P, Hill JE. Characterization of the upper respiratory tract microbiomes of patients with pandemic H1N1 influenza. PloS One 2013;8(7):e69559.

[192] Yi H, Yong D, Lee K, Cho YJ, Chun J. Profiling bacterial community in upper respiratory tracts. BMC Infect Dis 2014;14:583.

[193] Oh JZ, Ravindran R, Chassaing B, Carvalho FA, Maddur MS, Bower M, et al. TLR5-mediated sensing of gut microbiota is necessary for antibody responses to seasonal influenza vaccination. Immunity 2014;41(3):478–92.

[194] Finkelstein EA, Khavjou OA, Thompson H, Trogdon JG, Pan L, Sherry B, et al. Obesity and severe obesity forecasts through 2030. Am J Prev Med 2012;42(6):563–70.

[195] Breda J, Jewell J, Webber L, Galea G. In: WHO projections in adults to 2030. 22nd Eur Congress Obes (ECO2015); Vol. 6 to 9 May; Prague, Czech Republic; 2015.

GLOSSARY

Adjuvant   any pharmacological or immunological agent added to a vaccine or other preparation to boost response

Antigenic drift   gradual accumulation of mutations in influenza antigenic sites over time

Antigenic shift   sudden change in influenza virus resulting from a reassortment or major mutation event

Antiviral resistance   viral mutation or other factor that confers ability to influenza to resist antiviral drugs

CD4+ T cell   also known as helper T cells; coordinate immune responses, especially involved with the adaptive response

CD8+ T cell   also known as cytotoxic T cells; seek out and destroy infected cells

Cytokine   factors released by cells of many different bodily systems to act as signals and to coordinate functions

Hemagglutinantion inhibition   standard assay in influenza vaccine response testing that measures the ability of antibodies in serum to neutralize hemagglutination ability of an influenza virus

Immune correlate of protection   measurable quality that indicates that a specific host is protected against infection or developing a disease

Malnutrition   lack of proper nutrition resulting from inadequate or overadequate food consumption

Memory T cell   T cells that develop and persist after having already encountered a particular antigen from a pathogen and can respond rapidly to a subsequent encounter with the same pathogen

Pandemic   outbreak of a virus in several communities throughout the world

Primary immune response   immune response that occurs during the first time the host has experienced a specific pathogen

Prophylactic strategy   any strategy (i.e., vaccination) used to prevent infection or spread of disease

Regulatory T cell   also known as suppressor T cells; help to maintain immune function and regulate immune processes, mainly by immunosuppression

Seasonal human influenza virus   influenza virus groups that circulate seasonally within the human population

Secondary immune response   immune response that occurs upon reinfection with a pathogen previously experienced by the host

Secondary microbial infection   any infection (viral, bacterial, fungal) that occurs after, and possibly as a result of, a primary influenza infection

Seroprotective   protection obtained as a result of inflection or vaccination; usually measured by antibody titers in the serum

Therapeutic strategy   any strategy (i.e., antiviral medication) used to treat or limit infection once it has occurred in the host

Vaccine effectiveness   the measure of the ability of a vaccine to protect against infection

Zoonotic transmission   transmission of a pathogen from animals to humans

Obesity and the acute respiratory distress syndrome

11

Renee D. Stapleton, Benjamin T. Suratt

Department of Medicine, Division of Pulmonary and Critical Care Medicine, University of Vermont, Burlington, VT, United States

ABBREVIATIONS

APACHE	acute physiology and chronic health evaluation
ARDS	acute respiratory distress syndrome
BMI	body mass index
ICU	intensive care unit
IL	interleukin
KCLIP	King County Lung Injury Project
LOS	length of stay
NIS	Nationwide Inpatient Sample
OR	odds ratio
PaO$_2$/FiO$_2$	partial pressure of arterial oxygen to the fraction of inspired oxygen ratio
PEEP	positive end expiratory pressure
RCT	randomized control trial
RR	relative risk
SAPS	simplified acute physiology score
TNF	tumor necrosis factor
USCIITG-LIPS	U.S. Critical Illness and Injury Trials Group-Lung Injury Prevention Study
VILI	ventilator-induced lung injury

INTRODUCTION

Acute respiratory distress syndrome (ARDS) is a clinical syndrome, with an updated definition in 2012 that includes new acute hypoxemic respiratory failure, bilateral pulmonary infiltrates not fully explained by heart failure, and severity from mild to severe based on the ratio of partial pressure of arterial oxygen to the fraction of inspired oxygen (PaO$_2$/FiO$_2$) on at least a positive-end expiratory pressure (PEEP) of 5 cm H$_2$O [1]. ARDS is common, with 200,000 cases annually in the United States and a case fatality of 30%–40% [2–4]. For the past few decades, the prevalence of

Mechanisms and Manifestations of Obesity in Lung Disease. https://doi.org/10.1016/B978-0-12-813553-2.00011-7

obesity has consistently increased, especially in developed countries, with important public health consequences including increased cardiac disease, diabetes, and all-cause mortality [5–7]. However, in critically ill patients with ARDS, nearly 20 years of evidence has found that obese patients have lower mortality than their normal weight counterparts. Although some literature suggests that obese critically ill patients may have increased morbidity such as longer intensive care unit (ICU) and hospital lengths of stay (LOS), multiple studies have shown that survival of obese patients is at least as good as if not better than lean critically ill patients. The mechanisms for these findings are not clear, but early biologic data may provide some preliminary explanations.

In this chapter, we will review the current literature on the epidemiology and outcomes of critically ill patients with obesity and ARDS. Biologic mechanisms that might explain these outcomes are then discussed.

OBESITY AND ARDS—CLINICAL OUTCOMES

ARDS results in severe hypoxemic respiratory failure as a complication of direct pulmonary or extrapulmonary injury [4,8,9]. This syndrome is common, with an annual U.S. incidence of approximately 200,000 cases. Sepsis is the most common risk factor for ARDS, accounting for approximately 80% of all cases of ARDS, and patients whose risk is sepsis generally have worse outcomes than those with other risk factors [2,4,10]. ARDS has a tremendous public health effect, with > 3.6 million hospital days per year. When it was first described in the 1960s, ARDS mortality was $> 60\%$ [11], but mortality has decreased over the past few decades and is now 30%–40% [2,3,12,13]. Although more ARDS patients are surviving, this improved survivorship comes at a cost, as ARDS survivors suffer substantial morbidity including reduced quality of life and physical, cognitive, and psychological impairments that may persist for decades after critical illness [14–17].

ARDS results from massive activation of the host response through complex pro-inflammatory cascades, ultimately leading to severe end organ damage [9,18,19]. Myriad cytokines including interleukin (IL)-1β, IL-6, IL-8, and tumor necrosis factor (TNF)-α are released from activated macrophages, which in turn activate neutrophils that release oxidants, lipid mediators, and proteases that continue to injure the lung and perpetuate the ARDS cycle [9]. Few ARDS-specific therapies have ever been found to reduce mortality. In a landmark randomized controlled trial (RCT) published in 2000, low tidal volume ventilation decreased mortality with an absolute reduction of 8.8%, as is now standard of care [20]. Early prone positioning for at least 16 hours a day has also been shown to improve survival (16% in intervention group versus 32.8% in control group, $P < 0.001$) [21]. Finally, another large RCT demonstrated that neuromuscular blockade in early ARDS reduced mortality (hazard ratio for death at 90 days in cisatracurium group compared with placebo group = 0.68, 95% confidence interval [CI], 0.48–0.98, $P = 0.04$) [22]. However, many other clinical trials of therapies once considered hopeful to improve survival have found no difference compared with placebo. Therefore greater understanding of ARDS pathogenesis is urgently needed to enable progress of research into novel therapies for this common and devastating syndrome.

Over the past 3 decades in developed countries, the prevalence of obesity has steadily increased. This is particularly true for extreme obesity (BMI $\geq$ 40 kg/m^2) [5,7]. >35% of Americans are now obese, and 6.3% are extremely obese [23]. The resultant public health consequences of the obesity epidemic are enormous, as obesity is associated with increased all-cause mortality in both men and women [5,7].

PRIOR STUDIES OF OBESITY IN ARDS

Nearly 2 decades ago, emerging evidence suggested that, although obesity might be associated with increased risk of developing ARDS and other morbidities of the critically ill, survival in obese ICU patients was "paradoxically" improved. Since that time, multiple additional observational studies and several metaanalyses, several of which are discussed later, have generally confirmed these findings.

Several observational studies (mostly larger investigations) have reported that the risk of developing ARDS is associated with increasing BMI [24–32], whereas other studies (mostly smaller investigations) have found no association [33–38], and one study reported decreased risk of ARDS development with worsening obesity [39]. The largest of these studies is a recent secondary data analysis of data obtained prospectively for the U.S. Critical Illness and Injury Trials Group-Lung Injury Prevention Study (USCIITG-LIPS) [36]. This study examined 5585 patients with at least one risk factor for ARDS in 22 U.S. hospitals and found that body mass index (BMI; modeled as a linear variable) was associated with subsequent development of ARDS (OR = 1.02, 95% CI 1.00–1.04, P = 0.03). Another large study investigated 3032 patients in France during the 2009–10 H1N1 influenza epidemic and also reported an increase in the odds of developing ARDS among obese patients with BMI $\geq$ 30 kg/m^2 compared with normal weight patients (OR = 1.8, 95% CI 1.1–3.0) [25]. Additionally, a study of 2046 nontrauma surgical patients reported that BMI $\geq$ 40 kg/m^2 independently predicted development of ARDS (OR = 1.57, 95% CI 1.01–2.45) [32]. And similarly, a study in a mixed medical-surgical population of 1795 patients found that the risk of ARDS increased with increasing BMI (OR = 1.66, 95% CI 1.21–2.28 for BMI > 30 kg/m^2 compared with normal weight and OR = 1.78, 95% CI 1.12–2.92 for BMI > 40 kg/m^2) [30]. One criticism of studies finding that obesity is associated with a higher risk of developing ARDS is that bilateral infiltrates on chest imaging is a criterion for ARDS, and chest radiographs in obese patients are more difficult to interpret due to overlying soft tissue and might erroneously be interpreted as bilateral infiltrates. However, Gong *et al.* [30] found that obese patients in the cohort were not more likely to have bilateral infiltrates on chest radiograph (but were more likely than leans to have hypoxemic respiratory failure), and excluding severely obese patients from their regression models actually strengthened the association between ARDS and obesity, implying that their finding of association between increasing obesity and incidence of ARDS was not explained by the severely obese group receiving a false diagnosis of ARDS on the basis of difficult-to-interpret chest radiographs.

Beginning in 2004, nine observational studies have been reported examining survival in obese critically ill patients with ARDS. The majority of these studies have demonstrated that survival is counterintuitively improved in the obese compared with normal weight patients [40–44], whereas the others have reported that survival in the obese and extremely obese is as least as good as that of critically ill lean patients [30,45–47]. The first study on obesity in ARDS was a secondary analysis of 807 patients who were enrolled in a RCT of low versus traditional tidal volumes conducted by the National Heart, Lung, and Blood Institute ARDS Network (the ARMA and ALVEOLI studies) [45]. Severely obese patients (BMI > 40 kg/m^2) comprised a small percentage (4.7%) of the study group because the parent study excluded patients with a weight-to-height ratio of > 1.0 kg/cm. BMI data were also missing for 6.1% of participants. Although an association was found between rising BMI and decreased mortality, this was not significant after multivariate analysis. Additional morbidity outcomes, such as ICU length of stay, were not reported. Another more recent study also examined outcomes in obese patients enrolled in ARDS Network trials, and similar to the first study, found no significant differences [46]. Reasons why these two secondary analyses of ARDS Network data have not found mortality differences, as opposed to other studies discussed later, are not clear but may be partially explained by inclusion and exclusion criteria into the parent RCTs, as well as the limited cohort sizes.

Another study analyzed data from 825 participants in the King County Lung Injury Project (KCLIP), a landmark study of ARDS epidemiology [47]. Because this study analyzed population-based data, the results may be more generalizable to all ARDS patients. Adjusted analyses revealed no significant mortality differences between overweight, obese, or extremely obese patients compared with normal weight patients. Severely obese patients (BMI > 40 kg/m^2) who survived had longer duration of mechanical ventilation, ICU LOS, and hospital LOS than normal weight patients, and they were more likely to be discharged to a skilled nursing facility or to a rehabilitation facility than to home.

In contrast, a study analyzing data from Project Impact® (a subscription database operated by the Society of Critical Care Medicine designed to measure and describe the care of ICU patients) explored outcomes in 1488 patients with ARDS [40] and found that unadjusted hospital mortality was significantly associated with BMI. The highest mortality was found in underweight patients (54.6%) and the lowest in extremely obese patients (29.0%). In adjusted regression analyses, patients in the overweight, obese, and morbidly obese categories had decreased risk of death compared with normal weight patients. However, this decreased risk was only statistically significant in patients in the obese category (BMI 30–39.9 kg/m^2). ICU and hospital LOS and discharge location did not differ by BMI category.

Finally, a recent very large epidemiologic study analyzing data from the Nationwide Inpatient Sample (NIS) [43] investigated outcomes in 9,149,030 patients undergoing hospital admission for surgical procedures. Among these patients,

5.48% were obese and 1.82% of those developed ARDS during their hospitalization. Mortality was 11% among obese ARDS patients and 25% among normal weight patients with ARDS ($P < 0.0001$).

Beyond the previously discussed individual studies, several metaanalyses examining outcomes in obese critically ill patients have been published [48–51]. The most recent, and therefore comprehensive, metaanalysis evaluated 9,187,248 patients in 24 studies and confirmed results that, although obese critically ill patients may be at increased risk of developing ARDS, their survival appears to be better than lean patients [51]. In this metaanalysis, pooled data from 16 studies evaluating ARDS risk found that obesity (BMI $\geq$ 30 kg/m^2) was associated with increased risk of ARDS development compared with normal weight patients (odds ratio [OR] 1.75, 95% confidence interval [CI] = 1.42–2.15). Aggregate data from nine studies investigating the association of obesity with mortality demonstrated reduced mortality (OR = 0.63, 95% CI 0.41–0.89 for BMI $\geq$ 30 kg/m^2 compared with BMI $\leq$ 30 kg/m^2). One significant limitation of this metaanalysis, however, is the high heterogeneity between studies, which may reflect that the individual investigations utilized varying definitions of ARDS, patient populations, and definitions of obesity, and occurred over more than a decade. Another metaanalysis included 62,045 patients from 14 individual studies and found significantly lower hospital mortality in obese (BMI $\geq$ 30 kg/m^2) compared with normal weight ARDS patients (risk ratio [RR] 0.83, 95% CI 0.74–0.92). However, both duration of mechanical ventilation and ICU LOS were significantly longer in obese patients (1.48 days, 95% CI 0.07–2.89, $P = 0.04$ for ventilation and 1.08 days, 95% CI 0.27–1.88, $P = 0.009$ for ICU LOS) [48]. A third metaanalysis included 88,051 patients from 22 studies and reported a significantly decreased hospital mortality associated with obesity (RR = 0.76 with 95% CI 0.59–0.92), but no association between obesity and ICU mortality, duration of mechanical ventilation, or ICU LOS [49]. Differences in the results of these various metaanalyses highlight the variability of the included studies and definitions, but nonetheless their findings add to an increasingly robust body of literature supporting the finding that obese ARDS patients have a survival advantage compared with lean ARDS patients.

These same findings that obesity might confer a survival advantage in ARDS patients have also been recapitulated in general critical illness. One important study not included in the previously mentioned metaanalyses, as it did not focus solely on critically ill patients with ARDS, is a comprehensive secondary analysis of prospective data from a multicenter international observational ICU nutrition practices survey in 355 ICUs in 33 countries during 2007–09 [52]. Included patients were adults $\geq$ 18 years old receiving mechanical ventilation for $\geq$ 72 hours. Total sample size was 8813 patients, with 3490 of normal weight, 2604 overweight (BMI 25–29.9 kg/m^2), 1772 obese (BMI 30–39.9 kg/m^2), 348 with BMI 40–49.9 kg/m^2, 118 with BMI 50–59.9 kg/m^2, and 58 with BMI $\geq$ 60 kg/m^2. Mortality in both overweight (OR = 0.80, 95% CI 0.71–0.90) and obese (OR = 0.73, 95% CI 0.64–0.84) patients was significantly lower than in normal weight patients. Interestingly, there was also a trend toward lower mortality in the extremely obese group (BMI $\geq$ 40 kg/m^2), with odds of death = 0.87 (95% CI 0.69–1.09, $P = 0.07$) compared with normal weight patients. Additionally, obese and extremely obese survivors had longer durations of mechanical ventilation and ICU LOS than lean patients. Another more recent study in general critical illness investigated 154,308 patients in Dutch ICUs and concluded that greater BMI was significantly associated with improved hospital survival, with patients in the overweight group (BMI 30–39.9 kg/m^2) having the lowest risk of death (OR = 0.86, 95% CI 0.83–0.90) [53].

To summarize, available evidence suggests that, although overweight, obese, and extremely obese patients appear to have an increased risk of developing ARDS, these same patients, once developing ARDS, also have improved survival compared with normal weight patients. Obese patients with ARDS, however, may experience greater morbidity, including greater duration of mechanical ventilation and longer ICU length of stay.

LIMITATIONS OF PRIOR STUDIES

All investigations of outcomes in critically ill patients with ARDS have limitations. First, BMI is typically used to quantify obesity, but this measure may not accurately reflect adiposity or obesity-related syndromes compared with other measurements, such as waist circumference [54]. Furthermore, body weight may be altered by receipt of intravenous fluid or diuretics in the ICU, whereas measurement of height in supine critically ill patients may be inaccurate, thus leading to inaccuracy in determining BMI [55]. Second, there is no measure to assess severity of illness in obese patients specifically, and current tools, such as APACHE and SAPS [56,57], may inaccurately reflect mortality risk in obese patients due to unknown factors specific to this population. Third, care processes for obese patients may greatly vary by hospital and ICU, and may bias results, either toward improved or worse outcomes for obese patients [40,58]. Finally, it can be difficult to assess the degree of critical illness in extremely obese patients (e.g., measuring noninvasive blood pressure [59] or interpreting chest radiographs), thus producing misclassification and incorrect case ascertainment; however, many studies previously discussed report improved outcomes in overweight and obese groups, in addition to the extremely obese group, where misclassification is less likely [42]. Furthermore, as discussed earlier, one study found that the increased risk of ARDS development was not caused by greater incidence of bilateral pulmonary infiltrates but instead by greater incidence of hypoxemic respiratory failure in obese patients [30]. It should also be noted that all studies of obesity in ARDS thus far have used the former American-European Consensus Conference definition of ARDS [8] rather than the newer Berlin definition [1], and whether or not results would differ with the new definition is unclear.

POSSIBLE EXPLANATION OF FINDINGS

The previously discussed body of literature has led to a new interest in understanding the mechanisms by which obesity might alter development of ARDS as well as outcomes in patients who have ARDS. The increased duration of mechanical ventilation and ICU LOS observed in obese patients in some studies might be explained by physiologic factors that might lead to longer time in the ICU but do not increase mortality. For example, abdominal and chest wall weight might lead to lung derecruitment that leads to longer duration of mechanical ventilation due to prolonged hypoxemia [60]. However, explanations for why risk of ARDS is greater in obese critically ill patients, yet survival in obese patients with ARDS is improved, or at least as good as, survival in normal weight patients with ARDS, are perhaps not as clear. Mechanisms underlying these counterintuitive findings have not been elucidated but are being actively researched. Recent studies in animals have suggested that obesity may be associated with "priming" the individual for ARDS development but, at the same time, ARDS once acquired is less severe in obesity due to an abnormal neutrophil chemoattractant response among other factors [61,62]. Additionally, obese patients with ARDS, compared with their lean counterparts, have lower circulating levels of several proinflammatory cytokines (IL-6, IL-8) known to be both increased in ARDS and associated with increased mortality [42]. These findings suggest that obesity may alter both innate immunity and the acute inflammatory response.

BIOLOGIC RELATIONSHIP OF OBESITY AND ARDS
OBESITY AND THE PATHOGENESIS OF ARDS

Despite decades of research, the pathogenesis of ARDS remains incompletely understood. Following an inciting event, alveolar macrophage activation, pulmonary recruitment of neutrophils, and alveolar endothelial and epithelial injury are thought to be central factors in both the onset and progression of this syndrome, which is also then promulgated by ventilator-induced lung injury (VILI) [9,19,63–66]. In accordance with such a schema, increases in both airspace neutrophilia and plasma cytokine levels, including TNFα, IL-1β, IL-6, and IL-8, have been shown to correlate with increased morbidity and mortality from this disease [64,65,67–69].

It is increasingly recognized that ARDS pathogenesis and hence outcome may be influenced by host factors, including genetic polymorphisms and comorbid conditions [9,66]. In this light, the clinical evidence that obesity may have an ameliorative effect on ARDS and critical care outcomes suggests obesity may be one such factor. Such an interaction would be surprising, as obesity is itself believed to be an inflammatory state with baseline-increased circulating neutrophil levels [70,71]; elevations in blood TNFα, IL-1β, IL-6, and IL-8 [72,73]; and innate immune cell activation [74–76] with endothelial injury [77–79], perhaps predictive of inflammatory synergy between the obese state and ARDS. Nevertheless, it has recently been reported that plasma IL-6 and IL-8 *fall* with rising BMI in ARDS patients [42], indicating that, although obesity may increase the risk of *developing* ARDS [30], perhaps through preexisting vascular injury, it may paradoxically have an *attenuating* effect on ARDS-associated inflammation and hence the progression of the disease.

Although human studies examining the effects of obesity on ARDS pathophysiology are scarce, recent reports in animal models suggest that such models may recapitulate the clinical effects of obesity, allowing further dissection of the underlying mechanisms. Most animal studies examining obesity-associated effects on pulmonary immunity and inflammation have focused on models of asthma and bacterial pneumonia, and although some forms of airway inflammation appear to be amplified by obesity [80–83], the response to pneumonia is blunted [62,84–86], suggesting that the inflammatory response in the alveoli (the site of ARDS) is impaired, particularly over time. In the majority of published reports examining obesity's effects on acute lung injury models, obese mice and rats demonstrate reduced inflammation, lung injury, and mortality from LPS-, hyperoxia-, and ozone-induced ARDS [87–91], although in the case of ozone exposure, findings are mixed and appear to vary with the acuity of exposure and the timing of examination [90,92,93], such that there appears to be a reversal of obesity's effects on lung injury as the duration (and perhaps severity) of precipitating exposure increases: for example, acute (3 hours) vs. subacute (72 hours) ozone exposures yield diametrically opposite effects (increased and reduced, respectively) in the same model of obesity [92,94]. Furthermore, the timing of examination following injury appears to influence the witnessed effects of obesity on injury. This is particularly clear in the setting of LPS injury, in which early time points (2–6 hours) after injury appear to manifest a proinflammatory effect of obesity, but by 24 hours the effect is opposite [61], a finding that appears to recapitulate the human findings of obesity-associated increased risk for developing ARDS but subsequently reduced inflammation and improved survival once ARDS has developed.

OBESITY, THE METABOLIC SYNDROME, AND THE INFLAMMATORY RESPONSE IN ARDS

Given the systemic abnormalities associated with obesity and the accompanying metabolic syndrome, obesity's effects on the pathogenesis of ARDS almost certainly reflect interaction between multiple facets of the obese state (Figure 11.1). Although few reports focus on obesity itself, growing literature examines the effects of the metabolic syndrome on ARDS pathogenesis and outcome. The most extensively investigated element of the metabolic syndrome in this regard is diabetes.

Diabetes has been shown to be associated with a reduced risk of developing ARDS in three large clinical studies of high-risk patients including those with sepsis, aspiration, trauma, and massive transfusion [95–97], with an adjusted odds ratio ranging from 0.33 to 0.58 [98]. Recently, diabetes has also been shown to have a salutary effect on survival in those with ARDS due to direct injury to the lung (e.g., pneumonia, aspiration) [99]. Although this protective effect is reproducible in animal models of diabetes and lung injury [87,91,100–102], the underlying mechanisms remain unclear. Diabetes is associated with impaired innate immune response [103,104], which, although believed to drive the increased risk of infection in diabetics [105], might conversely attenuate inappropriate inflammatory states such as ARDS. Evidence supporting roles for either hyperglycemia or insulin resistance in

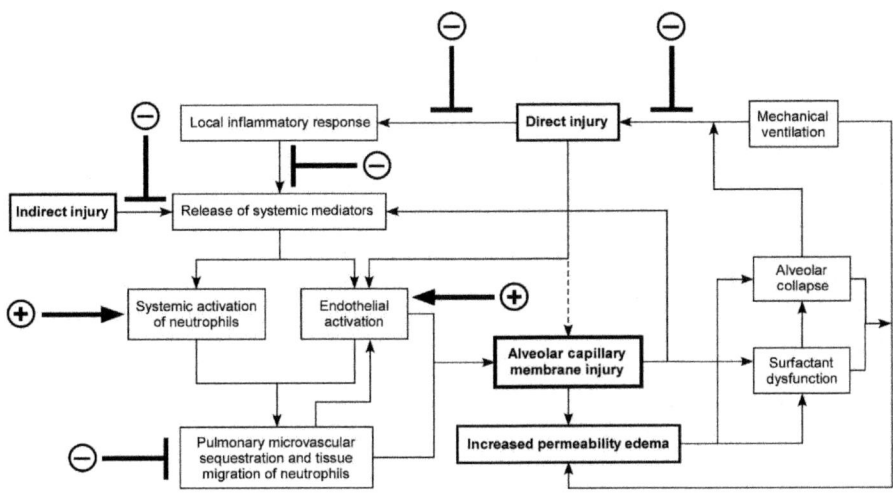

FIGURE 11.1

Pathophysiologic mechanisms of acute lung injury and the possible effects of obesity. In this schematized view of the pathophysiologic pathways of ALI/ARDS, direct injuries to the lung damage the alveolar capillary membrane (ACM) and initiate local and subsequently systemic inflammatory cascades, which may be dampened by the effects of obesity and the metabolic syndrome ($\perp$). Indirect injuries initiate the pathophysiologic pathways of ALI/ARDS primarily through release of systemic cytokines. Following both direct and indirect initiators of ALI/ARDS, the release of systemic inflammatory mediators activates circulating neutrophils and the vascular endothelium of the lung, leading to pulmonary microvascular sequestration of neutrophils and inflammatory injury to the ACM. Although obesity may promote vascular injury in ALI, it appears to attenuate cytokine release and neutrophil activation. This injury is critical in the failure of alveolar/capillary membrane barrier function and flooding of the alveoli with proteinaceous edema fluid. Both ACM injury and alveolar edema cause surfactant loss and dysfunction, which promote alveolar instability and collapse, driving further edema formation and alveolar injury, particularly in the setting of mechanical ventilation, which also may be attenuated in the setting of obesity.

Used with permission from Stapleton & Suratt: Obesity and nutrition in acute respiratory distress syndrome.
Clin Chest Med 2014;35(4):655–71.

the attenuation of ARDS is conflicting, and a examination of the possibility that this protective effect might be related to the systemic actions of diabetic *therapies*, such as exogenous insulin and PPAR-γ (gamma) agonists, showed no such effects of these treatments [106].

Comparable studies investigating dyslipidemia and its effects on ARDS risk and pathogenesis have not yet been published. However, as with obesity in general, dyslipidemia is associated with baseline elevations in circulating neutrophil levels in both humans and mouse models, often in the absence of accompanying obesity [107–110]. Persistent activation of both monocytes and neutrophils are described in dyslipidemic states, accompanied by endothelial injury [111], and may be driven by direct effects of lipid species on leukocytes [112,113]. Yet, in this setting there appear to be defects in neutrophil and monocyte function [114,115], and recently animal models of hypercholesterolemia without obesity have suggested that LPS-induced ARDS is blunted [116]. Whether such defects might reflect tonic activation of the innate immune system with both "priming" of the pulmonary endothelium and simultaneous "desensitization" of innate immune cells to acute stimuli or other effects of the dyslipidemic state is not yet known [117].

Another significant feature of the metabolic syndrome and obesity in general is the dysregulation of adipokine release and response. Although initially described as hormone-like signaling molecules released by adipose tissue and involved in metabolic homeostasis, adipokines such as leptin, adiponectin, and visfatin have recently been shown to have quite protean effects including modulation of both innate and adaptive immune systems [118,119]. The best studied of these molecules is leptin, which was originally described as a regulator of appetite. Leptin has been shown to be important in the marrow development of the myelomonocytic lineages [120,121], and to serve as an activation and survival signal for neutrophils in the periphery [122–124]. Interestingly, leptin also appears to act as a neutrophil chemoattractant [125–127] and may be released by the injured lung [127–129], whereas serum levels of leptin are elevated in critical illness [130–132], and leptin administered to the lung augments LPS-induced lung injury in mice [127], suggesting a possible role for leptin in the development of ARDS.

How leptin's effects on innate immune function may be altered in obesity is poorly understood. Obesity is typically accompanied by a state of hyperleptinemic leptin resistance in which leptin response is blunted despite high circulating levels of this cytokine, presumably due to receptor desensitization. Elevated leptin levels in patients with end-stage renal disease have been implicated in the neutrophil dysfunction that accompanies that state [133], whereas animal models of aleptinemia, leptin-resistance, and lean hyperleptinemia suggest that the development of ARDS is blunted in this setting [88,89,134]. Whether hyperleptinemia and leptin resistance may affect the development of human ARDS has not yet been addressed. Little is known of the potential roles for other adipokines in the pathogenesis of ARDS with the exception of adiponectin, deficiency of which (as seen in obesity) appears to increase the risk of ARDS in animal models [91].

In reviewing the literature describing the many effects of the metabolic syndrome, it is important to emphasize that both human and, with rare exceptions, animal studies examining "discrete" elements of the metabolic syndrome have not examined these in isolation of obesity or the other facets of the syndrome. For instance, few of the reported clinical studies on diabetes and ARDS included BMI as a confounding variable, and the db/db mouse model, although used in various studies to specifically examine leptin resistance, diabetes, or obesity, is also noted to be extremely dyslipidemic. Thus it remains unclear which elements of obesity and the metabolic syndrome may be operative in the majority of reported findings.

OBESE PULMONARY MECHANICS: VILI ATTENUATED?

One further possibility that must be considered when examining the potential interaction between obesity and ARDS focuses on the biomechanical effects of obesity. As detailed elsewhere in this volume (Chapter 3), obese individuals manifest altered pulmonary mechanics compared with lean individuals at baseline and when mechanically ventilated. Obese patients with ARDS demonstrate similar changes, with a combination of reduced chest wall and lung compliance, leading to a lower FRC and consequently atelectasis, increased airways resistance and closure, and ventilation/perfusion mismatch [135,136]. Although these changes likely underlie obesity-associated delays in liberation from mechanical ventilation, how such alterations might alter ARDS pathogenesis are unclear.

Given the biomechanical effects of obesity, there may be an inherent tendency toward progressive derecruitment, particularly in the supine position [137,138], and it has been shown that obese patients in the ICU are at greater risk of respiratory failure, independent of ARDS [139]. Thus such derecruitment and proclivity for respiratory failure may contribute to the increased risk of clinical ARDS in the obese. Once intubated with ARDS, how obese biomechanics may improve survival is less clear. It is possible that the combination of lower respiratory system compliance and higher airways resistance, which yield atelectasis and higher static and dynamic airway pressures for a given tidal volume, may prompt clinicians to selectively increase PEEP and decrease tidal volumes in the obese, thus mimicking or accentuating a protective low tidal volume strategy. However, it has been shown that obese ARDS patients are typically ventilated at *higher* tidal volumes (cc/kg ideal body weight) than normal weight patients [45], [47], indicating that, in light of comparable to improved survival in the obese, mechanical ventilation (even at higher tidal volumes) may be better tolerated in these patients. Furthermore, although overall mortality in the RCT of low tidal volume ventilation in ARDS (ARMA) was not different between lean and obese patients [45], data from this study suggest that obese patients may have tolerated higher tidal volumes (12 cc/kg IBW) better than lean patients. The relative reduction in mortality attributable to lower tidal volumes (6 cc/kg IBW) in the overall cohort was 30%, yet when stratified into normal, overweight, and obese BMI categories, the relative reductions in mortality between high and low tidal volume arms of the study were 42%, 27%, and 12%, respectively, although this finding did not reach statistical significance. Thus obese patients may be less susceptible to VILI. Whether this might be related to the mechanical interaction between obese patients and ventilation or an additional manifestation of attenuated inflammatory response in obese ARDS has yet to be examined.

CONCLUSIONS

Survival in the general population is J-shaped, with increased mortality in underweight people, lowest mortality in patients with a BMI near 25 kg/m^2, and increasing mortality rates in overweight, obese, and extremely obese patients [5].

Evidence in critically ill patients, however, suggests that overweight, obese, and extremely obese patients have lower mortality compared with normal weight patients. The limited studies of obese patients with ARDS published to date show that rising BMI may increase the risk for the development of ARDS [30] but paradoxically does not increase mortality from this disease, and may, in fact, be protective [40,45,47]. Health professionals may assume that obese patients have worse survival and morbidity due to presumed difficulties of caring for such critically ill patients including transport, body positioning, intravascular access, diagnostic imaging, and ventilator weaning. Although the literature does suggest that obese patients may have longer durations of ventilation and ICU LOS, their survival is at least as good as normal weight patients. This information is important for clinicians to recognize when discussing prognosis and expectations with critically ill patients and their families, and remains a critical area for research to better define the underlying mechanisms of these findings.

REFERENCES

[1] Force ADT, Ranieri VM, Rubenfeld GD, Thompson BT, Ferguson ND, Caldwell E, et al. Acute respiratory distress syndrome: the Berlin definition. JAMA 2012;307(23):2526–33.

[2] Rubenfeld GD, Caldwell E, Peabody E, Weaver J, Martin DP, Neff M, et al. Incidence and outcomes of acute lung injury. N Engl J Med 2005;353(16):1685–93.

[3] Stapleton RD, Wang BM, Hudson LD, Rubenfeld GD, Caldwell ES, Steinberg KP. Causes and timing of death in patients with ARDS. Chest 2005;128(2):525–32.

[4] Fan E, Brodie D, Slutsky AS. Acute respiratory distress syndrome: advances in diagnosis and treatment. JAMA 2018;319(7):698–710.

[5] Adams KF, Schatzkin A, Harris TB, Kipnis V, Mouw T, Ballard-Barbash R, et al. Overweight, obesity, and mortality in a large prospective cohort of persons 50 to 71 years old. N Engl J Med 2006;355(8):763–78.

[6] Heymsfield SB, Mechanisms WTA. Pathophysiology, and management of obesity. N Engl J Med 2017;376(15):1492.

[7] Health, United States, 2015. With special feature on racial and ethnic health disparities. Hyattsville (MD): Health, United States; 2016.

[8] Bernard GR, Artigas A, Brigham KL, Carlet J, Falke K, Hudson L, et al. The American-European consensus conference on ARDS. Definitions, mechanisms, relevant outcomes, and clinical trial coordination. Am J Respir Crit Care Med 1994;149(3 Pt 1):818–24.

[9] Ware LB, Matthay MA. The acute respiratory distress syndrome. N Engl J Med 2000;342(18):1334–49.

[10] Hudson LD, Milberg JA, Anardi D, Maunder RJ. Clinical risks for development of the acute respiratory distress syndrome. Am J Respir Crit Care Med 1995;151(2 Pt 1):293–301.

[11] Ashbaugh DG, Bigelow DB, Petty TL, Levine BE. Acute respiratory distress in adults. Lancet 1967;2(7511):319–23.

[12] Milberg JA, Davis DR, Steinberg KP, Hudson LD. Improved survival of patients with acute respiratory distress syndrome (ARDS): 1983–1993. JAMA 1995;273(4):306–9.

[13] Abel SJ, Finney SJ, Brett SJ, Keogh BF, Morgan CJ, Evans TW. Reduced mortality in association with the acute respiratory distress syndrome (ARDS). Thorax 1998;53(4):292–4.

[14] Rubenfeld GD, Herridge MS. Epidemiology and outcomes of acute lung injury. Chest 2007;131(2):554–62.

[15] Dowdy DW, Eid MP, Dennison CR, Mendez-Tellez PA, Herridge MS, Guallar E, et al. Quality of life after acute respiratory distress syndrome: a meta-analysis. Intensive Care Med 2006;32(8):1115–24.

[16] Herridge MS, Tansey CM, Matte A, Tomlinson G, Diaz-Granados N, Cooper A, et al. Functional disability 5 years after acute respiratory distress syndrome. N Engl J Med 2011;364(14):1293–304.

[17] Fan E, Dowdy DW, Colantuoni E, Mendez-Tellez PA, Sevransky JE, Shanholtz C, et al. Physical complications in acute lung injury survivors: a two-year longitudinal prospective study. Crit Care Med 2014;42(4):849–59.

[18] Weiland JE, Davis WB, Holter JF, Mohammed JR, Dorinsky PM, Gadek JE. Lung neutrophils in the adult respiratory distress syndrome. Clinical and pathophysiologic significance. Am Rev Respir Dis 1986;133(2):218–25.

[19] Tate RM, Repine JE. Neutrophils and the adult respiratory distress syndrome. Am Rev Respir Dis 1983;128(3):552–9.

[20] Acute Respiratory Distress Syndrome Network, Brower RG, Matthay MA, Morris A, Schoenfeld D, Thompson BT, Wheeler A. Ventilation with lower tidal volumes as compared with traditional tidal volumes for acute lung injury and the acute respiratory distress syndrome. The Acute Respiratory Distress Syndrome Network. N Engl J Med 2000;342(18):1301–8.

[21] Guerin C, Reignier J, Richard JC, Beuret P, Gacouin A, Boulain T, et al. Prone positioning in severe acute respiratory distress syndrome. N Engl J Med 2013;368(23):2159–68.

[22] Papazian L, Forel JM, Gacouin A, Penot-Ragon C, Perrin G, Loundou A, et al. Neuromuscular blockers in early acute respiratory distress syndrome. N Engl J Med 2010;363(12):1107–16.

[23] Flegal KM, Carroll MD, Ogden CL, Curtin LR. Prevalence and trends in obesity among US adults, 1999–2008. JAMA 2010;303(3):235–41.

[24] Lei H, Minghao W, Xiaonan Y, Ping X, Ziqi L, Qing X. Acute lung injury in patients with severe acute pancreatitis. Turk J Gastroenterol 2013;24(6):502–7.

[25] Bonmarin I, Belchior E, Bergounioux J, Brun-Buisson C, Megarbane B, Chappert JL, et al. Intensive care unit surveillance of influenza infection in France: the 2009/10 pandemic and the three subsequent seasons. Euro Surveill 2015;20(46).

[26] Palakshappa JA, Anderson BJ, Reilly JP, Shashaty MG, Ueno R, Wu Q, et al. Low plasma levels of adiponectin do not explain acute respiratory distress syndrome risk: a prospective cohort study of patients with severe sepsis. Crit Care 2016;20:71.

[27] Weinlein JC, Deaderick S, Murphy RF. Morbid obesity increases the risk for systemic complications in patients with femoral shaft fractures. J Orthop Trauma 2015;29(3):e91–5.

[28] Elmer J, Hou P, Wilcox SR, Chang Y, Schreiber H, Okechukwu I, et al. Acute respiratory distress syndrome after spontaneous intracerebral hemorrhage. Crit Care Med 2013;41(8):1992–2001.

[29] Newell MA, Bard MR, Goettler CE, Toschlog EA, Schenarts PJ, Sagraves SG, et al. Body mass index and outcomes in critically injured blunt trauma patients: weighing the impact. J Am Coll Surg 2007;204(5):1056–61 [discussion 62-4].

[30] Gong MN, Bajwa EK, Thompson BT, Christiani DC. Body mass index is associated with the development of acute respiratory distress syndrome. Thorax 2010;65(1):44–50.

[31] Hagau N, Slavcovici A, Gonganau DN, Oltean S, Dirzu DS, Brezoszki ES, et al. Clinical aspects and cytokine response in severe H1N1 influenza A virus infection. Crit Care 2010;14(6):R203.

[32] Towfigh S, Peralta MV, Martin MJ, Salim A, Kelso R, Sohn H, et al. Acute respiratory distress syndrome in nontrauma surgical patients: a 6-year study. J Trauma 2009;67(6):1239–43.

[33] Bramley AM, Dasgupta S, Skarbinski J, Kamimoto L, Fry AM, Finelli L, et al. Intensive care unit patients with 2009 pandemic influenza A (H1N1pdm09) virus infection—United States, 2009. Influenza Other Respi Viruses 2012;6(6):e134–42.

[34] Dhakal B, Eastwood D, Sukumaran S, Hassler G, Tisol W, Gasparri M, et al. Morbidities of lung cancer surgery in obese patients. J Thorac Cardiovasc Surg 2013;146(2):379–84.

[35] Duchesne JC, Schmieg Jr RE, Simmons JD, Islam T, McGinness CL, McSwain Jr NE. Impact of obesity in damage control laparotomy patients. J Trauma 2009;67(1):108–12 [discussion 12-4].

[36] Karnatovskaia LV, Lee AS, Bender SP, Talmor D, Festic E, Illness USC, et al. Obstructive sleep apnea, obesity, and the development of acute respiratory distress syndrome. J Clin Sleep Med 2014;10(6):657–62.

[37] Kumar MA, Chanderraj R, Gant R, Butler C, Frangos S, Maloney-Wilensky E, et al. Obesity is associated with reduced brain tissue oxygen tension after severe brain injury. Neurocrit Care 2012;16(2):286–93.

[38] Yao S, Mao T, Fang W, Xu M, Chen W. Incidence and risk factors for acute lung injury after open thoracotomy for thoracic diseases. J Thorac Dis 2013;5(4):455–60.

[39] Dossett LA, Heffernan D, Lightfoot M, Collier B, Diaz JJ, Sawyer RG, et al. Obesity and pulmonary complications in critically injured adults. Chest 2008;134(5):974–80.

[40] O'Brien Jr JM, Phillips GS, Ali NA, Lucarelli M, Marsh CB, Lemeshow S. Body mass index is independently associated with hospital mortality in mechanically ventilated adults with acute lung injury. Crit Care Med 2006;34(3):738–44.

[41] De Jong A, Molinari N, Sebbane M, Prades A, Futier E, Jung B, et al. Feasibility and effectiveness of prone position in morbidly obese patients with ARDS: a case-control clinical study. Chest 2013;143(6):1554–61.

[42] Stapleton RD, Dixon AE, Parsons PE, Ware LB, Suratt BT. The association between BMI and plasma cytokine levels in patients with acute lung injury. Chest 2010;138(3):568–77.

[43] Memtsoudis SG, Bombardieri AM, Ma Y, Walz JM, Chiu YL, Mazumdar M. Mortality of patients with respiratory insufficiency and adult respiratory distress syndrome after surgery: the obesity paradox. J Intensive Care Med 2012;27(5):306–11.

[44] Soto GJ, Frank AJ, Christiani DC, Gong MN. Body mass index and acute kidney injury in the acute respiratory distress syndrome. Crit Care Med 2012;40(9):2601–8.

[45] O'Brien Jr JM, Welsh CH, Fish RH, Ancukiewicz M, Kramer AM. Excess body weight is not independently associated with outcome in mechanically ventilated patients with acute lung injury. Ann Intern Med 2004;140(5):338–45.

[46] Soubani AO, Chen W, Jang H. The outcome of acute respiratory distress syndrome in relation to body mass index and diabetes mellitus. Heart Lung 2015;44(5):441–7.

[47] Morris AE, Stapleton RD, Rubenfeld GD, Hudson LD, Caldwell E, Steinberg KP. The association between body mass index and clinical outcomes in acute lung injury. Chest 2007;131(2):342–8.

[48] Akinnusi ME, Pineda LA, El Solh AA. Effect of obesity on intensive care morbidity and mortality: a meta-analysis. Crit Care Med 2008;36(1):151–8.

[49] Hogue Jr CW, Stearns JD, Colantuoni E, Robinson KA, Stierer T, Mitter N, et al. The impact of obesity on outcomes after critical illness: a meta-analysis. Intensive Care Med 2009;35(7):1152–70.

[50] Oliveros H, Villamor E. Obesity and mortality in critically ill adults: a systematic review and meta-analysis. Obesity (Silver Spring) 2008;16(3):515–21.

[51] Zhi G, Xin W, Ying W, Guohong X, Shuying L. "Obesity paradox" in acute respiratory distress syndrome: asystematic review and meta-analysis. PLoS One 2016;11(9):e0163677.

[52] Martino JL, Stapleton RD, Wang M, Day AG, Cahill NE, Dixon AE, et al. Extreme obesity and outcomes in critically ill patients. Chest 2011;.

[53] Pickkers P, de Keizer N, Dusseljee J, Weerheijm D, van der Hoeven JG, Peek N. Body mass index is associated with hospital mortality in critically ill patients: an observational cohort study. Crit Care Med 2013;41(8):1878–83.

[54] Clinical guidelines on the identification, evaluation, and treatment of overweight and obesity in adults—The evidence report. National Institutes of Health. Obes Res 1998;6(Suppl 2):51S–209S.

[55] McCallister JW, Adkins EJ, O'Brien Jr JM. Obesity and acute lung injury. Clin Chest Med 2009;30(3):495–508 [viii].

[56] Knaus WA, Draper EA, Wagner DP, Zimmerman JE. APACHE II: a severity of disease classification system. Crit Care Med 1985;13(10):818–29.

[57] Le Gall JR, Lemeshow S, Saulnier F. A new simplified acute physiology score (SAPS II) based on a European/North American multicenter study. JAMA 1993;270(24):2957–63.

[58] O'Brien Jr JM, Philips GS, Ali NA, Aberegg SK, Marsh CB, Lemeshow S. The association between body mass index, processes of care, and outcomes from mechanical ventilation: a prospective cohort study. Crit Care Med 2012;40(5):1456–63.

[59] Maxwell MH, Waks AU, Schroth PC, Karam M, Dornfeld LP. Error in blood-pressure measurement due to incorrect cuff size in obese patients. Lancet 1982;2(8288):33–6.

[60] Walz JM, Zayaruzny M, Heard SO. Airway management in critical illness. Chest 2007;131(2):608–20.

[61] Kordonowy LL, Burg E, Lenox CC, Gauthier LM, Petty JM, Antkowiak M, et al. Obesity is associated with neutrophil dysfunction and attenuation of murine acute lung injury. Am J Respir Cell Mol Biol 2012;47(1):120–7.

[62] Ubags ND, Burg E, Antkowiak M, Wallace AM, Dilli E, Bement J, et al. A comparative study of lung host defense in murine obesity models. Insights into neutrophil function. Am J Respir Cell Mol Biol 2016;55(2):188–200.

[63] Moraes TJ, Chow CW, Downey GP. Proteases and lung injury. Crit Care Med 2003;31(4 Suppl):S189–94.

[64] Baughman RP, Gunther KL, Rashkin MC, Keeton DA, Pattishall EN. Changes in the inflammatory response of the lung during acute respiratory distress syndrome: prognostic indicators. Am J Respir Crit Care Med 1996;154(1):76–81.

[65] Goodman RB, Strieter RM, Martin DP, Steinberg KP, Milberg JA, Maunder RJ, et al. Inflammatory cytokines in patients with persistence of the acute respiratory distress syndrome. Am J Respir Crit Care Med 1996;154(3 Pt 1):602–11.

[66] Suratt BT, Parsons PE. Mechanisms of acute lung injury/acute respiratory distress syndrome. Clin Chest Med 2006;27(4):579–89 [abstract viii].

[67] Meduri GU, Headley S, Kohler G, Stentz F, Tolley E, Umberger R, et al. Persistent elevation of inflammatory cytokines predicts a poor outcome in ARDS. Plasma IL-1 beta and IL-6 levels are consistent and efficient predictors of outcome over time. Chest 1995;107(4):1062–73.

[68] Parsons PE, Eisner MD, Thompson BT, Matthay MA, Ancukiewicz M, Bernard GR, et al. Lower tidal volume ventilation and plasma cytokine markers of inflammation in patients with acute lung injury. Crit Care Med 2005;33(1):1–6. [discussion 230-2].

[69] Ware LB. Prognostic determinants of acute respiratory distress syndrome in adults: impact on clinical trial design. Crit Care Med 2005;33(3 Suppl):S217–22.

[70] Kim JA, Park HS. White blood cell count and abdominal fat distribution in female obese adolescents. Metabolism 2008;57(10):1375–9.

[71] Desai MY, Dalal D, Santos RD, Carvalho JA, Nasir K, Blumenthal RS. Association of body mass index, metabolic syndrome, and leukocyte count. Am J Cardiol 2006;97(6):835–8.

[72] Ramos EJ, Xu Y, Romanova I, Middleton F, Chen C, Quinn R, et al. Is obesity an inflammatory disease? Surgery 2003;134(2):329–35.

[73] Yudkin JS. Adipose tissue, insulin action and vascular disease: inflammatory signals. Int J Obes Relat Metab Disord 2003;27(Suppl 3):S25–8.

[74] Cottam DR, Schaefer PA, Shaftan GW, Velcu L, Angus LD. Effect of surgically-induced weight loss on leukocyte indicators of chronic inflammation in morbid obesity. Obes Surg 2002;12(3):335–42.

[75] Cottam DR, Schaefer PA, Fahmy D, Shaftan GW, Angus LD. The effect of obesity on neutrophil Fc receptors and adhesion molecules (CD16, CD11b, CD62L). Obes Surg 2002;12(2):230–5.

[76] Nijhuis J, Rensen SS, Slaats Y, van Dielen FM, Buurman WA, Greve JW. Neutrophil activation in morbid obesity, chronic activation of acute inflammation. Obesity (Silver Spring) 2009;17(11):2014–8.

[77] Blann AD, Bushell D, Davies A, Faragher EB, Miller JP, McCollum CN. von Willebrand factor, the endothelium and obesity. Int J Obes Relat Metab Disord 1993;17(12):723–5.

[78] van Harmelen V, Eriksson A, Astrom G, Wahlen K, Naslund E, Karpe F, et al. Vascular peptide endothelin-1 links fat accumulation with alterations of visceral adipocyte lipolysis. Diabetes 2008;57(2):378–86.

[79] Pontiroli AE, Frige F, Paganelli M, Folli F. In morbid obesity, metabolic abnormalities and adhesion molecules correlate with visceral fat, not with subcutaneous fat: effect of weight loss through surgery. Obes Surg 2009;19(6):745–50.

[80] Shore SA. Obesity and asthma: lessons from animal models. J Appl Physiol 2007;102(2):516–28.

[81] Calixto MC, Lintomen L, Schenka A, Saad MJ, Zanesco A, Antunes E. Obesity enhances eosinophilic inflammation in a murine model of allergic asthma. Br J Pharmacol 2010;159(3):617–25.

[82] Kim HY, Lee HJ, Chang YJ, Pichavant M, Shore SA, Fitzgerald KA, et al. Interleukin-17-producing innate lymphoid cells and the NLRP3 inflammasome facilitate obesity-associated airway hyperreactivity. Nat Med 2014;20(1):54–61.

[83] Ather JL, Chung M, Hoyt LR, Randall MJ, Georgsdottir A, Daphtary NA, et al. Weight loss decreases inherent and allergic methacholine hyperresponsiveness in mouse models of diet-induced obese asthma. Am J Respir Cell Mol Biol 2016;55(2):176–87.

[84] Ikejima S, Sasaki S, Sashinami H, Mori F, Ogawa Y, Nakamura T, et al. Impairment of host resistance to listeria monocytogenes infection in liver of db/db and ob/ob mice. Diabetes 2005;54(1):182–9.

[85] Hsu A, Aronoff DM, Phipps J, Goel D, Mancuso P. Leptin improves pulmonary bacterial clearance and survival in ob/ob mice during pneumococcal pneumonia. Clin Exp Immunol 2007;150(2):332–9.

[86] Mancuso P, Gottschalk A, Phare SM, Peters-Golden M, Lukacs NW, Huffnagle GB. Leptin-deficient mice exhibit impaired host defense in Gram-negative pneumonia. J Immunol 2002;168(8):4018–24.

[87] Wright JK, Nwariaku FN, Clark J, Falck JC, Rogers T, Turnage RH. Effect of diabetes mellitus on endotoxin-induced lung injury. Arch Surg 1999;134(12):1354–8 [discussion 8-9].

[88] Bellmeyer A, Martino JM, Chandel NS, Scott Budinger GR, Dean DA, Mutlu GM. Leptin resistance protects mice from hyperoxia-induced acute lung injury. Am J Respir Crit Care Med 2007;175(6):587–94.

[89] Barazzone-Argiroffo C, Muzzin P, Donati YR, Kan CD, Aubert ML, Piguet PF. Hyperoxia increases leptin production: a mechanism mediated through endogenous elevation of corticosterone. Am J Physiol 2001;281(5):L1150–6.

[90] Shore SA, Lang JE, Kasahara DI, Lu FL, Verbout NG, Si H, et al. Pulmonary responses to subacute ozone exposure in obese vs. lean mice. J Appl Physiol 2009;107(5):1445–52.

[91] Shah D, Romero F, Duong M, Wang N, Paudyal B, Suratt BT, et al. Obesity-induced adipokine imbalance impairs mouse pulmonary vascular endothelial function and primes the lung for injury. Sci Rep 2015;5:11362.

[92] Lu FL, Johnston RA, Flynt L, Theman TA, Terry RD, Schwartzman IN, et al. Increased pulmonary responses to acute ozone exposure in obese db/db mice. Am J Physiol 2006;290(5):L856–65.

[93] Johnston RA, Theman TA, Lu FL, Terry RD, Williams ES, Shore SA. Diet-induced obesity causes innate airway hyperresponsiveness to methacholine and enhances ozone-induced pulmonary inflammation. J Appl Physiol 2008;104(6):1727–35.

[94] Shore SA, Lang JE, Kasahara DI, Lu FL, Verbout NG, Si H, et al. Pulmonary responses to subacute ozone exposure in obese vs lean mice. J Appl Physiol (1985) 2009;107(5):1445–52.

[95] Moss M, Guidot DM, Steinberg KP, Duhon GF, Treece P, Wolken R, et al. Diabetic patients have a decreased incidence of acute respiratory distress syndrome. Crit Care Med 2000;28(7):2187–92.

[96] Gong MN, Thompson BT, Williams P, Pothier L, Boyce PD, Christiani DC. Clinical predictors of and mortality in acute respiratory distress syndrome: potential role of red cell transfusion. Crit Care Med 2005;33(6):1191–8.

[97] Iscimen R, Cartin-Ceba R, Yilmaz M, Khan H, Hubmayr RD, Afessa B, et al. Risk factors for the development of acute lung injury in patients with septic shock: an observational cohort study. Crit Care Med 2008;36(5):1518–22.

[98] Honiden S, Gong MN. Diabetes, insulin, and development of acute lung injury. Crit Care Med 2009;37(8):2455–64.

[99] Luo L, Shaver CM, Zhao Z, Koyama T, Calfee CS, Bastarache JA, et al. Clinical predictors of hospital mortality differ between direct and indirect ARDS. Chest 2017;151(4):755–63.

[100] Boichot E, Sannomiya P, Escofier N, Germain N, Fortes ZB, Lagente V. Endotoxin-induced acute lung injury in rats. Role of insulin. Pulm Pharmacol Ther 1999;12(5):285–90.

[101] Alba-Loureiro TC, Martins EF, Landgraf RG, Jancar S, Curi R, Sannomiya P. Role of insulin on PGE2 generation during LPS-induced lung inflammation in rats. Life Sci 2006;78(6):578–85.

[102] de Oliveira Martins J, Meyer-Pflug AR, Alba-Loureiro TC, Melbostad H, Costa da Cruz JW, Coimbra R, et al. Modulation of lipopolysaccharide-induced acute lung inflammation: Role of insulin. Shock 2006;25(3):260–6.

[103] Alba-Loureiro TC, Munhoz CD, Martins JO, Cerchiaro GA, Scavone C, Curi R, et al. Neutrophil function and metabolism in individuals with diabetes mellitus. Braz J Med Biol Res 2007;40(8):1037–44.

[104] Moreno-Navarrete JM, Fernandez-Real JM. Antimicrobial-sensing proteins in obesity and type 2 diabetes: the buffering efficiency hypothesis. Diabetes Care 2011;34(Suppl 2):S335–41.

[105] Moutschen MP, Scheen AJ, Lefebvre PJ. Impaired immune responses in diabetes mellitus: analysis of the factors and mechanisms involved. Relevance to the increased susceptibility of diabetic patients to specific infections. Diabete Metab 1992;18(3):187–201.

[106] Yu S, Christiani DC, Thompson BT, Bajwa EK, Gong MN. Role of diabetes in the development of acute respiratory distress syndrome. Crit Care Med 2013;41(12):2720–32.

[107] Huang ZS, Chien KL, Yang CY, Tsai KS, Wang CH. Peripheral differential leukocyte counts in humans vary with hyperlipidemia, smoking, and body mass index. Lipids 2001;36(3):237–45.

[108] Giugliano G, Brevetti G, Lanero S, Schiano V, Laurenzano E, Chiariello M. Leukocyte count in peripheral arterial disease: a simple, reliable, inexpensive approach to cardiovascular risk prediction. Atherosclerosis 2010;210(1):288–93.

[109] Gomes AL, Carvalho T, Serpa J, Torre C, Dias S. Hypercholesterolemia promotes bone marrow cell mobilization by perturbing the SDF-1:CXCR4 axis. Blood 2010;115(19):3886–94.

[110] Drechsler M, Megens RT, van Zandvoort M, Weber C, Soehnlein O. Hyperlipidemia-triggered neutrophilia promotes early atherosclerosis. Circulation 2010;122(18):1837–45.

[111] Hansson GK, Libby P. The immune response in atherosclerosis: A double-edged sword. Nat Rev Immunol 2006;6(7):508–19.

[112] Kopprasch S, Leonhardt W, Pietzsch J, Kuhne H. Hypochlorite-modified low-density lipoprotein stimulates human polymorphonuclear leukocytes for enhanced production of reactive oxygen metabolites, enzyme secretion, and adhesion to endothelial cells. Atherosclerosis 1998;136(2):315–24.

[113] Lehr HA, Krombach F, Munzing S, Bodlaj R, Glaubitt SI, Seiffge D, et al. In vitro effects of oxidized low density lipoprotein on CD11b/CD18 and L-selectin presentation on neutrophils and monocytes with relevance for the in vivo situation. Am J Pathol 1995;146(1):218–27.

[114] Porreca E, Sergi R, Baccante G, Reale M, Orsini L, Febbo CD, et al. Peripheral blood mononuclear cell production of interleukin-8 and IL-8-dependent neutrophil function in hypercholesterolemic patients. Atherosclerosis 1999;146(2):345–50.

[115] Stragliotto E, Camera M, Postiglione A, Sirtori M, Di Minno G, Tremoli E. Functionally abnormal monocytes in hypercholesterolemia. Arterioscler Thromb 1993;13(6):944–50.

[116] Madenspacher JH, Draper DW, Smoak KA, Li H, Griffiths GL, Suratt BT, et al. Dyslipidemia induces opposing effects on intrapulmonary and extrapulmonary host defense through divergent TLR response phenotypes. J Immunol 2010;185(3):1660–9.

[117] Suratt BT. Mouse modeling of obese lung disease. Insights and caveats. Am J Respir Cell Mol Biol 2016;55(2):153–8.

[118] Fantuzzi G. Adipose tissue, adipokines, and inflammation. J Allergy Clin Immunol 2005;115(5):911–9 [quiz 20].

[119] La Cava A, Matarese G. The weight of leptin in immunity. Nat Rev Immunol 2004;4(5):371–9.

[120] Umemoto Y, Tsuji K, Yang FC, Ebihara Y, Kaneko A, Furukawa S, et al. Leptin stimulates the proliferation of murine myelocytic and primitive hematopoietic progenitor cells. Blood 1997;90(9):3438–43.

[121] Claycombe K, King LE, Fraker PJ. A role for leptin in sustaining lymphopoiesis and myelopoiesis. Proc Natl Acad Sci U S A 2008;105(6):2017–21.

[122] Moore SI, Huffnagle GB, Chen GH, White ES, Mancuso P. Leptin modulates neutrophil phagocytosis of *Klebsiella pneumoniae*. Infect Immun 2003;71(7):4182–5.

[123] Caldefie-Chezet F, Poulin A, Tridon A, Sion B, Vasson MP. Leptin: a potential regulator of polymorphonuclear neutrophil bactericidal action? J Leukoc Biol 2001;69(3):414–8.

[124] Bruno A, Conus S, Schmid I, Simon HU. Apoptotic pathways are inhibited by leptin receptor activation in neutrophils. J Immunol 2005;174(12):8090–6.

[125] Ottonello L, Gnerre P, Bertolotto M, Mancini M, Dapino P, Russo R, et al. Leptin as a uremic toxin interferes with neutrophil chemotaxis. J Am Soc Nephrol 2004;15(9):2366–72.

[126] Caldefie-Chezet F, Poulin A, Vasson MP. Leptin regulates functional capacities of polymorphonuclear neutrophils. Free Radic Res 2003;37(8):809–14.

[127] Ubags ND, Vernooy JH, Burg E, Hayes C, Bement J, Dilli E, et al. The role of leptin in the development of pulmonary neutrophilia in infection and acute lung injury. Crit Care Med 2014;42(2):e143–51.

[128] Bruno A, Chanez P, Chiappara G, Siena L, Giammanco S, Gjomarkaj M, et al. Does leptin play a cytokine-like role within the airways of COPD patients? Eur Respir J 2005;26(3):398–405.

[129] Vernooy JH, Drummen NE, van Suylen RJ, Cloots RH, Moller GM, Bracke KR, et al. Enhanced pulmonary leptin expression in patients with severe COPD and asymptomatic smokers. Thorax 2009;64(1):26–32.

[130] Torpy DJ, Bornstein SR, Chrousos GP. Leptin and interleukin-6 in sepsis. Horm Metab Res 1998;30(12):726–9.

[131] Arnalich F, Lopez J, Codoceo R, Jim nez M, Madero R, Montiel C. Relationship of plasma leptin to plasma cytokines and human survivalin sepsis and septic shock. J Infect Dis 1999;180(3):908–11.

[132] Yousef AA, Amr YM, Suliman GA. The diagnostic value of serum leptin monitoring and its correlation with tumor necrosis factor-alpha in critically ill patients: a prospective observational study. Crit Care 2010;14(2):R33.

[133] Montecucco F, Bianchi G, Gnerre P, Bertolotto M, Dallegri F, Ottonello L. Induction of neutrophil chemotaxis by leptin: crucial role for p38 and Src kinases. Ann N Y Acad Sci 2006;1069:463–71.

[134] Ubags ND, Stapleton RD, Vernooy JH, Burg E, Bement J, Hayes CM, et al. Hyperleptinemia is associated with impaired pulmonary host defense. JCI Insight 2016;1(8).

[135] Gattinoni L, Chiumello D, Carlesso E, Valenza F. Bench-to-bedside review: chest wall elastance in acute lung injury/acute respiratory distress syndrome patients. Crit Care 2004;8(5):350–5.

[136] Hess DR, Bigatello LM. The chest wall in acute lung injury/acute respiratory distress syndrome. Curr Opin Crit Care 2008;14(1):94–102.

[137] Watson RA, Pride NB. Postural changes in lung volumes and respiratory resistance in subjects with obesity. J Appl Physiol (1985) 2005;98(2):512–7.

[138] Burns SM, Egloff MB, Ryan B, Carpenter R, Burns JE. Effect of body position on spontaneous respiratory rate and tidal volume in patients with obesity, abdominal distension and ascites. Am J Crit Care 1994;3(2):102–6.

[139] Adler D, Pepin JL, Dupuis-Lozeron E, Espa-Cervena K, Merlet-Violet R, Muller H, et al. Comorbidities and subgroups of patients surviving severe acute hypercapnic respiratory failure in the intensive care unit. Am J Respir Crit Care Med 2017;196(2):200–7.

Obesity and the microbiome: Big changes on a small scale?

12

Niki D.J. Ubags*, Benjamin J. Marsland*,†

Faculty of Biology and Medicine, University of Lausanne, Epalinges, Switzerland* Department of Immunology and Pathology, Monash University, Melbourne, VIC, Australia†

ABBREVIATIONS

BMI	body mass index
DOHAD	developmental origins of health and disease
EFF	excess free fructose
NHANES	National Health and Nutrition Examination Survey
pHA	postnatal hyperalimentation

The human body is colonized with a diverse array of microbes (bacteria, fungi, and viruses) whose dynamic composition is dependent on the body site and can be influenced by both host and environmental factors. Microbial colonization takes place at body sites exposed to the external environment, including the skin [1], the oral and nasal cavity [2, 3], and the respiratory [4] and gastrointestinal tracts [5, 6]. The predominant bacterial colonizers of these body sites are *Firmicutes, Bacteroidetes, Proteobacteria, Actinobacteria, Fusobacteria* and *Cyanobacteria* [1–6]. However, the total bacterial load and the relative abundance of these phyla may differ depending on the body site [7]. In addition, the composition of the bacterial microbiota on the genus and species level may greatly vary between body sites and individuals. Although studied to a much lesser extent, commensal fungi (mycobiome) have been suggested to be critical players in human health and disease, as well. The main commensal fungi detected in healthy subjects belong to the *Aspergillus, Candida, Cladrosporium, Malassezia* and *Saccharomyces* genera [8, 9]. Human and murine studies have demonstrated that long-term antibiotic treatment decreases the bacterial load in the intestine, leading to an expansion of fungal communities (mostly *Candida*) in the murine intestine [10] and a predisposition to develop fungal infections in humans [11, 12]. Even though, at the genomic level, the mycobiome is only a minor component of the entire microbiota, these studies suggest that its interrelationship with the bacterial microbial composition is crucial in maintaining health.

The microbiota exerts a fundamental role in the maturation and education of the immune system. Immune responses in germ-free mice (mice with a complete

Mechanisms and Manifestations of Obesity in Lung Disease. https://doi.org/10.1016/B978-0-12-813553-2.00012-9

absence of any microbiota) can be seen as "naïve" to the maturation and education provided by both beneficial and pathogenic microbes. The immune defects observed in germ-free mice are most pronounced in the gut, in which underdevelopment of gut-associated lymphoid tissue (GALT) is observed [13]. In addition, a decrease in secretory IgA production by B cells, essential for maintaining mucosal barrier integrity, has been observed in germ-free mice [14, 15]. Germ-free mice also exhibit alterations in T-cell populations. T-helper cell subsets in the spleen of germ-free mice are skewed toward a T-helper 2 (Th2) phenotype, which has been associated with increased allergic responses [16]. Furthermore, circulating T-helper 17 (Th17) cell and regulatory T-cell (Treg) numbers are decreased in germ-free mice [17, 18]. Recolonization of germ-free mice with a conventional microbiota rescues the Th2-Th1 imbalance and is able to restore CD4$^+$ T-cell populations in these mice, indicating the necessity of a microbiota for T-cell maintenance.

Dysbiosis (disturbances in microbial colonization and composition) caused by the use of antimicrobials, dietary changes, or other strategies has been shown to exert profound effects on immune function [19]. The vast majority of external factors influencing the microbiome composition are highly modifiable. Early life exposures such as mode of delivery (vaginal delivery versus caesarean section) and feeding (breastfeeding versus formula) can alter the infant microbiome. In addition, the use of antibiotics and direct or indirect changes induced by the environment and lifestyle factors, such as dietary intake, have the potential to influence the microbiome throughout life. This plasticity of the microbiome could be targeted for future therapies or preventive strategies.

The prevalence of overweight and obesity has increased markedly in the past 30 years [20], which is, at least in part, attributable to drastic changes in dietary habits in the developed world [21]. An increased consumption of a diet consisting of energy-dense and processed foods (also called the "Western diet") coincided with an increase in communicable diseases, as well as inflammatory (e.g., allergies and asthma) and autoimmune diseases [22, 23]. The effects of obesity and alterations in dietary composition on gut microbial composition have been well described, and investigations of the link between obesity and lung diseases are an emerging area of research. However, thus far there are only a few published studies linking obesity-induced dysbiosis to obesity-induced effects on the development and progression of respiratory diseases. In this chapter, we review the current understanding of obesity-associated alterations on gut microbiota composition and its consequent immunomodulatory properties. Furthermore, potential mechanisms that link obesity-induced dysbiosis to initiation and progression of respiratory diseases will be discussed and priority areas for future studies will be highlighted.

OBESITY AND THE MICROBIOME: LINKING DYSBIOSIS TO DISEASE

Changes in dietary habits, such as consumption of a "Western diet," which is generally low in fiber content yet high in animal protein, digestible sugar, and starch and fat content [21], are important contributors to the obesity epidemic. Alterations in nutrient intake have also been shown to influence gut microbial composition [24–30] (Table 12.1). Furthermore, changes in the gut microbiota are associated with the predisposition to and development of obesity. The effects of obesity and dietary composition on the gut microbiome may also have immunomodulatory consequences. However, studies dissecting the potential mechanisms that link obesity-induced dysbiosis to initiation and progression of respiratory diseases remain scarce.

GUT MICROBIOTA AND OBESITY

Germ-free rodents require 30% more calories to maintain their body weight compared with conventional rodents (which possess a normal microbiota) [48]. Interestingly, despite increased food intake, germ-free mice were reported to have significantly less total body fat compared with conventional mice, and recolonization with a conventional microbiota could restore body fat within 2 weeks [49]. A mechanism that may explain the observed differences in body fat is an increase in energy harvest from the ingested food in conventional compared with germ-free rodents, because germ-free rodents have been shown to excrete more calories into their feces [48]. Not only can the gut microbiota influence the nutritional value of a diet, nutrition itself can also shape the composition of the microbiota. Obesity and high-fat diet consumption have been shown to alter microbial composition in both the human and murine intestine, and one of the most profound changes is an increase in the *Firmicutes* to *Bacteroidetes* ratio [37, 38]. This shift has been associated with increased gut permeability and increased circulating levels of lipopolysaccharides (LPS), which can induce systemic inflammation [50, 51]. The *Firmicutes* to *Bacteroidetes* ratio is a rough indicator of shifts in the microbial composition. An increase in this ratio can reflect three different scenarios: an overall increase in the abundance of *Firmicutes*; a decrease in the abundance of *Bacteroidetes*; or increases in both *Firmicutes* and *Bacteroidetes* abundance, with *Firmicutes* increasing faster. Following high-fat diet consumption, the increased abundance of *Firmicutes* accounts for the increases seen in the *Firmicutes* to *Bacteroidetes* ratio. It is postulated that *Firmicutes* promote increased energy harvest from a given energy source and thereby further obesity development.

Changes in the gut microbiota are not only *induced by* obesity but may also then *promote the development of* obesity. Studies in gnotobiotic (mice populated with a defined microbiota) and germ-free mice have shown that such mice do not become obese when fed a high-fat diet but do after bacterial reconstitution with feces from conventionally raised mice [38]. Moreover, increased weight-gain is observed in germ-free mice colonized with fecal contents from conventionally raised obese mice compared with fecal contents from lean mice [52]. Ridaura and colleagues colonized germ-free mice with fecal contents of an obese human subject versus her lean monozygotic twin and observed similar effects [53]. Interestingly, when they cohoused germ-free mice harboring the obese twin's microbiota with mice that received the fecal contents of her lean counterpart, this cohousing prevented the development of increased body mass and associated metabolic phenotypes in the mice colonized with the obese microbiota. These data indicate that a "lean" microbiota can dominate or outcompete an "obese" microbiota.

Table 12.1 Effects of Dietary Composition on the Microbiome, Immune Modulation, and Respiratory Diseases

Dietary Composition	Influence on Microbiome	Immunomodulatory Effects	Effect on Respiratory Diseases
High fructose content	• Increased *Firmicutes/Bacteroidetes* ratio in murine gut [31] • *Bacteroides* and *Erysipelotrichi* increased in murine gut [32]	• Increased circulating uric acid [33], which has potential to stimulate Th2 immunity by activation of inflammatory dendritic cells [34]	• Associated with increased asthma in children [35] and increased prevalence and odds of chronic bronchitis in adults [36]
High fat content	• Increased *Firmicutes/Bacteroidetes* ratio in human and murine intestine [37, 38] • Increase in intestinal *Erysipelotrichi* [38]	• Elevated systemic inflammation upon high-fat meal challenge in human [39] • Dysfunctional monocyte maturation [40] • Neutrophil dysfunction [41]	• Exacerbation of allergic airway responses [42] • Increased influenza mortality [43, 44] • Increased pneumonia severity [41] • Attenuation of acute lung injury response [41, 45]
Low fiber content	• Increased *Firmicutes/Bacteroidetes* ratio murine intestine [46] • Increase in intestinal *Erysipelotrichi* [46]	• Reduced short-chain fatty acid levels (gut and circulation) [46] • Altered dendritic cell hematopoiesis and function [46]	• Associated with decreased lung function [47] • Exaggerated allergic airway inflammation [46]

It is suggested that not only obesity but also the composition of the food itself drives the observed alterations in gut microbial communities. Ravussin and colleagues demonstrated that mice fed a high-fat diet, but in limited quantities to reduce body weight-gain, manifested a gut bacterial community structure similar to that seen in mice fed a high-fat diet *ad libitum*. However, the weights of the high-fat diet-restricted mice were more similar to those seen in mice on a low-fat diet [54]. The type of dietary fat used (either lard or fish-oil-based) can lead to divergent changes in the gut microbiome and differential activation of immune responses, including Toll-like receptor (TLR) activation and consequent increased adipose tissue inflammation following a lard-based fat diet [55]. High-fat diet-induced alterations in the gut microbiota have been linked to overexpansion of endotoxin-producing bacteria, thereby increasing gut permeability [50] and promoting translocation of live gram-negative bacteria through the intestinal mucosa to the circulation and mesenteric adipose tissue [56], which may also contribute to the increased systemic and adipose tissue inflammation seen in obesity.

DIETARY COMPOSITION, MICROBIOME, AND LUNG DISEASE

Long-term dietary patterns have a powerful influence on the gut microbiota composition [57], and they are associated with three gut microbiota profile clusters, also known as enterotypes [58]. The enterotypes are divided according to the predominance of *Bacteroides* (associated with high protein and animal fat diets) [26], *Prevotella* (associated with carbohydrate-rich diets) [26], or *Ruminococcus* (associated with a diet high in resistant starch) [59]; these enterotypes may correlate with individual health status. The Western diet is generally high in added sugars, such as glucose and fructose. Exposure to high levels of fructose (which is mainly found in fruit juice and sweetened beverages) has been suggested to influence asthmatic responses in children. High fructose intake, in the form of consumption of excess free fructose (EFF)-containing beverages, is significantly associated with asthma in children aged 2–9 years [35] and with increased odds for the occurrence of chronic bronchitis in adults aged 20–55 years [36]. Aeberli *et al.* reported that dietary sugar, especially fructose, increases circulating C-reactive protein levels [60], which may suggest low-grade systemic inflammation. Fructose can be metabolized to uric acid, and high fructose intake is associated with increased circulating uric acid levels [33]. In addition, fructose can also stimulate uric acid synthesis from amino acid precursors [61]. Interestingly, uric acid has been suggested to function as an essential initiator and amplifier of Th2 immunity and allergic inflammation through activation of inflammatory dendritic cells [34]. These data suggest that the association between consumption of high fructose-containing beverages and allergic airway responses and asthma may be mediated by uric acid-induced Th2 responses. However, additional studies are required to establish such a link.

High fructose consumption has been suggested to alter gut microbial composition. Consumption of drinking water supplemented with fructose increased the *Firmicutes* to *Bacteroidetes* ratio in the gut and endotoxin translocation in mice fed a normal chow diet [31]. More specifically, *Bacteroides* (a genus within the *Bacteroidetes*) and

Erysipelotrichi (a class of bacteria within the *Firmicutes*) appeared to be increased in the murine intestine upon consumption of fructose-supplemented drinking water [32]. Interestingly, an increase in *Erysipelotrichi* abundance in the mouse gut has also been demonstrated following high-fat diet [38] and low-fiber diet [46]. Descriptive studies have linked alterations in the gut microbiome to respiratory diseases. Although these studies suggest the potential involvement of the gut microbiome in the association between fructose intake and airways disease, a mechanistic link between alterations in dietary sugar intake, gut dysbiosis, and lung disease remains to be established.

Another characteristic of the Western diet is a high dietary fat content [21]. Comparison of lard-derived versus a fish-oil-derived fat in a high-fat diet revealed that *Bacteroides* and *Bilophila* were increased in the gut of lard-fed mice, whereas *Actinobacteria* (specifically *Bifidobacterium* and *Adlercreutzia*), *Lactobacillus*, *Streptococcus* (lactic acid bacteria), and *Verumicrobia* (*Akkermansia muciniphila*) were increased in fish-oil-fed mice [55]. As discussed in the previous section, this was shown to differentially activate immune responses and induce adipose tissue inflammation. In line with those studies, gut microbial changes have been shown to control metabolic endotoxemia-induced inflammation in high-fat diet-fed mice, which was suggested to be mediated by increased gut permeability [50]. In addition, treatment of high-fat diet mice with *A. muciniphila*, which is typically decreased in this setting, reversed endotoxemia and adipose tissue inflammation [51], suggesting potential probiotic applications in obesity.

Consumption of a high-fat diet in rodents has been shown to increase allergic airway inflammation [42], influenza mortality [43, 44], and bacterial pneumonia severity [41], whereas pulmonary inflammation was attenuated in a model of experimentally induced acute lung injury [41, 45]. Although experimental evidence linking these obesity-induced effects on respiratory outcomes to microbial dysbiosis is scarce, involvement of dysbiosis on immune regulation and metabolite production will be a key focus for further investigation.

In recent years, the interest in the potential beneficial effects of a high-fiber diet has increased. Fiber is an important component of our diet [62] and is a substrate for intestinal microbes and short-chain fatty acid production [63]. Fiber-free diet consumption has been reported to promote the expansion of mucus-degrading bacteria in the colon of mice, which consequently increases the susceptibility to infections [64]. High-fiber diet feeding in mice leads to increased short-chain fatty acid levels and is accompanied by an increase in the *Bacteroidetes* to *Firmicutes* ratio in both the intestine [46, 65] and the lung [46]. Low-fiber intake has been associated with lower lung function in an analysis of the National Health and Nutrition Examination Survey (NHANES) database [47]. Moreover, increased dietary fiber intake has been associated with a decrease in airway inflammation (by sputum) and improved lung function in asthmatic subjects [66], and a decrease in respiratory-related deaths [67, 68]. The beneficial effects of high-fiber intake on allergic airway inflammation are mediated by changes in short-chain fatty acid levels and consequent alterations in dendritic cell hematopoiesis [46]. Adequate fiber consumption, and its effects on microbial composition in both the gut and the lung, thus has considerable relevance for lung health.

PARENTAL EFFECTS OF OBESITY ON NEONATAL HEALTH: THE MICROBIOME AS A KEY PLAYER?

The increasing prevalence of obesity may not only affect the health of the overweight or obese individual but may also result in major health concerns for their offspring. The prevalence of obesity in women of childbearing age is of great concern. Results from the 2002 U.S. National Survey of Family Growth demonstrated that 24.5% of nonpregnant women (20–44 years of age) were overweight (BMI 25.0–29.9 kg/m^2) and 23.0% were obese (BMI ≥ 30.0 kg/m^2) [69]. Among obese women, 10.3% were classified as having class II or III obesity (BMI ≥ 35.0 kg/m^2). In line with this data, analysis of the NHANES database reported that the overall prevalence of overweight and obesity in women (20–39 years of age) in 2003–04 was 51.7%, of which 28.9% were obese (BMI ≥ 30.0 kg/m^2) and 8.0% extremely obese (BMI ≥ 40.0 kg/m^2) [70]. Moreover, the overall prevalence of overweight and obesity in men of the same age was 62.2%, of which 28.0% were obese and 3.1% were extremely obese [70]. The influence of maternal weight and gestational weight gain on both perinatal health and transgenerational health are subjects of much study, and the effects of paternal obesity on offspring is gaining interest. Furthermore, the first microbial contact of neonates is the maternal microbiota during pregnancy, birth, and lactation. The contribution of the microbiome to maternal obesity-associated effects on offspring health and disease is highlighted in Figure 12.1.

PARENTAL DIETARY INTAKE AND NEONATAL HEALTH: MAJOR CONCERNS FOR DEVELOPMENTAL PROGRAMMING

Based on the Barker's hypothesis (theory of developmental origins of health and disease or "DOHAD"), it may be suggested that the nutrition of both parents prior to conception is important in the developmental programming of the future child [71, 72]. In an era in which women become pregnant while overweight or obese, it is important to consider the effects of maternal (and paternal) obesity on pregnancy outcome and the developmental programming of offspring health. Maternal obesity during pregnancy is associated with adverse events that may directly influence maternal health and pregnancy outcome [73]. The relative risk of childhood obesity at 4 years of age is associated with maternal obesity during the first trimester [74], and birth weight is strongly correlated with BMI later in life [75]. Differences in maternal gut microbiome between normal weight versus obese pregnant women suggest that the microbiome might be important in weight management during pregnancy. Pregnant women who have greater abundance of *Lactobacillus* appear to be protected against excessive gestational weight gain, and their infants appear less likely to be within the higher percentiles of the expected size for their gestational age at birth [76, 77]. Moreover, maternal obesity increases the risk for developing preeclampsia [78], which is in turn a risk factor for intrauterine growth restriction [79], allergic rhinitis, and increased bronchial responsiveness in childhood [80].

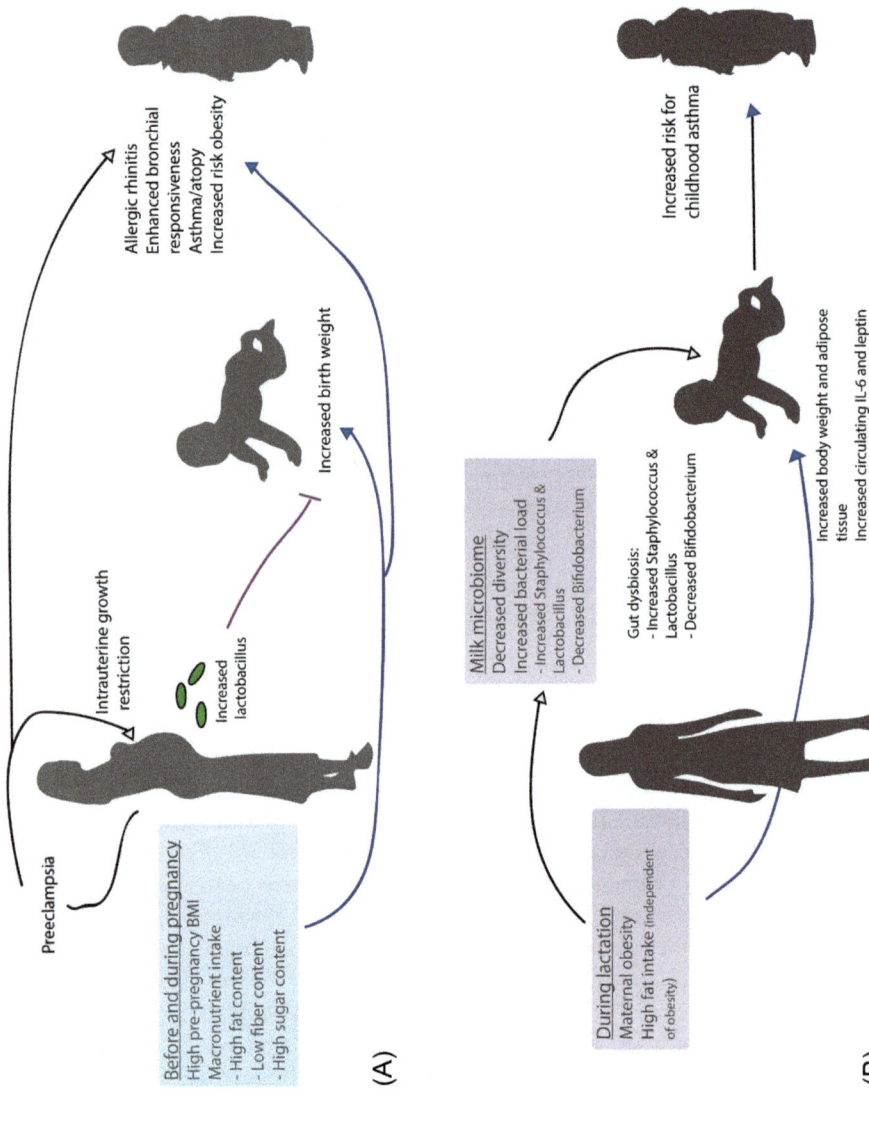

Broadney and colleagues studied the effects of parental obesity on offspring immune programming. They reported a positive association between increasing maternal BMI and neonatal IgM levels and a negative association between increasing paternal weight and neonatal IgM levels [81], which can be suggestive of an increased susceptibility to infections. The net effect of parental BMI on neonatal IgM levels is unclear as maternal and paternal BMI was correlated in this study, and it remains undefined whether one parental influence takes precedence. In line with these findings, experimental evidence in rodents suggests that pups born to high-fat diet-fed dams (prior to and during gestation) exhibit worse outcomes in response to bacterial infection (*Escherichia coli* sepsis model and Methicillin-resistant *Staphylococcus aureus* infection) and experimentally induced autoimmunity, despite being weaned onto a control diet [82]. *ex vivo* stimulation of the colon from high-fat diet-fed dam's offspring with LPS enhanced the production of IL-6, IL-1β, and IL-17A. This hyperinflammatory milieu in the colon resulted in increased systemic LPS exposure and reduced LPS responsiveness, which likely contributed to the observed immune dysregulation and disease susceptibility seen in these mice. These effects were shown to be mediated by alterations in gut microbiota, because cohousing of the pups from high-fat diet-fed dams with pups from low-fat diet-fed dams at the time of weaning reduced the colonic hyperinflammation and systemic LPS exposure, and the susceptibility of the offspring to *E. coli*-induced sepsis [82].

Various research groups have investigated the effects of specific dietary components, independent of obesity, on offspring microbiota and health. Maternal high-fat diet consumption during gestation and lactation, independent of obesity, was shown to alter the composition of the offspring gut microbiome in a nonhuman primate model [83]. These alterations were found to persist up to 1 year of age regardless of the infant's postweaning diet. In a population-based prospective cohort study, the effects of maternal fat intake during gestation, independent of obesity, on offspring gut microbial composition were studied. Fecal samples collected immediately after delivery (meconium) and at 6 weeks of age from infants born to mothers with a low-fat intake (24.4% fat) and mothers with a high-fat intake (43.1% fat) during gestation were analyzed using 16S rRNA gene sequencing. Exposure to a high-fat-containing diet was significantly associated with an enrichment of *Enterococcus* and a relative depletion of *Bacteroides* in meconium samples [84]. Similar to the observations in the nonhuman primate model [83], the neonatal gut microbiome at 6 weeks still possessed the same significant differences at both the phylum and operational taxonomic unit (OTU) levels, indicating a broad effect upon the infants' gut microbiome.

A recent study by Bédard and colleagues investigated the association between maternal free-sugar intake during pregnancy, and adverse respiratory and atopic outcomes in the offspring. Higher maternal intake of free-sugar was associated with an increased risk of atopy and asthma in their offspring (at 7 years of age) [85]. These associations were shown to be independent of sugar intake during early childhood.

Decreased fiber intake is also an important part of the Western diet. Maternal consumption of a high-fiber diet in mice (started at gestational age E13) was shown to attenuate allergic airway inflammation following house dust mite exposure through a regulatory T-cell-dependent mechanism [65]. Similar results were found after maternal consumption of acetate-supplemented drinking water, suggesting that these effects can be mediated by short-chain fatty acids and, in particular, acetate. Moreover, using litter-swap experiments, it was demonstrated that these effects on the offspring immune response were mediated *in utero* and were independent of the transfer of a specific microbiota [65].

MATERNAL NUTRIENT INTAKE, LACTATION, AND INFANT HEALTH

The World Health Organization guidelines on infant feeding practices recommend "exclusive human milk feeding during the first 6 months of infant life, after which nutritionally adequate and safe complimentary feeding should be started from 6 months of age with continued breastfeeding up to 2 years of age and beyond" [86]. Breast milk is a dynamic and complex biological fluid that provides nutrients and bioactive compounds, such as enzymes, proteins, hormones, oligosaccharides, and microbes, needed for infant health and development [87]. *Firmicutes* and *Proteobacteria* phyla are found with the highest relative abundance in breast milk [88, 89]. Maternal obesity decreases microbial diversity and microbiota composition in breast milk. More specifically, obesity and a high BMI were associated with increased total bacterial load, increased abundance of *Staphylococcus* and *Lactobacillus,* and a decrease in *Bifidobacterium* abundance [90]. Interestingly, allergic status of mothers has also been linked to altered human milk microbiota [91]. Allergic mothers were found to have lower concentrations of *Bifidobacteria* in their breast milk compared with nonallergic mothers. In addition, total fecal counts of *Bifidobacteria* were decreased in infants of allergic compared with nonallergic mothers. Because maternal obesity has been associated with increased asthma during childhood [92], it could be speculated that the maternal obesity induced alterations in breast milk microbiome may change the human milk-fed infants' gut microbiome and consequently increase the risk for developing childhood asthma.

Most studies have focused on the effects of maternal obesity through the duration of pregnancy and lactation on offspring health. However, a few studies have investigated the effects of postnatal hyperalimentation, independent of maternal obesity, on infant development. Rolls and colleagues published one of the first studies focusing on this subject in 1986. They determined the effects of cafeteria diet consumption (consisting of a variety of palatable foods: salami, cheese crackers, and chocolate chip cookies, in addition to the stock diet) during either pregnancy and lactation or lactation only in rats that were lean or obese before mating [93]. Cafeteria diet feeding during pregnancy and lactation reduced the proportion of medium-chain fatty acids and increased the proportion of long-chain fatty acids in the milk fat of these animals compared with that from normal chow fed controls. The greatest influence of diet on fatty acid composition was observed during the lactation period. Dinger and colleagues used a mouse model of postnatal hyperalimentation (pHA) to study the effects of maternal high-fat consumption during lactation on offspring health. In this model, dams were fed a high-fat diet starting on the day their offspring were born until the day of weaning, and their offspring was weaned onto a normal chow diet [94]. A significant increase in body weight and white adipose tissue was observed at postnatal day 21 for mice born to dams fed a high-fat diet during lactation, compared with offspring born to dams fed a normal chow diet, whereas body weight and white adipose tissue were similar to controls by postnatal day 70. Interestingly, circulating leptin and IL-6 levels were increased in the pHA offspring from weaning into adulthood [94], and increased circulating leptin and IL-6 levels have been associated with allergic airway inflammation and asthma [95, 96]. In line with this, pHA offspring exhibited increased airway resistance and airway hyperresponsiveness compared with control offspring [94], suggesting that early postnatal hyperalimentation, independent of obesity, leads to excessive postnatal weight gain and metabolic reprogramming of an asthma-like disease in adulthood. Although such changes may reflect the direct effects of diet on the offspring's immune system, data suggests a role for diet-driven changes in the microbiome as well. Furthermore, high-fat diet consumption influences maternal metabolism and can induce hormonal changes [97], which can consequently alter breast milk composition and, in turn, contribute to changes in the offspring's microbiome.

Alterations in the composition of maternal dietary intake may also affect breast milk composition independent of obesity. Hallam *et al.* reported that high-fiber or high-protein feeding of the dams (diet was started 1 week before breeding) affects predominance of certain oligosaccharides (sialic acid, D-glucose, D-galactose, and N-glucosamine) in the maternal milk compared with regular chow-fed controls [98]. These differences were found to be associated with gut microbial composition in the offspring, in particular, increases in *Bifidobacterium* spp. in offspring from high-fiber-fed dams, and stability (no alterations over time even after feeding of a high-fat diet to the offspring) of *Lactobacillus* spp. in offspring of both high-fiber and high-protein-fed dams [98].

Taken together, these studies suggest that consumption of an "unhealthy" diet during pregnancy and lactation (e.g., high fat and low fiber content) can alter the microbial composition in breast milk and consequently the infant gut microbiome, thereby predisposing the infant to diseases such as asthma. This may explain, at least in part, the increased incidence of childhood asthma in children born to obese mothers and mothers with unhealthy dietary intake during lactation. Although additional studies linking the breast milk microbiome to childhood asthma are required to establish causality, such studies may provide crucial insight into the potential dangers of unhealthy diet during lactation for the infant. Recommendations and guidelines on infant feeding practices may require revisions to protect infant health in this era of the obesity epidemic.

WEIGHT LOSS INTERVENTIONS IN OBESE ASTHMA: MICROBIAL INTERACTIONS CONNECTING THE DOTS?

In the current era of the obesity epidemic, a major focus has been on weight reduction in order to minimize obesity-associated comorbidities. Weight loss can be established in various ways. Weight loss counseling with increased exercise combined with a calorie-restricted diet is frequently used and can be an effective intervention [99]. Furthermore, gastric-bypass surgery can be performed. Gastric-bypass surgery is usually performed in extremely obese subjects or obese subjects with a serious weight-related health problem, such as type 2 diabetes, high blood pressure, or severe sleep apnea. As previously discussed, obesity leads to profound alterations in the gut microbial composition. Several studies have indicated that the gut microbiome from obese individuals after weight loss by either dietary intervention or bariatric surgery becomes more similar to that of lean controls [37, 100–103]. The *Firmicutes* to *Bacteroidetes* ratio is decreased during weight loss, and this appears to be mostly due to an increase in *Bacteroides/Prevotella* [101]. In addition, functional analysis has revealed that the abundance of obesity-associated Kyoto Encyclopedia of Genes and Genomes (KEGG) pathways (including pathways involved in carbohydrate fermentation, citrate cycle, glycosaminoglycan degradation, and LPS synthesis) also becomes more similar to that seen in lean subjects 3 months postbariatric surgery [100]. Bariatric surgery goes hand-in-hand with alterations in dietary intake. However it remains to be determined whether the observed effects of bariatric surgery-induced weight loss on gut microbial composition are solely due to the surgery-induced weight loss itself, or are mediated by dietary changes.

Weight reduction has been suggested as a potential therapy to improve control of asthma in severely obese subjects. A recent position paper on weight-loss interventions in asthma concluded that the evidence for weight reduction on asthma is weak [104]. However, only studies using weight loss induced by dietary interventions were included in this analysis. Although the effects of both weight loss through dietary intervention and bariatric surgery on comorbidities may be equal, the sustained effect of weight loss through dietary intervention is small because the likelihood of regaining weight is much higher. Bariatric surgery-induced weight loss significantly reduces the need for inhaled corticosteroids as well as levels of exhaled nitric oxide, and thus decreases bronchial inflammation, in asthmatics [105]. Furthermore, airway responsiveness, small airway function, and asthma severity and control markedly improved with weight loss following bariatric surgery in severely obese patients [106–108]. Dixon and colleagues also observed that obese asthmatics did not report asthma exacerbations in the 12 months following bariatric surgery [108].

Although the effects of bariatric surgery on asthma control and severity are promising, our current understanding of the mechanisms underlying these outcomes is limited. Weight loss is associated with a decrease in adipose tissue mass and a more antiinflammatory activation profile of adipose tissue macrophages. Fat mass reduction following bariatric surgery is associated with decreased numbers of subcutaneous adipose tissue macrophages and a reduction in crown-like structures [109], implicated in the obesity-associated inflammatory state. The remaining adipose tissue macrophages displayed a more M2-like phenotype following weight reduction, compared with the M1-phenotype observed in obesity [109]. However, whether these changes in adipose tissue inflammation directly contribute to the decrease in systemic inflammation observed in obese asthmatics after bariatric surgery remains to be delineated.

Alterations in gut microbiome composition may also contribute to the association between weight reduction and improved asthma severity and control. Ather and colleagues published the first study that set out to determine such a mechanistic link. They used a murine model of diet-induced obesity after which weight loss was initiated by either low-fat diet feeding or gastric sleeve bariatric surgery. Both diet-induced and bariatric surgery-induced weight loss were shown to reduce obesity-associated inherent metacholine responsiveness to levels comparable with lean controls [110], which is in line with reports in human subjects as earlier. Analysis of the fecal microbiome in lean and obese mice with diet-induced weight loss revealed that the *Firmicutes* to *Bacteroidetes* ratio, which was increased in obese mice, had established a "lean" ratio following diet-induced weight loss. Although a clear mechanistic link between alterations in microbial composition and airway reactivity was not established in this work, analysis of metabolite levels, such as short-chain fatty acids and bile acids, in future similar studies could enhance our knowledge in this field and suggest new research avenues.

OUTLOOK

Descriptive studies have not only detailed the effects of obesity on the microbiome but also the effects of alterations in gut microbial composition on obesity in both humans and experimental animal models. Although mechanistic studies linking the microbiome and obesity are currently an area of ongoing investigation, most of these studies focus on the mechanisms underlying the microbiome and obesity-associated metabolic diseases. Other diseases that can be complicated by obesity, such as respiratory diseases, have been broadly suggested to be affected by changes in the microbiome, but mechanistic studies examining this potential link are currently lacking. To delineate the isolated effects of a specific component of obesity, such as dietary composition, diabetes, dyslipidemia, or hypertension, studies in experimental animal models are required, as they may allow the study of a single component's effects on both the microbiome and lung disease.

In light of the DOHAD theory, together with an increasing prevalence of obesity in woman of childbearing age, studies aiming to establish causal relations between maternal obesity and, for example, childhood respiratory diseases should be prioritized in the near future. Furthermore, given the emerging literature on the effects of high-fat diet consumption during lactation on the breast milk microbiome, and consequently the infant microbiome, it is important to expand our understanding of the long-term effects of perinatal exposures.

The ability of the microbial composition in the gut to shape the host immune system in early life and during adulthood may have therapeutic implications. Modulation of the gut microbiome with the addition of microbes (e.g., probiotics), their nutrients (e.g., probiotics/diet), or a combination of both (e.g., synbiotics) in obese subjects may not only contribute to weight reduction but may also ameliorate respiratory symptoms, and may even have the potential to replace invasive bariatric surgeries. Furthermore, when considering the potential negative side effects of bad dietary habits during lactation for infant gut microbial composition, restoration of a "healthy" infant gut microbiota using microbiota-based treatments may decrease childhood allergies. A better understanding of the effects of obesity-associated microbial dysbiosis on lung health can lead the way to the development of microbiota-based treatments.

REFERENCES

[1] Grice EA, Segre JA. The skin microbiome. Nat Rev Microbiol 2011;9(4):244–53.
[2] Frank DN, Feazel LM, Bessesen MT, Price CS, Janoff EN, Pace NR. The human nasal microbiota and *Staphylococcus aureus* carriage. PLoS One 2010;5(5):e10598.
[3] Lazarevic V, Whiteson K, Huse S, Hernandez D, Farinelli L, Osteras M, et al. Metagenomic study of the oral microbiota by Illumina high-throughput sequencing. J Microbiol Methods 2009;79(3):266–71.
[4] Charlson ES, Bittinger K, Haas AR, Fitzgerald AS, Frank I, Yadav A, et al. Topographical continuity of bacterial populations in the healthy human respiratory tract. Am J Respir Crit Care Med 2011;184(8):957–63.
[5] Maldonado-Contreras A, Goldfarb KC, Godoy-Vitorino F, Karaoz U, Contreras M, Blaser MJ, et al. Structure of the human gastric bacterial community in relation to helicobacter pylori status. ISME J 2011;5(4):574–9.
[6] Eckburg PB, Bik EM, Bernstein CN, Purdom E, Dethlefsen L, Sargent M, et al. Diversity of the human intestinal microbial flora. Science 2005;308(5728):1635–8.
[7] Cho I, Blaser MJ. The human microbiome: at the interface of health and disease. Nat Rev Genet 2012;13(4):260–70.
[8] Underhill DM, Iliev ID. The mycobiota: interactions between commensal fungi and the host immune system. Nat Rev Immunol 2014;14(6):405–16.
[9] Iliev ID, Leonardi I. Fungal dysbiosis: immunity and interactions at mucosal barriers. Nat Rev Immunol 2017;.
[10] Dollive S, Chen YY, Grunberg S, Bittinger K, Hoffmann C, Vandivier L, et al. Fungi of the murine gut: episodic variation and proliferation during antibiotic treatment. PLoS One 2013;8(8):e71806.
[11] Samonis G, Gikas A, Anaissie EJ, Vrenzos G, Maraki S, Tselentis Y, et al. Prospective evaluation of effects of broad-spectrum antibiotics on gastrointestinal yeast colonization of humans. Antimicrob Agents Chemother 1993;37(1):51–3.
[12] Mulligan ME, Citron DM, McNamara BT, Finegold SM. Impact of cefoperazone therapy on fecal flora. Antimicrob Agents Chemother 1982;22(2):226–30.
[13] Rhee KJ, Sethupathi P, Driks A, Lanning DK, Knight KL. Role of commensal bacteria in development of gut-associated lymphoid tissues and preimmune antibody repertoire. J Immunol 2004;172(2):1118–24.
[14] Cahenzli J, Koller Y, Wyss M, Geuking MB, McCoy KD. Intestinal microbial diversity during early-life colonization shapes long-term IgE levels. Cell Host Microbe 2013;14(5):559–70.
[15] Lecuyer E, Rakotobe S, Lengline-Garnier H, Lebreton C, Picard M, Juste C, et al. Segmented filamentous bacterium uses secondary and tertiary lymphoid tissues to induce gut IgA and specific T helper 17 cell responses. Immunity 2014;40(4):608–20.

[16] Mazmanian SK, Liu CH, Tzianabos AO, Kasper DL. An immunomodulatory molecule of symbiotic bacteria directs maturation of the host immune system. Cell 2005;122(1):107–18.

[17] Ivanov II, Frutos Rde L, Manel N, Yoshinaga K, Rifkin DB, Sartor RB, et al. Specific microbiota direct the differentiation of IL-17-producing T-helper cells in the mucosa of the small intestine. Cell Host Microbe 2008;4(4):337–49.

[18] Strauch UG, Obermeier F, Grunwald N, Gurster S, Dunger N, Schultz M, et al. Influence of intestinal bacteria on induction of regulatory T cells: lessons from a transfer model of colitis. Gut 2005;54(11):1546–52.

[19] Logan AC, Jacka FN, Prescott SL. Immune-microbiota interactions: dysbiosis as a Global Health Issue. Curr Allergy Asthma Rep 2016;16(2):13.

[20] Ng M, Fleming T, Robinson M, Thomson B, Graetz N, Margono C, et al. Global, regional, and national prevalence of overweight and obesity in children and adults during 1980-2013: a systematic analysis for the Global Burden of Disease Study 2013. Lancet 2014;384(9945):766–81.

[21] Cordain L, Eaton SB, Sebastian A, Mann N, Lindeberg S, Watkins BA, et al. Origins and evolution of the western diet: health implications for the 21st century. Am J Clin Nutr 2005;81(2):341–54.

[22] Eder W, Ege MJ, von Mutius E. The asthma epidemic. N Engl J Med 2006;355(21):2226–35.

[23] Bach JF. The effect of infections on susceptibility to autoimmune and allergic diseases. N Engl J Med 2002;347(12):911–20.

[24] David LA, Maurice CF, Carmody RN, Gootenberg DB, Button JE, Wolfe BE, et al. Diet rapidly and reproducibly alters the human gut microbiome. Nature 2014;505(7484):559–63.

[25] De Filippo C, Cavalieri D, Di Paola M, Ramazzotti M, Poullet JB, Massart S, et al. Impact of diet in shaping gut microbiota revealed by a comparative study in children from Europe and rural Africa. Proc Natl Acad Sci U S A 2010;107(33):14691–6.

[26] Wu GD, Chen J, Hoffmann C, Bittinger K, Chen YY, Keilbaugh SA, et al. Linking long-term dietary patterns with gut microbial enterotypes. Science 2011;334(6052):105–8.

[27] Cotillard A, Kennedy SP, Kong LC, Prifti E, Pons N, Le Chatelier E, et al. Dietary intervention impact on gut microbial gene richness. Nature 2013;500(7464):585–8.

[28] Kovatcheva-Datchary P, Nilsson A, Akrami R, Lee YS, De Vadder F, Arora T, et al. Dietary fiber-induced improvement in glucose metabolism is associated with increased abundance of Prevotella. Cell Metab 2015;22(6):971–82.

[29] Walker AW, Ince J, Duncan SH, Webster LM, Holtrop G, Ze X, et al. Dominant and diet-responsive groups of bacteria within the human colonic microbiota. ISME J 2011;5(2):220–30.

[30] Muegge BD, Kuczynski J, Knights D, Clemente JC, Gonzalez A, Fontana L, et al. Diet drives convergence in gut microbiome functions across mammalian phylogeny and within humans. Science 2011;332(6032):970–4.

[31] Volynets V, Louis S, Pretz D, Lang L, Ostaff MJ, Wehkamp J, et al. Intestinal barrier function and the gut microbiome are differentially affected in mice fed a western-style diet or drinking water supplemented with fructose. J Nutr 2017;147(5):770–80.

[32] Ferrere G, Leroux A, Wrzosek L, Puchois V, Gaudin F, Ciocan D, et al. Activation of Kupffer cells is associated with a specific Dysbiosis induced by fructose or high fat diet in mice. PLoS One 2016;11(1):e0146177.

[33] Johnson RJ, Nakagawa T, Sanchez-Lozada LG, Shafiu M, Sundaram S, Le M, et al. Sugar, uric acid, and the etiology of diabetes and obesity. Diabetes 2013;62(10):3307–15.

[34] Kool M, Willart MA, van Nimwegen M, Bergen I, Pouliot P, Virchow JC, et al. An unexpected role for uric acid as an inducer of T helper 2 cell immunity to inhaled antigens and inflammatory mediator of allergic asthma. Immunity 2011;34(4):527–40.

[35] DeChristopher LR, Uribarri J, Tucker KL. Intakes of apple juice, fruit drinks and soda are associated with prevalent asthma in US children aged 2–9 years. Public Health Nutr 2016;19(1):123–30.

[36] DeChristopher LR, Uribarri J, Tucker KL. Intake of high fructose corn syrup sweetened soft drinks is associated with prevalent chronic bronchitis in U.S. adults, ages 20–55 y. Nutr J 2015;14:107.

[37] Ley RE, Turnbaugh PJ, Klein S, Gordon JI. Microbial ecology: human gut microbes associated with obesity. Nature 2006;444(7122):1022–3.

[38] Turnbaugh PJ, Backhed F, Fulton L, Gordon JI. Diet-induced obesity is linked to marked but reversible alterations in the mouse distal gut microbiome. Cell Host Microbe 2008;3(4):213–23.

[39] Wood LG, Garg ML, Gibson PG. A high-fat challenge increases airway inflammation and impairs bronchodilator recovery in asthma. J Allergy Clin Immunol 2011;127(5):1133–40.

[40] Zhou Q, Leeman SE, Amar S. Signaling mechanisms involved in altered function of macrophages from diet-induced obese mice affect immune responses. Proc Natl Acad Sci U S A 2009;106(26):10740–5.

[41] Ubags ND, Burg E, Antkowiak M, Wallace AM, Dilli E, Bement J, et al. A comparative study of lung host defense in murine obesity models. Insights into neutrophil function. Am J Respir Cell Mol Biol 2016;55(2):188–200.

[42] Kim HY, Lee HJ, Chang YJ, Pichavant M, Shore SA, Fitzgerald KA, et al. Interleukin-17-producing innate lymphoid cells and the NLRP3 inflammasome facilitate obesity-associated airway hyperreactivity. Nat Med 2014;20(1):54–61.

[43] Smith AG, Sheridan PA, Harp JB, Beck MA. Diet-induced obese mice have increased mortality and altered immune responses when infected with influenza virus. J Nutr 2007;137(5):1236–43.

[44] Easterbrook JD, Dunfee RL, Schwartzman LM, Jagger BW, Sandouk A, Kash JC, et al. Obese mice have increased morbidity and mortality compared to non-obese mice during infection with the 2009 pandemic H1N1 influenza virus. Influenza Other Respi Viruses 2011;5(6):418–25.

[45] Kordonowy LL, Burg E, Lenox CC, Gauthier LM, Petty JM, Antkowiak M, et al. Obesity is associated with neutrophil dysfunction and attenuation of murine acute lung injury. Am J Respir Cell Mol Biol 2012;47(1):120–7.

[46] Trompette A, Gollwitzer ES, Yadava K, Sichelstiel AK, Sprenger N, Ngom-Bru C, et al. Gut microbiota metabolism of dietary fiber influences allergic airway disease and hematopoiesis. Nat Med 2014;20(2):159–66.

[47] Hanson C, Lyden E, Rennard S, Mannino DM, Rutten EP, Hopkins R, et al. The relationship between dietary fiber intake and lung function in the National Health and Nutrition Examination Surveys. Ann Am Thorac Soc 2016;13(5):643–50.

[48] Wostmann BS, Larkin C, Moriarty A, Bruckner-Kardoss E. Dietary intake, energy metabolism, and excretory losses of adult male germfree Wistar rats. Lab Anim Sci 1983;33(1):46–50.

[49] Backhed F, Ding H, Wang T, Hooper LV, Koh GY, Nagy A, et al. The gut microbiota as an environmental factor that regulates fat storage. Proc Natl Acad Sci U S A 2004;101(44):15718–23.

[50] Cani PD, Bibiloni R, Knauf C, Waget A, Neyrinck AM, Delzenne NM, et al. Changes in gut microbiota control metabolic endotoxemia-induced inflammation in high-fat diet-induced obesity and diabetes in mice. Diabetes 2008;57(6):1470–81.

[51] Cani PD, Possemiers S, Van de Wiele T, Guiot Y, Everard A, Rottier O, et al. Changes in gut microbiota control inflammation in obese mice through a mechanism involving GLP-2-driven improvement of gut permeability. Gut 2009;58(8):1091–103.

[52] Turnbaugh PJ, Ley RE, Mahowald MA, Magrini V, Mardis ER, Gordon JI. An obesity-associated gut microbiome with increased capacity for energy harvest. Nature 2006;444(7122):1027–31.

[53] Ridaura VK, Faith JJ, Rey FE, Cheng J, Duncan AE, Kau AL, et al. Gut microbiota from twins discordant for obesity modulate metabolism in mice. Science 2013;341(6150):1241214.

[54] Ravussin Y, Koren O, Spor A, LeDuc C, Gutman R, Stombaugh J, et al. Responses of gut microbiota to diet composition and weight loss in lean and obese mice. Obesity (Silver Spring) 2012;20(4):738–47.

[55] Caesar R, Tremaroli V, Kovatcheva-Datchary P, Cani PD, Backhed F. Crosstalk between gut microbiota and dietary lipids aggravates WAT inflammation through TLR signaling. Cell Metab 2015;22(4):658–68.

[56] Amar J, Chabo C, Waget A, Klopp P, Vachoux C, Bermudez-Humaran LG, et al. Intestinal mucosal adherence and translocation of commensal bacteria at the early onset of type 2 diabetes: molecular mechanisms and probiotic treatment. EMBO Mol Med 2011;3(9):559–72.

[57] Ursell LK, Clemente JC, Rideout JR, Gevers D, Caporaso JG, Knight R. The interpersonal and intrapersonal diversity of human-associated microbiota in key body sites. J Allergy Clin Immunol 2012;129(5):1204–8.

[58] Arumugam M, Raes J, Pelletier E, Le Paslier D, Yamada T, Mende DR, et al. Enterotypes of the human gut microbiome. Nature 2011;473(7346):174–80.

[59] Salonen A, Lahti L, Salojarvi J, Holtrop G, Korpela K, Duncan SH, et al. Impact of diet and individual variation on intestinal microbiota composition and fermentation products in obese men. ISME J 2014;8(11):2218–30.

[60] Aeberli I, Gerber PA, Hochuli M, Kohler S, Haile SR, Gouni-Berthold I, et al. Low to moderate sugar-sweetened beverage consumption impairs glucose and lipid metabolism and promotes inflammation in healthy young men: a randomized controlled trial. Am J Clin Nutr 2011;94(2):479–85.

[61] Emmerson BT. Effect of oral fructose on urate production. Ann Rheum Dis 1974;33(3):276–80.

[62] King DE, Mainous 3rd AG, Lambourne CA. Trends in dietary fiber intake in the United States, 1999–2008. J Acad Nutr Diet 2012;112(5):642–8.

[63] Sonnenburg ED, Sonnenburg JL. Starving our microbial self: the deleterious consequences of a diet deficient in microbiota-accessible carbohydrates. Cell Metab 2014;20(5):779–86.

[64] Desai MS, Seekatz AM, Koropatkin NM, Kamada N, Hickey CA, Wolter M, et al. A dietary fiber-deprived gut microbiota degrades the colonic mucus barrier and enhances pathogen susceptibility. Cell 2016;167(5). 1339-53.e21.

[65] Thorburn AN, McKenzie CI, Shen S, Stanley D, Macia L, Mason LJ, et al. Evidence that asthma is a developmental origin disease influenced by maternal diet and bacterial metabolites. Nat Commun 2015;6:7320.

[66] Halnes I, Baines KJ, Berthon BS, MacDonald-Wicks LK, Gibson PG, Wood LG. Soluble fibre meal challenge reduces airway inflammation and expression of GPR43 and GPR41 in asthma. Nutrients 2017;9(1).

[67] Chuang SC, Norat T, Murphy N, Olsen A, Tjonneland A, Overvad K, et al. Fiber intake and total and cause-specific mortality in the European prospective investigation into cancer and nutrition cohort. Am J Clin Nutr 2012;96(1):164–74.

[68] Park Y, Subar AF, Hollenbeck A, Schatzkin A. Dietary fiber intake and mortality in the NIH-AARP diet and health study. Arch Intern Med 2011;171(12):1061–8.

[69] Vahratian A. Prevalence of overweight and obesity among women of childbearing age: results from the 2002 National Survey of Family Growth. Matern Child Health J 2009;13(2):268–73.

[70] Ogden CL, Carroll MD, Curtin LR, McDowell MA, Tabak CJ, Flegal KM. Prevalence of overweight and obesity in the United States, 1999-2004. JAMA 2006;295(13):1549–55.

[71] Barker DJ, Osmond C. Infant mortality, childhood nutrition, and ischaemic heart disease in England and Wales. Lancet 1986;1(8489):1077–81.

[72] Wadhwa PD, Buss C, Entringer S, Swanson JM. Developmental origins of health and disease: brief history of the approach and current focus on epigenetic mechanisms. Semin Reprod Med 2009;27(5):358–68.

[73] Leddy MA, Power ML, Schulkin J. The impact of maternal obesity on maternal and fetal health. Rev Obstet Gynecol 2008;1(4):170–8.

[74] Whitaker RC. Predicting preschooler obesity at birth: the role of maternal obesity in early pregnancy. Pediatrics 2004;114(1):e29–36.

[75] Oken E, Gillman MW. Fetal origins of obesity. Obes Res 2003;11(4):496–506.

[76] Collado MC, Laitinen K, Salminen S, Isolauri E. Maternal weight and excessive weight gain during pregnancy modify the immunomodulatory potential of breast milk. Pediatr Res 2012;72(1):77–85.

[77] Gronlund MM, Grzeskowiak L, Isolauri E, Salminen S. Influence of mother's intestinal microbiota on gut colonization in the infant. Gut Microbes 2011;2(4):227–33.

[78] O'Brien TE, Ray JG, Chan WS. Maternal body mass index and the risk of preeclampsia: a systematic overview. Epidemiology 2003;14(3):368–74.

[79] Castro LC, Avina RL. Maternal obesity and pregnancy outcomes. Curr Opin Obstet Gynecol 2002;14(6):601–6.

[80] Stokholm J, Sevelsted A, Anderson UD, Bisgaard H. Preeclampsia associates with asthma, allergy, and eczema in childhood. Am J Respir Crit Care Med 2017;195(5):614–21.

[81] Broadney MM, Chahal N, Michels KA, McLain AC, Ghassabian A, Lawrence DA, et al. Impact of parental obesity on neonatal markers of inflammation and immune response. Int J Obes (Lond) 2017;41(1):30–7.

[82] Myles IA, Fontecilla NM, Janelsins BM, Vithayathil PJ, Segre JA, Datta SK. Parental dietary fat intake alters offspring microbiome and immunity. J Immunol 2013;191(6):3200–9.

[83] Ma J, Prince AL, Bader D, Hu M, Ganu R, Baquero K, et al. High-fat maternal diet during pregnancy persistently alters the offspring microbiome in a primate model. Nat Commun 2014;5:3889.

[84] Chu DM, Antony KM, Ma J, Prince AL, Showalter L, Moller M, et al. The early infant gut microbiome varies in association with a maternal high-fat diet. Genome Med 2016;8(1):77.

[85] Bédard A, Northstone K, Henderson AJ, Shaheen SO. Maternal intake of sugar during pregnancy and childhood respiratory and atopic outcomes. Eur Respir J 2017;50(1):1700073.

[86] Kramer MS, Kakuma R. The optimal duration of exclusive breastfeeding: a systematic review. Adv Exp Med Biol 2004;554:63–77.

[87] Ballard O, Morrow AL. Human milk composition: nutrients and bioactive factors. Pediatr Clin N Am 2013;60(1):49–74.

[88] Pannaraj PS, Li F, Cerini C, Bender JM, Yang S, Rollie A, et al. Association between breast milk bacterial communities and establishment and development of the infant gut microbiome. JAMA Pediatr 2017;171(7):647–54.

[89] Jost T, Lacroix C, Braegger C, Chassard C. Assessment of bacterial diversity in breast milk using culture-dependent and culture-independent approaches. Br J Nutr 2013;110(7):1253–62.

[90] Cabrera-Rubio R, Collado MC, Laitinen K, Salminen S, Isolauri E, Mira A. The human milk microbiome changes over lactation and is shaped by maternal weight and mode of delivery. Am J Clin Nutr 2012;96(3):544–51.

[91] Gronlund MM, Gueimonde M, Laitinen K, Kociubinski G, Gronroos T, Salminen S, et al. Maternal breast-milk and intestinal bifidobacteria guide the compositional development of the *Bifidobacterium* microbiota in infants at risk of allergic disease. Clin Exp Allergy 2007;37(12):1764–72.

[92] Forno E, Young OM, Kumar R, Simhan H, Celedon JC. Maternal obesity in pregnancy, gestational weight gain, and risk of childhood asthma. Pediatrics 2014;134(2):e535–46.

[93] Rolls BA, Gurr MI, van Duijvenvoorde PM, Rolls BJ, Rowe EA. Lactation in lean and obese rats: effect of cafeteria feeding and of dietary obesity on milk composition. Physiol Behav 1986;38(2):185–90.

[94] Dinger K, Kasper P, Hucklenbruch-Rother E, Vohlen C, Jobst E, Janoschek R, et al. Early-onset obesity dysregulates pulmonary adipocytokine/insulin signaling and induces asthma-like disease in mice. Sci Rep 2016;6:24168.

[95] Yokoyama A, Kohno N, Fujino S, Hamada H, Inoue Y, Fujioka S, et al. Circulating interleukin-6 levels in patients with bronchial asthma. Am J Respir Crit Care Med 1995;151(5):1354–8.

[96] Shore SA, Schwartzman IN, Mellema MS, Flynt L, Imrich A, Johnston RA. Effect of leptin on allergic airway responses in mice. J Allergy Clin Immunol 2005;115(1):103–9.

[97] Fields DA, Demerath EW. Relationship of insulin, glucose, leptin, IL-6 and TNF-alpha in human breast milk with infant growth and body composition. Pediatr Obes 2012;7(4):304–12.

[98] Hallam MC, Barile D, Meyrand M, German JB, Reimer RA. Maternal high-protein or high-prebiotic-fiber diets affect maternal milk composition and gut microbiota in rat dams and their offspring. Obesity (Silver Spring) 2014;22(11):2344–51.

[99] Martins C, Strommen M, Stavne OA, Nossum R, Marvik R, Kulseng B. Bariatric surgery versus lifestyle interventions for morbid obesity—changes in body weight, risk factors and comorbidities at 1 year. Obes Surg 2011;21(7):841–9.

[100] Liu R, Hong J, Xu X, Feng Q, Zhang D, Gu Y, et al. Gut microbiome and serum metabolome alterations in obesity and after weight-loss intervention. Nat Med 2017;23(7):859–68.

[101] Furet JP, Kong LC, Tap J, Poitou C, Basdevant A, Bouillot JL, et al. Differential adaptation of human gut microbiota to bariatric surgery-induced weight loss: links with metabolic and low-grade inflammation markers. Diabetes 2010;59(12):3049–57.

[102] Kong LC, Tap J, Aron-Wisnewsky J, Pelloux V, Basdevant A, Bouillot JL, et al. Gut microbiota after gastric bypass in human obesity: increased richness and associations of bacterial genera with adipose tissue genes. Am J Clin Nutr 2013;98(1):16–24.

[103] Turnbaugh PJ, Hamady M, Yatsunenko T, Cantarel BL, Duncan A, Ley RE, et al. A core gut microbiome in obese and lean twins. Nature 2009;457(7228):480–4.

[104] Moreira A, Bonini M, Garcia-Larsen V, Bonini S, Del Giacco SR, Agache I, et al. Weight loss interventions in asthma: EAACI evidence-based clinical practice guideline (part I). Allergy 2013;68(4):425–39.

[105] Lombardi C, Gargioni S, Gardinazzi A, Canonica GW, Passalacqua G. Impact of bariatric surgery on pulmonary function and nitric oxide in asthmatic and non-asthmatic obese patients. J Asthma 2011;48(6):553–7.

[106] Boulet LP, Turcotte H, Martin J, Poirier P. Effect of bariatric surgery on airway response and lung function in obese subjects with asthma. Respir Med 2012;106(5):651–60.

[107] van Huisstede A, Rudolphus A, Castro Cabezas M, Biter LU, van de Geijn GJ, Taube C, et al. Effect of bariatric surgery on asthma control, lung function and bronchial and systemic inflammation in morbidly obese subjects with asthma. Thorax 2015;70(7):659–67.

[108] Dixon AE, Pratley RE, Forgione PM, Kaminsky DA, Whittaker-Leclair LA, Griffes LA, et al. Effects of obesity and bariatric surgery on airway hyperresponsiveness, asthma control, and inflammation. J Allergy Clin Immunol 2011;128(3). 508-15.e1-2.

[109] Cancello R, Henegar C, Viguerie N, Taleb S, Poitou C, Rouault C, et al. Reduction of macrophage infiltration and chemoattractant gene expression changes in white adipose tissue of morbidly obese subjects after surgery-induced weight loss. Diabetes 2005;54(8):2277–86.

[110] Ather JL, Chung M, Hoyt LR, Randall MJ, Georgsdottir A, Daphtary NA, et al. Weight loss decreases inherent and allergic methacholine hyperresponsiveness in mouse models of diet-induced obese asthma. Am J Respir Cell Mol Biol 2016;55(2):176–87.

GLOSSARY

Commensal   Different species living in close association, such that one species can benefit without harming the others

Dysbiosis   Disturbances in microbial colonization/composition

Germ-free   With a complete absence of any microbiota

Gnotobiotic   Populated with a defined microbiota

Microbiome   The entire collection of microorganisms and their collective genetic material present in or on the human body or in another environment

Microbiota   The community of microorganisms that reside either on the surface or in different cavities of the body: the skin, the mouth, the ears, the vagina, or the gastrointestinal tract, among others

Mycobiome   The entire fungal community and their collective genetic material present in the human body or in another environment

Relative abundance   The number of organisms of a particular kind (e.g., bacterial or fungal) as a percentage of the total number of organisms of a given community

Apolipoproteins as context-dependent regulators of lung inflammation

13

Debbie M. Figueroa, Elizabeth M. Gordon, Xianglan Yao, Stewart J. Levine

Laboratory of Asthma and Lung Inflammation, Pulmonary Branch, Division of Intramural Research, NHLBI, NIH, Bethesda, MD, United States

Apolipoproteins are the protein constituents of lipoprotein particles that play a key role in transporting cholesterol, triglycerides, phospholipids, and fat-soluble vitamins between the intestine, liver, and peripheral tissues [1]. This allows cholesterol to be utilized for membrane synthesis and steroid hormone production, as well as free fatty acids to be processed as a fuel source [2]. However, it is becoming increasingly recognized that apolipoproteins are also expressed and have local biological functions in the lung, where they modulate the pathogenesis and severity of lung inflammation.

The two main apolipoproteins that have been identified as regulators of lung inflammation are apolipoprotein E and apolipoprotein A-I [3]. Throughout this chapter, proteins for apolipoprotein E and apolipoprotein A-I are abbreviated as apoE and apoA-I, respectively; the corresponding human genes are denoted as *APOE* and *APOA1*; and the murine genes are denoted as *Apoe* and *Apoa1*. ApoE is primarily a component of chylomicron remnants, but is also present on very low-density lipoproteins (VLDL), intermediate-density lipoproteins (IDL), and high-density lipoproteins (HDL). A main function of apoE is to mediate the uptake of cholesterol and lipids into cells by binding to the low-density lipoprotein (LDL) receptor, which is a process termed receptor-mediated endocytosis (Figure 13.1) [4]. This facilitates the clearance of chylomicrons, which transport dietary cholesterol, triglycerides, fatty acids, phospholipids, and retinyl esters from the small intestine to peripheral cells and the liver, as well as VLDL, which transports lipids from the liver to peripheral cells [1]. In contrast, a main function of apoA-I is to transport cholesterol out of cells, which is a process termed reverse cholesterol transport [5–8]. In particular, apoA-I interacts with the ATP-binding cassette subfamily A member 1 (ABCA1) to efflux unesterified cholesterol and phospholipids from peripheral cells. This forms nascent HDL particles that can acquire additional cholesterol by interactions with the ABCG1 transporter and scavenger receptor class B member 1 (SRB1) [8]. HDL can then deliver esterified cholesterol to the liver via interactions with SRB1 [9]. Interestingly, apoE-containing HDL particles also mediate reverse cholesterol transport from macrophages [8]. Clusterin, which is also known as apolipoprotein J, is another apolipoprotein component of HDL that may modulate lung inflammation

Mechanisms and Manifestations of Obesity in Lung Disease. https://doi.org/10.1016/B978-0-12-813553-2.00013-0
2019 Published by Elsevier Inc.

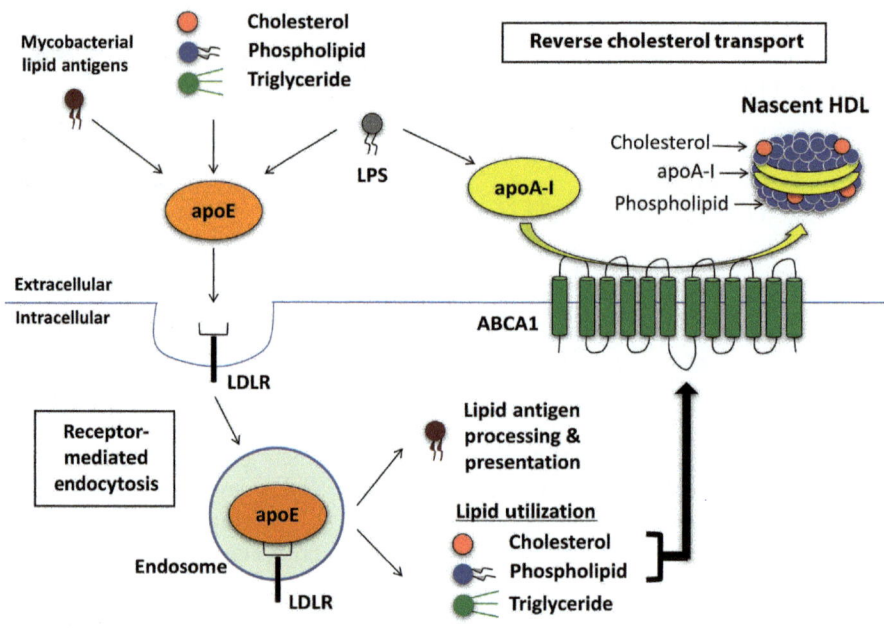

FIGURE 13.1

Modulation of lipid trafficking in the lung by apolipoprotein E and apolipoprotein A-I. Apolipoprotein E (apoE) mediates the cellular uptake of cholesterol, triglycerides, and phospholipids within lipoprotein particles by interacting with low-density lipoprotein receptors (LDLR) on the cell surface. LDLRs are then internalized into cells within clathrin-coated endocytic vesicles by a process termed receptor-mediated endocytosis. ApoE also mediates the LDLR-dependent internalization of mycobacterial lipid antigens by antigen-presenting cells. Lipid-free and lipid-poor apolipoprotein A-I (apoA-I) can interact with ATP-binding cassette subfamily A member 1 (ABCA1) transporters on the cell surface to efflux cholesterol and phospholipids out of cells and thereby create nascent HDL particles. Both apoE and apoA-I can directly bind and neutralize LPS.

Reprinted with permission of the American Thoracic Society. Copyright © 2017 American Thoracic Society. Yao X, Gordon EM, Figueroa DM, Barochia AV, Levine SJ. Emerging roles of apolipoprotein E and apolipoprotein A-I in the pathogenesis and treatment of lung disease. Am J Respir Cell Mol Biol 2016;55(2):159–69. The American Journal of Respiratory Cell and Molecular Biology is an official Journal of the American Thoracic Society.

in the setting of asthma and acute lung injury. Here, the context-dependent roles of apoE, apoA-I, and clusterin in regulating lung inflammation will be reviewed.

APOLIPOPROTEIN E

ApoE is a 34-kDa protein comprised of 299 amino acids and is highly expressed in the liver and brain [10]. ApoE is also expressed in the lung by several cell types, including alveolar macrophages, alveolar epithelial cells (both type I and type II), and pulmonary artery smooth muscle cells [11–14]. The apoE amino-terminus contains four helical bundles, one of which represents the LDL-receptor binding domain, whereas the carboxy-terminus functions as the lipid-binding domain. Apolipoprotein E binds to receptors belonging to the LDL receptor family, which includes the LDL receptor (LDLR), the very low-density lipoprotein receptor (VLDLR), ApoE receptor 2 (which is also known as LDL receptor-related protein 8), and multiple LDL receptor-related proteins (i.e., LRP1, LRP1B, LRP2 (megalin), LRP4 (MEGF7), LRP5, LRP6, and SORLA (sortilin-related receptor with A-type repeats)) [15].

The human *APOE* gene is polymorphic with three common alleles, ε2, ε3, and ε4, which have significant structural and functional differences [10, 16, 17]. Although these polymorphisms do not involve the receptor- or lipid-binding domains of apoE, they significantly alter its structural and functional properties. *APOE* ε3, which encodes for a cysteine at position 112 and an arginine at position 158, is the most common allele, whereas the ε2 allele encodes cysteines and the ε4 allele encodes arginines at both sites. The apoE2 protein has markedly diminished LDL-receptor binding activity, which is less than 2% of the apoE3 protein, and results in type III hyperlipoproteinemia. Although apoE4 has normal LDL receptor binding, it has altered lipid binding characteristics, with preferential binding to triglyceride-rich VLDL and chylomicron remnants. Consequently, the apoE4 protein is associated with elevated plasma levels of LDL, hyperlipidemia, and increased cardiovascular disease risk. In addition, the *APOE* ε4 allele is the major genetic risk factor for Alzheimer's disease and is also associated with diminished longevity [10, 18–20]. Furthermore, macrophages from humanized knock-in mice expressing the *APOE* ε4 allele display reduced efferocytosis, as well as augmented apoptosis and endoplasmic reticulum stress that may promote inflammation [21].

APOE KNOCKOUT MICE AS A MODEL OF DIET-INDUCED INFLAMMATION-DEPENDENT LUNG REMODELING

Apolipoprotein *E*-containing lipoprotein particles, such as chylomicrons, VLDL, and IDL, mediate the clearance of lipids and cholesterol from plasma into cells. Consistent with this, mice genetically deficient in *Apoe* (*Apoe*$^{-/-}$ mice) display a phenotype of hypercholesterolemia and spontaneous atherosclerosis that can be exacerbated when fed an atherogenic diet. Murine studies using *Apoe* knockout mice fed a high-fat diet have shown that hypercholesterolemia can also induce lung inflammation with consequent pathologic remodeling responses (Figure 13.2). For example, *Apoe*$^{-/-}$ mice fed a high-fat diet for 12 weeks develop both systemic and pulmonary inflammation manifested by elevated serum and bronchoalveolar lavage fluid (BALF) levels of TNF-α and IFN-γ, as well as increased lung inflammation with elevated numbers of BALF alveolar macrophages, matrix metalloproteinase 9 (MMP9) activity, and connective tissue deposition [22]. Interestingly, the TNF-mediated systemic and lung inflammation in this model may be driven by oxidized LDL. In addition, *Apoe*$^{-/-}$

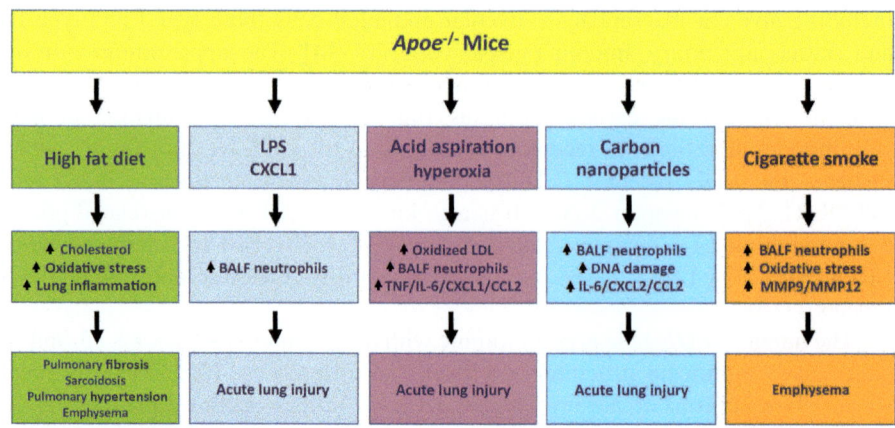

FIGURE 13.2

Apoe$^{-/-}$ mice display a phenotype of augmented lung inflammation and pulmonary disease in response to multiple proinflammatory stimuli. *Apoe$^{-/-}$* mice develop enhanced pulmonary inflammatory responses to a variety of heterogeneous stimuli that directly induce the development of lung disease.

mice fed a cholate-containing high-fat diet develop a phenotype of granulomatous inflammation that resembles pulmonary sarcoidosis and progresses to pulmonary fibrosis over time [23]. Similar to human granulomas found in sarcoidosis, the granulomas in cholate-fed *Apoe$^{-/-}$* mice contain both CD4$^+$ and CD8$^+$ T cells. Similarly, *Apoe$^{-/-}$* mice fed a high-fat Western diet develop a phenotype of increased lung cholesterol content and pulmonary inflammation with associated granuloma formation, as well as alveolar septal thickening [24]. In contrast, *Apoe$^{-/-}$* mice fed a normal diet develop minimal pulmonary inflammation with very few granulomas, smaller increases in lung cholesterol content, and normal alveolar septal thickness. Silencing TLR4 expression by administration of a lentiviral short hairpin RNA construct attenuated the phenotype of lung inflammation and alveolar septal thickening, which suggests the involvement of a TLR4/Myd88-dependent pathway.

When fed a high-fat Paigen diet, which is predominantly comprised of saturated fatty acids derived from butter fat or cocoa butter, *Apoe$^{-/-}$* mice develop pulmonary hypertension that is dependent on IL-1 signaling and could be prevented by administration of an IL-1 receptor antagonist [25, 26]. This suggests a role for proinflammatory IL-1 receptor signaling in the mechanism of diet-induced lung vascular remodeling with consequent pulmonary arterial hypertension. Feeding a high-fat diet to *Apoe$^{-/-}$* mice has also been shown to induce a phenotype of airspace enlargement suggestive of emphysema, with associated increases in lung macrophages, as well as augmented MMP9 activity and increased MMP12 expression [27]. This phenotype may be mediated by signaling pathways involving TLR4 and IL-1 receptor-associated kinase (IRAK). Collectively, these murine studies suggest that hypercholesterolemia in *Apoe$^{-/-}$* mice fed a high-fat diet induces pulmonary inflammation and oxidative stress that results in remodeling responses that recapitulate several human lung diseases, including pulmonary fibrosis, sarcoidosis, pulmonary arterial hypertension, and emphysema.

APOLIPOPROTEIN E: REGULATORY MECHANISMS OF IMMUNE AND INFLAMMATORY CELL FUNCTION

Apolipoprotein E mediates context-dependent effects on innate and adaptive immune responses via several mechanisms (Figure 13.3). Both the apoE holoprotein, as well as an apoE mimetic peptide, can be internalized into cells via a receptor-independent pathway, where they bind the carboxy-terminal domain of the SET protein, which is a physiological binding partner and inhibitor of protein phosphatase 2A (PP2A) [28]. The apoE-SET interaction thereby liberates PP2A, which can reduce inflammatory responses by dephosphorylating lipopolysaccharide (LPS)-induced signaling kinases, such as p38 mitogen activated kinase and Akt, as well as by decreasing nitric oxide generation. Thus the ability of apoE to bind SET and liberate PP2A may suppress inflammatory responses to systemic and pulmonary infections. Additionally, binding of SET by an apoE mimetic peptide in cancer cell lines increases PP2A activity, which suggests that the apoE-SET pathway may be relevant for the treatment of cancer [29].

Apolipoprotein E can also inhibit macrophage inflammatory responses via a pathway that involves the increased expression of the transcription factor, PU.1

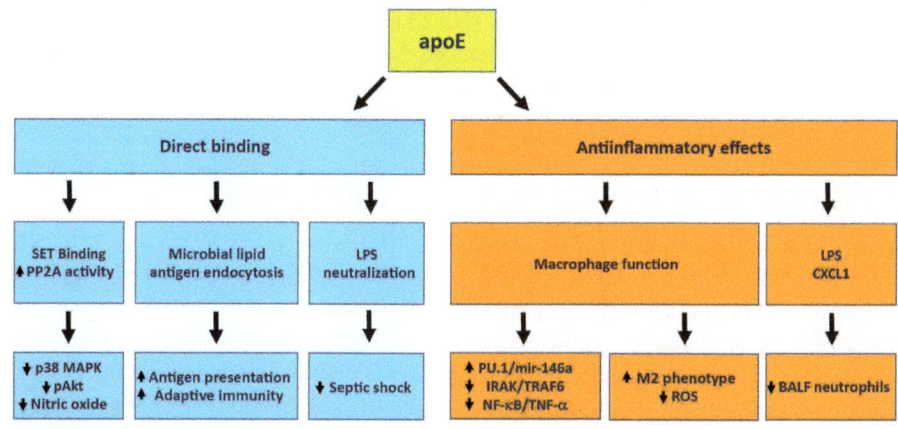

FIGURE 13.3

Context-dependent effects of apolipoprotein E on inflammation. Apolipoprotein E (apoE) ameliorates inflammatory responses either by the direct binding of proteins (i.e., SET) and lipids (i.e., bacterial lipopolysaccharide (LPS)), or, alternatively, by attenuating macrophage and neutrophil functions. In contrast, the binding of apoE to mycobacterial lipid antigens mediates receptor-mediated endocytosis and antigen presentation that promotes adaptive immune responses.

(purine-rich PU-box-binding protein 1) [30]. PU.1 subsequently upregulates the expression of mir-146a, which is a key negative modulator of NF-κB. This suppresses the expression of proinflammatory genes, including IRAK1 and TRAF6, which reduces TNF-α mRNA levels during hyperlipidemia. This pathway, which was identified as attenuating atherosclerosis, may also be relevant in the context of other inflammatory conditions. ApoE has also been reported to convert the murine macrophage RAW264.7 cell line from a proinflammatory M1 phenotype to an anti-inflammatory M2 phenotype, as characterized by the reduced expression of IL-12, CCL3, and inducible nitric oxide synthase, and the increased expression of arginase 1, SOCS3, and IL-1 receptor antagonist [31]. Furthermore, apoE inhibited the proinflammatory responses of macrophages to IFN-γ and the TLR3 agonist, poly I:C, as well as attenuated ROS generation. Similarly, intratracheal administration of the alkylating agent, nitrogen mustard, to rats induced an increase in apoE expression by M2-biased CD11b$^+$/CD43$^-$ infiltrating macrophages in the lung [32, 33]. The role of apoE in modulating the phenotype of human macrophages has not yet been demonstrated.

In contrast, apoE can initiate adaptive immune responses to lipid antigens expressed by *Mycobacterium tuberculosis* [34]. In particular, apoE binds mycobacterial lipid antigens and thereby delivers them into the endosomal compartments of antigen-presenting cells, such as dendritic cells (DCs), by receptor-mediated endocytosis. ApoE allows the lipid antigens to interact with CD1, which facilitates their presentation to lipid-antigen reactive T cells. This represents a novel mechanism by which apoE-mediated lipid transport pathways can enhance adaptive immune responses to pulmonary pathogens. Similarly, an apoE-LDLR-dependent pathway mediates lipid antigen uptake and presentation by B cells to natural killer T (NKT) cells [35]. Macrophages from *Apoe$^{-/-}$* mice have also been shown to have enhanced antigen-presenting capabilities that may result from the increased expression of major histocompatibility complex (MHC) class II and costimulatory proteins, CD40 and CD80 [36].

APOLIPOPROTEIN E: AN ENDOGENOUS MODULATOR OF INNATE HOST DEFENSE AND SEPSIS

ApoE, via its ability to bind and neutralize bacterial LPS, can directly attenuate systemic inflammatory responses to infection (Figure 13.3) [37–39]. In particular, the ability of apoE-LPS complexes to be endocytosed by liver parenchymal cells, and subsequently secreted into the bile where it is inactivated, can suppress systemic LPS-mediated inflammation, cytokine production, and mortality secondary to septic shock [37–40]. Furthermore, LPS can induce increases in serum apoE levels via a mechanism that involves reduced plasma clearance, which in turn enhances host defense [37, 39]. Consistent with this, apoE levels in serum and cerebrospinal fluid have been reported to be increased in pediatric patients in the setting of bacterial infections [41, 42].

Murine studies investigating the role of apoE in the pathogenesis of septic shock have produced context-dependent results based upon whether LPS or bacterial-based model systems were utilized. For example, LPS-challenged $Apoe^{-/-}$ mice have increased production of proinflammatory cytokines (i.e., TNF-α, IL-6, IL-12, and IFN-γ), whereas administration of the apoE protein or apoE mimetic peptides that correspond to the LDLR binding domain protected against LPS-induced lethality in rodents [39, 43–45]. Similarly, $Apoe^{-/-}$ mice have enhanced susceptibility to infections caused by *Klebsiella pneumoniae*, *Listeria monocytogenes*, and *Candida albicans*, as well as impaired immune responses manifested by increased circulating levels of TNF-α [46–49]. In contrast, in a rat model of septic peritonitis caused by cecal ligation and puncture, the serial administration of apoE was associated with increased mortality secondary to enhanced presentation of the endogenous glycosphingolipid antigen, iGb3, to NKT cells [50, 51].

The common polymorphic *APOE* genotypes have also been shown to differentially modify inflammatory responses during sepsis. For example, mice expressing the human *APOE* ε4 allele displayed a phenotype of hypothermia, acute liver dysfunction, and increased TNF-α and IL-6 levels in response to systemic LPS challenge compared with mice expressing the human *APOE* ε3 allele [44, 52]. Furthermore, mice expressing the human *APOE* ε4 allele had increased lethality following cecal ligation and puncture as compared with mice expressing the human *APOE* ε3 allele [45]. Similarly, *ex vivo* stimulation of whole blood from healthy *APOE* ε3/ε4 subjects with TLR2, TLR4, and TLR5 agonists induced larger increases in TNF-α production than blood from ε3/ε3 subjects [52]. The mechanism by which TNF-α expression is increased in *APOE* ε3/ε4 subjects may involve increased lipid raft assembly in blood monocytes. Healthy volunteer carriers of the *APOE* ε3/ε4 genotype also had increased body temperature and plasma TNF-α levels after intravenous LPS administration compared with carriers of the ε3/ε3 genotype [52]. In addition, among a cohort of European American patients with severe sepsis, carriage of the *APOE* ε4 allele was associated with an increase in coagulation system failure [52]. In contrast, carriage of the *APOE* ε3 allele has been associated with a reduced incidence of severe sepsis and shorter ICU stay in 343 subjects following major elective noncardiac surgery [53]. Collectively, these studies suggest that carriage of the *APOE* ε4 allele may be associated with more severe inflammatory responses during sepsis.

Apolipoprotein E also participates in the pathogenesis of *Chlamydophila* (*Chlamydia*) *pneumoniae* infection, which is an obligate, intracellular, human respiratory pathogen that causes community-acquired pneumonia and may contribute to the pathogenesis of reactive arthritis and atherosclerosis [54–56]. The apoE4 protein enhances the attachment of the elementary body, which is the infectious extracellular form of the bacteria, to host eukaryotic cells [54]. The *APOE* ε4 allele has also been associated with an increased risk of synovial *C. pneumoniae* infection [55]. Furthermore, $Apoe^{-/-}$ mice develop a type 2 immune response to *C. pneumoniae* infection associated with a reduction in vascular infection [57]. Interestingly, apoE4 has also been associated with increased pathogenesis of viral infections, such as those caused by human immunodeficiency virus (HIV-1), hepatitis C, hepatitis E, and herpes simplex virus [58–60]. However, it is not known whether the apoE4 protein modifies the risk or severity of viral pneumonia.

APOLIPOPROTEIN E AS AN ENDOGENOUS MODULATOR OF LUNG INFLAMMATION

$Apoe^{-/-}$ mice have also been used to identify a protective role for apoE in lung inflammation caused by a variety of insults. For example, in an experimental model of bacterial pneumonia, inhalation of LPS, as well as tracheal inoculation with CXCL1, caused significant increases in BALF neutrophils in $Apoe^{-/-}$ mice (Figure 13.2) that were suppressed by pretreatment with an apoE mimetic peptide (Figure 13.3) [61]. In a model of experimental acute lung injury secondary to acid aspiration and hyperoxia, BALF neutrophils, cytokines (e.g., TNF-α and IL-6), and chemokines (e.g., CXCL1 (KC) and CCL2 (MCP-1)), were increased in $Apoe^{-/-}$ mice compared with WT mice [62]. The increased susceptibility to acute lung injury was proposed to be mediated by increased vascular permeability due to elevated levels of lipids, such as oxidized LDL. Similarly, the inhalation of carbon nanoparticles, such as carbon black, which results from incomplete combustion of hydrocarbons, induced significant increases in BALF neutrophils, DNA damage in BALF cells, as well as lung expression of IL-6, CXCL2, and CCL2 in $Apoe^{-/-}$ mice compared with WT mice (Figure 13.2) [63].

Studies using $Apoe^{-/-}$ mice have similarly demonstrated a protective effect of apoE on pulmonary inflammation in a variety of experimental lung diseases. For example, in a model of cigarette smoke-induced emphysema, $Apoe^{-/-}$ mice developed an exaggerated inflammatory response in the lung with increases in BALF neutrophils, oxidative stress, and matrix metalloproteinases (i.e., MMP-9 and MMP-12), which were associated with enhanced airspace enlargement (Figure 13.2) [64]. Although $Apoe^{-/-}$ mice did not have increased lung inflammation in a model of house dust mite-induced allergic asthma, administration of an apoE mimetic peptide, corresponding to the apoE LDL receptor-binding domain, suppressed increases in BALF eosinophils and macrophages, as well as serum IgE levels, mucous cell metaplasia, and airway hyperresponsiveness [12]. Interestingly, ovalbumin-challenged $Gata5^{-/-}$ mice have reduced pulmonary expression of apoE and a phenotype similar to $Apoe^{-/-}$ mice with increased airway hyperreactivity and mucous cell metaplasia, without associated alterations in BALF inflammatory cell numbers [65]. Furthermore, knock-in mice expressing the human $APOE$ ε3 allele had a reduction in house dust mite-induced airway inflammation mediated by decreases in BALF eosinophils and lymphocytes compared with mice expressing either the human $APOE$ ε2 or ε4 alleles [66].

APOLIPOPROTEIN A-I

Apolipoprotein A-I (apoA-I) is a 28-kDa protein comprised of 243 amino acids and is primarily synthesized by the liver and intestine but is also produced by lung cells, including alveolar epithelial cells and alveolar macrophages [67–71]. Expression of apoA-I can be detected in the developing murine lung by gestational day 15.5 with levels that exceed those found in the adult lung, which suggests that apoA-I modulates lipid metabolism during embryogenesis [72]. The structure of apoA-I is comprised of 10 conserved amphipathic α-helices consisting of a polar face, which interacts with water, and a hydrophobic face, which interacts with lipids [70, 71]. The ability of apoA-I to form stable micellar complexes with cholesterol, phospholipids, triglycerides, and cholesteryl esters is dependent upon its amphipathic α-helices. ApoA-I is the major structural protein component of HDL particles where it comprises approximately 70% of the protein mass [70]. After apoA-I is secreted into the blood, it associates with cholesterol and phospholipids to form nascent HDL particles that can acquire additional cholesterol and phospholipids from peripheral cells and tissues via interactions with ABCA1 and ABCG1 to mediate reverse cholesterol transport [6]. ApoA-I-mediated reverse cholesterol transport is a key component of normal lipid homeostasis, which facilitates the clearance of lipids from the body via hepatic uptake and subsequent excretion into the bile [6, 9].

The negative association between HDL-cholesterol (HDL-C) and the risk of cardiovascular disease has supported the concept of developing apoA-I mimetic peptides that mimic the class A α-helical structure for therapeutic use [6, 7, 9, 71]. This is based upon the ability of apoA-I mimetic peptides to mediate reverse cholesterol transport out of cells, as well as to exert antiinflammatory and antioxidant effects. Two different apoA-I mimetic peptides, 4F and 5A, have been utilized in murine studies that demonstrated their ability to attenuate lung inflammation. Furthermore, the 4F peptide, which can be synthesized using either L-peptides or D-peptides, has already been advanced to human clinical trials in cardiovascular disease [73, 74]. Mammalian proteins are typically comprised of L-amino acids. When the 4F apoA-I mimetic peptide is synthesized using L-amino acids (termed L-4F), it is too unstable to be given via an oral route. In contrast, when the 4F apoA-I mimetic peptide is synthesized using D-amino acids (termed D-4F), it is resistant to trypsin digestion in the gastrointestinal tract and can be administered orally [75]. The 4F peptide consists of 18 amino acids that share features with the apoA-I α-helical segment, whereas the 5A peptide is bihelical, with two amphipathic helices linked by a proline. One of the 5A helices is a high lipid affinity helix, whereas the other helix has five alanine substitutions that results in low lipid affinity and reduced cellular toxicity [76, 77].

APOLIPOPROTEIN A-I: MECHANISMS OF IMMUNE AND INFLAMMATORY CELL MODULATION

A primary mechanism of action by which apoA-I exerts its anti-inflammatory effects is by mediating reverse cholesterol transport from cells, which reduces the number of lipid raft microdomains in the plasma membrane and the endosomal system, which thereby decreases signaling via cytokine receptors, Toll-like receptors, T-cell receptors, and B-cell receptors (Figure 13.4) [78, 79]. Consistent with this, mice with genetic deletions of both *Abca1* and *Abcg1* have a phenotype of increased bone marrow myelopoiesis [80]. This phenotype is mediated by increased hematopoietic stem cell proliferation caused by enhanced cytokine signaling via the common β-subunit of the IL-3, IL-5, and GM-CSF receptors, which is localized to lipid raft microdomains

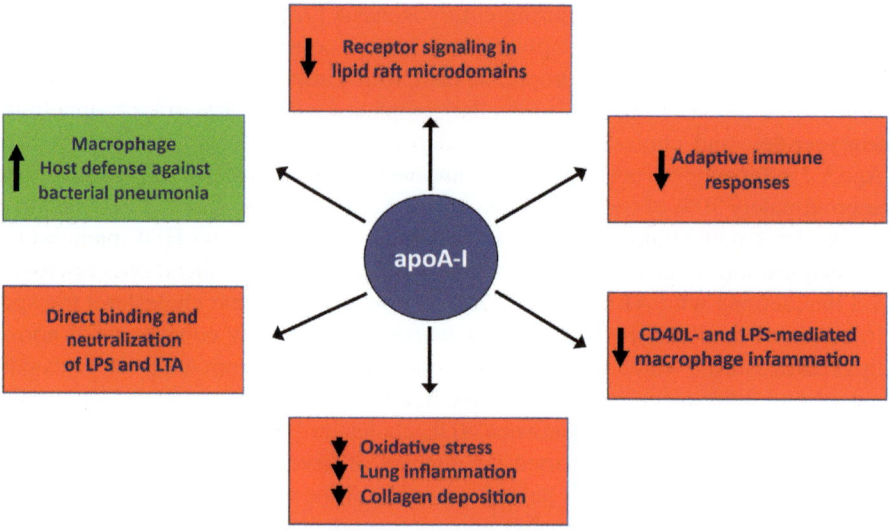

FIGURE 13.4

Context-dependent mechanisms by which apolipoprotein A-I modulates inflammation. Apolipoprotein A-I (apoA-I) attenuates inflammation by multiple mechanisms, including reductions in receptor signaling within lipid raft microdomains, oxidative stress, adaptive immune responses, and macrophage function. Apolipoprotein A-I can also reduce inflammation by directly binding and neutralizing both bacterial lipopolysaccharide (LPS) and lipoteichoic acid (LTA). Alternatively, apoA-I can augment macrophage-mediated host defense against bacterial pneumonia. (Boxes in red indicate antiinflammatory effects of apoA-I, whereas the green box indicates its proinflammatory effect.)

in the setting of cholesterol enrichment. Similarly, mice with a cell-specific deletion of *Abca1* and *Abcg1* in macrophages have increased atherosclerosis and enhanced production of proinflammatory C-C chemokines within atherosclerotic plaques [81].

Apolipoprotein A-I can suppress adaptive immunity by decreasing the differentiation and function of antigen-presenting cells. For example, apoA-I can reduce the antigen presentation function of murine DCs, B cells, and macrophages, which impairs their ability to induce T-cell activation [82]. Furthermore, apoA-I-mediated decreases in cholesterol and major histocompatibility class II protein within lipid raft microdomains is a mechanism by which adaptive immune responses are attenuated. Apolipoprotein A-I can similarly inhibit the differentiation and maturation of human monocytes into antigen-presenting DCs via the induction of IL-10 and PGE2 [83]. Apolipoprotein A-I also acts via ABCA1 to attenuate CD11b expression on activated human neutrophils, as well as to reduce neutrophil recruitment, adhesion, spreading, and migration [84]. Lipidated apoA-I can also inhibit oxidative stress in the human neutrophil-like HL-60 cell line by interrupting the redistribution of NADPH oxidase subunits into lipid raft microdomains [85].

Apolipoprotein A-I has been shown to modulate the pathogenesis of autoimmune disease in murine models. For example, increased cholesterol accumulation by cells activates the liver X receptors α and β (LXRα and LXRβ), which induces the expression of *Abca1* and *Abcg1* and promotes cholesterol efflux to apoA-I and HDL, respectively [86]. In contrast, cholesterol accumulation by CD11c$^+$ antigen-presenting cells in lipid X receptor (LXR)-deficient mice promotes autoimmune disease by causing B-cell proliferation and T-cell priming, whereas *Apoe$^{-/-}$Lxrb$^{-/-}$* mice that overexpress human apoA-I in the liver are protected from developing autoimmunity caused by cholesterol overload [86]. This suggests a role for cholesterol imbalance in the development of autoimmune disease, which can be reversed by apoA-I-mediated reverse cholesterol transport. Similarly, *Ldlr$^{-/-}$Apoa1$^{-/-}$* mice, when fed an atherogenic diet, display a phenotype of increased T-cell activation and proliferation, autoantibody production, and enlargement of cholesterol-enriched lymph nodes [87]. Subcutaneous administration of apoA-I decreased the number of lymph node cells and reduced the percentage of effector/effector memory T cells via a mechanism mediated by the expansion of CD4$^+$CD25$^+$FoxP3$^+$ regulatory T cells (Tregs) within the broader CD4$^+$ T-cell population. *Apoa1$^{-/-}$* mice also have increased Th1 and Th17 autoimmune responses in a murine model of antigen-induced arthritis that can be attenuated by administration of reconstituted HDL comprised of lipidated apoA-I [88]. Furthermore, the antiinflammatory effects of lipidated apoA-I involved the reduced expression of MHCII and costimulatory molecules by DCs, which was dependent upon reverse cholesterol transport via ABCA1 and SR-B1. In normocholesterolemic systemic lupus erythematous (SLE)-prone mice, apoA-I also suppressed the activation of CD4$^+$ T and B cells by a mechanism that was independent of cholesterol transport [89].

Apolipoprotein A-I has been shown to have context-dependent effects in macrophages. For example, apoA-I can attenuate proinflammatory functions of the human monocyte cell line, THP-1, following phorbol ester-induced differentiation into macrophages, by suppressing the ability of soluble CD40 ligand (CD40L) to induce NF-κB activation [90]. This is mediated by the decreased recruitment of tumor necrosis factor receptor-associated factor 6 (TRAF-6) to CD40 in lipid raft microdomains, which occurs secondary to ABCA1-dependent cholesterol efflux. Apolipoprotein A-I can also suppress LPS-induced cytokine production in macrophages by the posttranscriptional regulatory process of mRNA destabilization [91]. Apolipoprotein A-I mediates this effect by increasing the expression of the zinc-finger protein, tristetraprolin, which destabilizes mRNAs of several proinflammatory cytokines, such as TNF-α, which contain class II AU-rich elements in their 3′-untranslated regions. The apoA-I-mediated upregulation of tristetraprolin occurs via a JAK2/STAT3-dependent pathway. Furthermore, apoA-I can disrupt lipid raft microdomains in monocytes, which reduces chemotaxis by attenuating the PI3K/Akt signaling pathway that induces reorganization of the actin cytoskeleton [92]. In contrast, HDL-mediated cholesterol efflux from murine and human macrophages augments TLR-induced, proinflammatory responses in a macrophage-specific fashion that enhances host defense against bacterial infection via the activation of a PKC-NF-κB/STAT1-IRF1 signaling axis (Figure 13.4) [93]. Consistent with this, bacterial titers in BALF were significantly lower in human *APOA1* transgenic mice than *Apoa1* knockout mice after *in vivo* infection with *Pseudomonas aeruginosa*. This suggests that apoA-I and HDL support normal macrophage immune responses in the setting of bacterial infection.

Interestingly, proteases derived from inflammatory cells can cleave and thereby inactivate apoA-I as a feedback mechanism that amplifies innate inflammatory responses. For example, granule-associated proteases from mast cells, such as chymase, tryptase, cathepsin G, carboxypeptidases, and granzyme B, can cleave apoA-I, thereby generating a truncated form that cannot bind to vascular endothelial cells with high affinity [94]. Chymase was the most effective mast cell-derived protease that generated a C-terminally truncated and functionally inactive apoA-I. Chymase also cleaved and inactivated the L-4F apoA-I mimetic peptide so that it no longer possessed antiinflammatory properties, whereas a protease-resistant D-4F was not cleaved. Similarly, human macrophage-derived cathepsin B cleaves apoA-I between Ser228 and Phe229 to generate a C-terminally truncated form that has impaired ABCA1-dependent cholesterol efflux and antiatherogenic capabilities [95]. In contrast, lipidated forms and HDL-associated forms of apoA-I are protected from cleavage.

APOLIPOPROTEIN A-I AS AN ENDOGENOUS INHIBITOR OF LUNG INFLAMMATION

Mice with a genetic deletion of the *Apoa1* gene have provided insights into the normally anti-inflammatory effects of apoA-I. For example, HDL from *Apoa1*$^{-/-}$ mice oxidizes at a faster rate than HDL from wild type mice, which indicates that, in the absence of apoA-I, HDL is converted to a proinflammatory particle, whereas activity of the HDL-associated antioxidant protein, paraoxonase 1, is decreased in plasma [96]. Furthermore, under basal conditions, the *Apoa1*$^{-/-}$ mice display a phenotype of increased perialveolar and perivascular inflammation with increased numbers of lung neutrophils and eosinophils, which is associated with increased collagen deposition and oxidative stress in the lung, as well as augmented airway hyperresponsiveness (Figure 13.4). Collectively, these findings are consistent with the conclusion that apoA-I plays an important role to limit basal levels of systemic and pulmonary inflammation, as well as oxidative stress.

APOLIPOPROTEIN A-I: AN ENDOGENOUS MODULATOR OF LPS-MEDIATED ACUTE LUNG INJURY

In addition to attenuating basal levels of lung inflammation, apoA-I also suppresses LPS-mediated lung and systemic inflammation via several mechanisms, including its ability to directly bind and neutralize LPS, a cell wall component of gram-negative bacteria, which is mediated by domains located in the carboxy-terminal half of apoA-I (Figure 13.4) [97–100]. Furthermore, the mechanism by which apoA-I binds and neutralizes LPS may involve an initial interaction with the LPS-binding protein (LBP) in plasma, which then transfers LPS to apoA-I [101]. Similarly, apoA-I can directly bind, and potentially neutralize, lipoteichoic acid (LTA), a cell wall component of gram-positive bacteria [102].

Studies utilizing $Apoa1^{-/-}$ mice have suggested a protective role for apoA-I in LPS-induced lung inflammation and acute lung injury (Figure 13.5). For example, $Apoa1^{-/-}$ mice display a phenotype of enhanced lung inflammation caused by the increased recruitment of both neutrophils and alveolar macrophages in response to inhalation of LPS [61]. Similarly, $Apoa1^{-/-}$ mice have an increased influx of neutrophils into the airway in response to intratracheal inoculation of the chemokine CXCL1 (KC), which was inhibited by intravenous administration of the L-4F apoA-I mimetic peptide. Furthermore, administration of L-4F suppressed the LPS-mediated increases in BALF neutrophils and macrophages. Similarly, in *ex vivo* experiments, L-4F suppressed the chemotaxis of murine bone marrow-derived neutrophils.

Administration or overexpression of apoA-I has also protected against LPS-induced septic shock and acute lung injury in multiple rodent model systems. Following systemic LPS administration, apoA-I reduced TNF-α, IL-1β, and IL-6 in BALF and decreased mortality in recipient mice [99, 103, 104]. ApoA-I also attenuated lung injury,

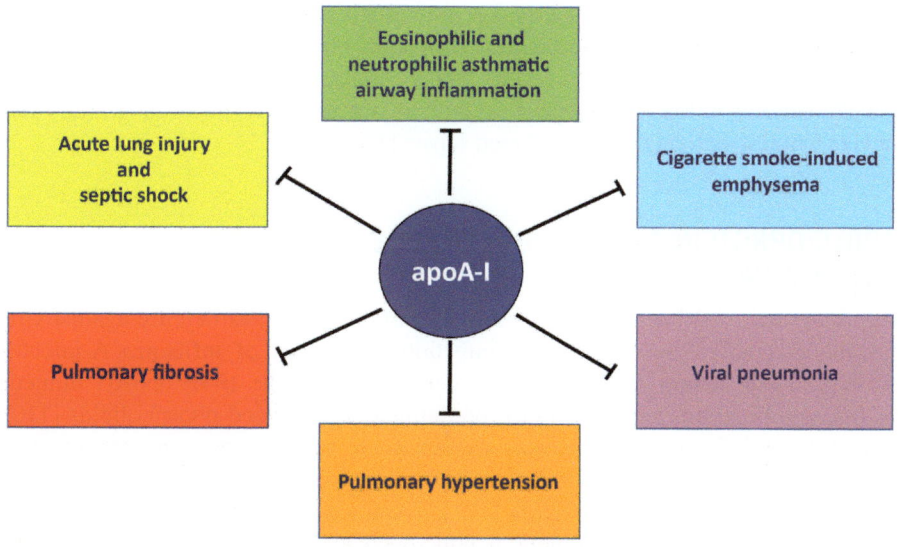

FIGURE 13.5

Apolipoprotein A-I has protective effects in multiple experimental murine models of lung disease. Apolipoprotein A-I (apoA-I) has antiinflammatory properties that attenuate disease severity in murine models of acute lung injury and septic shock, eosinophilic and neutrophilic asthmatic airway inflammation, cigarette smoke-induced emphysema, pulmonary fibrosis, pulmonary hypertension, and viral pneumonia.

as well as TNF-α and IL-1β levels in BALF, following systemic LTA administration [102]. Similarly, mice overexpressing human apoA-I had reductions in LPS-induced acute lung and kidney injury and BALF levels of TNF-α, IL-1β, and IL-6 [105], whereas systemic administration of an adenoviral vector expressing human apoA-I suppressed LPS-induced neutrophilic inflammation, lung edema, and mortality [100]. LPS-challenged rats that received the L-4F apoA-I mimetic peptide had improved survival associated with reductions in acute lung injury and systemic inflammation, as well as NF-κB activation, adhesion molecule expression, and lung myeloperoxidase activity [106, 107].

The protective effects of apoA-I in rodent models of LPS-induced septic shock and acute lung injury may be relevant for humans. For example, *ex vivo* experiments that stimulated peripheral blood cells with LPS showed that treatment with the L-4F apoA-I mimetic peptide reduced both LPS binding to leukocytes and endotoxin activity [106]. In addition, L-4F attenuated LPS-induced increases in IL-6 production and reductions in paraoxonase activity by peripheral blood cells. L-4F also inhibited neutrophil activation and superoxide formation by leukocytes induced by exposure to serum from patients with the acute respiratory distress syndrome (ARDS). Polymorphisms in the *APOA1* gene have also been associated with susceptibility to acute lung injury. In subjects who underwent cardiopulmonary bypass surgery, the −75 G/A single nucleotide polymorphism (SNP), which is located 75 base pairs upstream from the *APOA1* transcription start site, was associated with the risk of developing acute lung injury. In particular, individuals carrying the AA genotype have been found to have an increased risk of both acute lung injury and 30-day mortality after cardiopulmonary bypass surgery [108]. Similarly, another study showed that genetic variants in the *APOA1* gene might be associated with acute lung injury risk in Han Chinese patients who developed sepsis [109].

APOLIPOPROTEIN A-I AND HOST DEFENSE AGAINST VIRAL PNEUMONIA

Apolipoprotein A-I has also been shown to participate in host defense against viral pneumonia caused by influenza A infection. Treatment of influenza A-infected A549 cells (a human alveolar type II cell line) with the D-4F apoA-I mimetic peptide inhibited the production of proinflammatory oxidized phospholipids [110]. Furthermore, HDL particles became dysfunctional and lost their antiinflammatory properties in influenza A-infected mice, which may enhance disease severity and also increase the risk for atherosclerosis [111]. Treatment of influenza-infected $Ldlr^{-/-}$ mice with the D-4F apoA-I mimetic peptide decreased lung viral titers by 50% and also reduced lung IL-6 levels, which is consistent with an antiviral effect of apoA-I (Figure 13.5) [112]. Furthermore, D-4F increased HDL levels and paraoxonase activity, with an associated decrease in LDL-induced monocyte chemotactic activity. Thus the D-4F peptide displayed antiviral properties and allowed HDL to retain its antiinflammatory functions.

APOLIPOPROTEIN A-I ATTENUATES AIRWAY INFLAMMATION IN OBSTRUCTIVE LUNG DISEASES, SUCH AS ASTHMA AND EMPHYSEMA

Apoa1$^{-/-}$ mice display a phenotype of increased disease severity in experimental models of asthma and emphysema (Figure 13.5). For example, following sensitization and challenge with ovalbumin, *Apoa1*$^{-/-}$ mice have augmented neutrophilic airway inflammation that was primarily mediated by the increased expression of granulocyte colony-stimulating factor (G-CSF) in the lung, with concomitant increases in proinflammatory cytokines (IL-17 and TNF-α), CXCL5, and vascular cell adhesion molecule-1 [68]. Additional evidence regarding a protective role of apoA-I in asthmatic lung inflammation has been provided by studies that administered either apoA-I mimetic peptides or the whole apolipoprotein A-I protein in murine models of allergic airway inflammation. For example, intranasal administration of the 5A apoA-I mimetic peptide suppressed the phenotype of augmented neutrophilic airway inflammation in ovalbumin-challenged *Apoa1*$^{-/-}$ mice [68]. Similarly, administration of the 5A apoA-I mimetic peptide to wild type mice sensitized and challenged with house dust mite attenuated the three cardinal manifestations of asthma: airway inflammation, airway hyperresponsiveness, and airway remodeling responses, including mucous cell metaplasia and collagen expression [113]. Furthermore, the reductions in inflammation reflected a decrease in BALF neutrophils, eosinophils, and lymphocytes, as well as type 2 and type 17 cytokines, C-C chemokines, and alternatively activated macrophages. Similarly, administration of the D-4F apoA-I mimetic peptide to wild type mice that had been sensitized and challenged with ovalbumin suppressed airway hyperresponsiveness, eosinophilic airway inflammation, collagen deposition, and oxidative stress in the lung [114]. Administration of apoA-I to house dust mite-challenged mice also suppressed lung levels of IL-25, IL-33, and thymic stromal lymphopoietin, as well as increased the expression of the epithelial tight junction proteins, Zo-1 and occludin, which was mediated by increased expression of the proresolution mediator, lipoxin A4 [115].

Studies have also begun to demonstrate a role for apoA-I in modulating disease severity in asthmatic subjects. For example, the concentration of apoA-I in BALF from asthmatics has been reported to be five-fold lower than that from nonasthmatic subjects [115]. In addition, higher serum apoA-I levels have been found to be correlated with less severe airflow obstruction in atopic asthmatics, as indicated by larger FEV_1 values [116]. In contrast, an association was not found between serum apoA-I levels and biomarkers of type 2 inflammation (i.e., blood eosinophil counts or serum periostin levels) in atopic asthmatics [117].

ApoA-I has also been implicated as having a protective role in the pathogenesis of cigarette smoke-induced emphysema. ApoA-I levels are significantly decreased in both the lungs and sputum of subjects with emphysema as compared to normal control subjects, whereas mice that had been exposed to cigarette smoke also had decreased apoA-I expression [118, 119]. Furthermore, transgenic mice that conditionally expressed the human *APOA1* transgene in alveolar epithelial cells were protected from developing cigarette smoke-induced lung inflammation and emphysema. In particular, following exposure to cigarette smoke, *APOA1* transgenic mice had decreases in BALF neutrophils, proinflammatory cytokines (i.e., TNF-α and IL-1β) and chemokines (i.e., CXCL1 (KC)), as well as reduced oxidative stress, lung cell apoptosis, and metalloproteinase activation (i.e., MMP-2 and MMP-9).

ANTI-INFLAMMATORY EFFECTS OF APOLIPOPROTEIN A-I IN INTERSTITIAL AND VASCULAR LUNG DISEASE

ApoA-I also suppresses inflammatory responses in murine models of idiopathic pulmonary fibrosis (IPF) and pulmonary hypertension. For example, apoA-I was found to be decreased in BALF from subjects with IPF compared with healthy control subjects [69]. Furthermore, in a bleomycin-induced murine model of IPF, apoA-I administration significantly reduced both BALF inflammatory cells (i.e., neutrophils, macrophages, and lymphocytes) and collagen deposition. Similarly, transgenic mice that overexpressed the human *APOA1* gene in alveolar epithelial cells had a reduction in silica-induced pulmonary fibrosis associated with decreases in BALF inflammatory cells, as well as cytokines (i.e., TNF-α and IL-1β) and chemokines (i.e., CXCL1 (KC), CXCL2 (MIP-2), and CCL2 (MCP-1)) [120]. In addition, *APOA1* transgenic mice had increased expression of lipoxin A4, which is a biologically active eicosanoid that has antiinflammatory properties.

In murine models of pulmonary hypertension caused by monocrotaline and hypoxia, administration of the L-4F apoA-I mimetic peptide suppressed elevated plasma levels of oxidized fatty acids, including multiple HETEs (5-hydroxyeicosatetranoic acid (HETE), 12-HETE, and 15-HETE), as well as HODEs (9-hydroxyoctadecadienoic acid (HODE) and 13-HODE) [121]. Furthermore, L-4F prevented the HETE- and HODE-mediated downregulation of miR-193, which has a protective effect in pulmonary hypertension, via its transcription factor, RXR-α. This suggests that apoA-I may have a protective effect in pulmonary hypertension by attenuating signaling pathways mediated by proinflammatory oxidized phospholipids. Although apoA-I has been shown to have protective roles in rodent models of pulmonary fibrosis and pulmonary hypertension, further studies will be required to assess the relevance of these findings to human subjects.

CLUSTERIN (APOLIPOPROTEIN J)

Clusterin, which is also termed apolipoprotein J, is encoded by the *CLU* gene on human chromosome 8 and is highly expressed in the liver, brain, ovary, and testis, with lower levels of expression found in the lung, heart, breast, and spleen [122]. Clusterin is synthesized as a polypeptide comprised of 427 amino acids, which is cleaved at an internal bond between Arg-205 and Ser-206 to generate two subunits that remain associated via disulfide bonds to form a heterodimer [122, 123]. The precursor form has a predicted molecular mass of 60 kDa, whereas the mature glycosylated form is secreted as a protein with a molecular mass of 75–80 kDa, termed secretory clusterin (sCLU) [123, 124]. The structure of clusterin is predicted to contain amphipathic helices, which may facilitate its association with HDL particles [122, 125]. Clusterin has been suggested to have a protective role during pathological stresses based upon its antioxidant and antiapoptotic properties, as well as its chaperone-like activity similar to small heat shock proteins. Furthermore, clusterin expression is a marker of oxidative stress, as multiple stress-induced transcription factors can bind and activate the *CLU* promoter [126]. Secretory clusterin has also been demonstrated to have a cytoprotective role in tumor cells via inhibition of apoptosis and enhanced resistance to chemotherapy and radiotherapy [124]. Of note, a nonglycosylated form of clusterin has also been identified primarily localized to the cell nucleus and, in contrast to sCLU, is proapoptotic [124].

The antiinflammatory and antioxidant properties of clusterin may also be relevant for lung disease. For example, house dust mite-challenged clusterin ($Clu^{-/-}$) knockout mice display a phenotype of increased eosinophilic airway inflammation with increases in CD11c$^+$/CD11b$^+$ DCs and Ly6C^{high} monocytes in the lung, as well as increases in IL-4 and C-C chemokines in BALF [126]. Oxidative stress was also increased in the lungs of $Clu^{-/-}$ knockout mice, whereas clusterin treatment decreased both reactive oxygen species generation and CCL20 secretion by house dust mite-stimulated airway epithelial cells. This suggests that clusterin may negatively regulate CCL20 production by airway epithelial cells, which thereby attenuates the recruitment of inflammatory DCs and suppresses allergic airway inflammation. In a cohort of 170 childhood asthmatics, sputum clusterin levels were higher in stable asthmatics than in healthy nonasthmatic subjects and were associated with disease severity and eosinophilic airway inflammation [127]. Interestingly, sputum clusterin levels were lower during asthma exacerbations than during periods of stable asthma. Serum clusterin levels have also been found to be higher in children with atopic dermatitis than in healthy controls [128].

Clusterin may also participate in the pathogenesis of acute lung injury, cigarette smoke-induced oxidative stress, and pulmonary hypertension. For example, in a rabbit model of leukocyte-induced lung injury, clusterin pretreatment suppressed pulmonary edema, as well as thromboxane A2 production and complement activation [129]. However, in a rat model of pulmonary arterial hypertension, sCLU expression was increased and levels were positively correlated with pulmonary hemodynamics, such as right ventricular and pulmonary artery pressures [130]. In addition, sCLU promoted the proliferation and migration of human pulmonary artery smooth muscle cells, whereas apoptosis secondary to oxidative stress was inhibited. Lastly, cigarette smoke has been shown to induce oxidative stress and clusterin expression in human lung fibroblasts, which suggests that it may have a protective role in this setting [123].

CONCLUSION

An emerging body of evidence is being generated to support the concept that apolipoproteins modify inflammation and adaptive immune responses, both systemically, as well as in the lung, in a context-dependent fashion. This not only provides an opportunity to advance our understanding of lung disease pathogenesis, but also to identify new therapeutic approaches, such as apoA-I mimetic peptides, for the treatment of inflammatory pulmonary disorders. Consistent with this, apoA-I mimetic peptides are currently being developed for future clinical trials of asthmatics [77]. It is likely that additional apolipoprotein-based therapeutic approaches will be identified that may also be utilized to attenuate lung inflammation. Thus, apolipoprotein-based therapies that modify inflammatory responses in the lung could represent a new approach for the treatment of pulmonary disease.

ACKNOWLEDGMENT

Financial support: Division of Intramural Research, National Heart, Lung, and Blood Institute, NIH, Bethesda, Maryland.

REFERENCES

[1] Rader DJ, Hobbs HH. Disorders of lipoprotein metabolism. In: Kasper D, Fauci A, Hauser S, Longo DL, Jameson JL, Loscalzo J, editors. Harrison's Principles of Internal Medicine, 19e. New York, NY: McGraw-Hill Education; 2017.

[2] Ridker PM. LDL cholesterol: controversies and future therapeutic directions. Lancet 2014;384(9943):607–17.

[3] Yao X, Gordon EM, Figueroa DM, Barochia AV, The LSJ. Emerging roles of apolipoprotein E and apolipoprotein A-I in the pathogenesis and treatment of lung disease. Am J Respir Cell Mol Biol 2016;55(2):159–69.

[4] Goldstein JL, Brown MS. The LDL receptor. Arterioscler Thromb Vasc Biol 2009;29(4):431–8.

[5] Fisher EA, Feig JE, Hewing B, Hazen SL, Smith JD. High-density lipoprotein function, dysfunction, and reverse cholesterol transport. Arterioscler Thromb Vasc Biol 2012;32(12):2813–20.

[6] Kingwell BA, Chapman MJ, Kontush A, Miller NE. HDL-targeted therapies: progress, failures and future. Nat Rev Drug Discov 2014;13(6):445–64.

[7] Rader DJ, Hovingh GK. HDL and cardiovascular disease. Lancet 2014;384(9943):618–25.

[8] Rosenson RS, Brewer Jr. HB, Davidson WS, Fayad ZA, Fuster V, Goldstein J, et al. Cholesterol efflux and atheroprotection: advancing the concept of reverse cholesterol transport. Circulation 2012;125(15):1905–19.

[9] Navab M, Reddy ST, Van Lenten BJ, Fogelman AM. HDL and cardiovascular disease: atherogenic and atheroprotective mechanisms. Nat Rev Cardiol 2011;8(4):222–32.

[10] Mahley RW, Weisgraber KH, Huang Y. Apolipoprotein E: structure determines function, from atherosclerosis to Alzheimer's disease to AIDS. J Lipid Res 2009;50(Suppl):S183–8.

[11] Lin CT, Xu YF, Wu JY, Chan L. Immunoreactive apolipoprotein E is a widely distributed cellular protein. Immunohistochemical localization of apolipoprotein E in baboon tissues. J Clin Invest 1986;78(4):947–58.

[12] Yao X, Fredriksson K, Yu ZX, Xu X, Raghavachari N, Keeran KJ, et al. Apolipoprotein E negatively regulates house dust mite-induced asthma via a low-density lipoprotein receptor-mediated pathway. Am J Respir Crit Care Med 2010;182(10):1228–38.

[13] Chen J, Chen Z, Chintagari NR, Bhaskaran M, Jin N, Narasaraju T, et al. Alveolar type I cells protect rat lung epithelium from oxidative injury. J Physiol 2006;572(Pt 3):625–38.

[14] Hansmann G, de Jesus Perez VA, Alastalo TP, Alvira CM, Guignabert C, Bekker JM, et al. An antiproliferative BMP-2/PPARgamma/apoE axis in human and murine SMCs and its role in pulmonary hypertension. J Clin Invest 2008;118(5):1846–57.

[15] Bu G. Apolipoprotein E and its receptors in Alzheimer's disease: pathways, pathogenesis and therapy. Nat Rev Neurosci 2009;10(5):333–44.

[16] Frieden C, Garai K. Concerning the structure of apoE. Protein Sci 2013;22(12):1820–5.

[17] Hatters DM, Peters-Libeu CA, Weisgraber KH. Apolipoprotein E structure: insights into function. Trends Biochem Sci 2006;31(8):445–54.

[18] Finch CE. Evolution in health and medicine Sackler colloquium: evolution of the human lifespan and diseases of aging: Roles of infection, inflammation, and nutrition. Proc Natl Acad Sci U S A 2010;107(Suppl 1):1718–24.

[19] Riedel BC, Thompson PM, Brinton RD. Age, APOE and sex: triad of risk of Alzheimer's disease. J Steroid Biochem Mol Biol 2016;160:134–47.

[20] Kulminski AM, Arbeev KG, Culminskaya I, Arbeeva L, Ukraintseva SV, Stallard E, et al. Age, gender, and cancer but not neurodegenerative and cardiovascular diseases strongly modulate systemic effect of the apolipoprotein E4 allele on lifespan. PLoS Genet 2014;10(1):e1004141.

[21] Cash JG, Kuhel DG, Basford JE, Jaeschke A, Chatterjee TK, Weintraub NL, et al. Apolipoprotein E4 impairs macrophage efferocytosis and potentiates apoptosis by accelerating endoplasmic reticulum stress. J Biol Chem 2012;287(33):27876–84.

[22] Naura AS, Hans CP, Zerfaoui M, Errami Y, Ju J, Kim H, et al. High-fat diet induces lung remodeling in ApoE-deficient mice: an association with an increase in circulatory and lung inflammatory factors. Lab Invest 2009;89(11):1243–51.

[23] Samokhin AO, Buhling F, Theissig F, Bromme D. ApoE-deficient mice on cholate-containing high-fat diet reveal a pathology similar to lung sarcoidosis. Am J Pathol 2010;176(3):1148–56.

[24] Ouyang Q, Huang Z, Lin H, Ni J, Lu H, Chen X, et al. Apolipoprotein E deficiency and high-fat diet cooperate to trigger lipidosis and inflammation in the lung via the Toll-like receptor 4 pathway. Mol Med Rep 2015;12(2):2589–97.

[25] Getz GS, Reardon CA. Diet and murine atherosclerosis. Arterioscler Thromb Vasc Biol 2006;26(2):242–9.

[26] Lawrie A, Hameed AG, Chamberlain J, Arnold N, Kennerley A, Hopkinson K, et al. Paigen diet-fed apolipoprotein E knockout mice develop severe pulmonary hypertension in an interleukin-1-dependent manner. Am J Pathol 2011;179(4):1693–705.

[27] Goldklang M, Golovatch P, Zelonina T, Trischler J, Rabinowitz D, Lemaitre V, et al. Activation of the TLR4 signaling pathway and abnormal cholesterol efflux lead to emphysema in ApoE-deficient mice. Am J Physiol Lung Cell Mol Physiol 2012;302(11):L1200–8.

[28] Christensen DJ, Ohkubo N, Oddo J, Van Kanegan MJ, Neil J, Li F, et al. Apolipoprotein E and peptide mimetics modulate inflammation by binding the SET protein and activating protein phosphatase 2A. J Immunol 2011;186(4):2535–42.

[29] Switzer CH, Cheng RY, Vitek TM, Christensen DJ, Wink DA, Vitek MP. Targeting SET/I(2)PP2A oncoprotein functions as a multi-pathway strategy for cancer therapy. Oncogene 2011;30(22):2504–13.

[30] Li K, Ching D, Luk FS, Raffai RL. Apolipoprotein E enhances microRNA-146a in monocytes and macrophages to suppress nuclear factor-kappaB-driven inflammation and atherosclerosis. Circ Res 2015;117(1):e1–11.

[31] Baitsch D, Bock HH, Engel T, Telgmann R, Muller-Tidow C, Varga G, et al. Apolipoprotein E induces antiinflammatory phenotype in macrophages. Arterioscler Thromb Vasc Biol 2011;31(5):1160–8.

[32] Venosa A, Malaviya R, Choi H, Gow AJ, Laskin JD, Laskin DL. Characterization of distinct macrophage subpopulations during nitrogen mustard-induced injury and fibrosis. Am J Respir Cell Mol Biol 2015;.

[33] Venosa A, Malaviya R, Gow AJ, Hall L, Laskin JD, Laskin DL. Protective role of spleen-derived macrophages in lung inflammation, injury, and fibrosis induced by nitrogen mustard. Am J Physiol Lung Cell Mol Physiol 2015;309(12):L1487–98.

[34] van den Elzen P, Garg S, Leon L, Brigl M, Leadbetter EA, Gumperz JE, et al. Apolipoprotein-mediated pathways of lipid antigen presentation. Nature 2005;437(7060):906–10.

[35] Allan LL, Hoefl K, Zheng DJ, Chung BK, Kozak FK, Tan R, et al. Apolipoprotein-mediated lipid antigen presentation in B cells provides a pathway for innate help by NKT cells. Blood 2009;114(12):2411–6.

[36] Tenger C, Zhou X. Apolipoprotein E modulates immune activation by acting on the antigen-presenting cell. Immunology 2003;109(3):392–7.

[37] Li L, Thompson PA, Kitchens RL. Infection induces a positive acute phase apolipoprotein E response from a negative acute phase gene: role of hepatic LDL receptors. J Lipid Res 2008;49(8):1782–93.

[38] Rensen PC, Oosten M, Bilt E, Eck M, Kuiper J, Berkel TJ. Human recombinant apolipoprotein E redirects lipopolysaccharide from Kupffer cells to liver parenchymal cells in rats in vivo. J Clin Invest 1997;99(10):2438–45.

[39] Van Oosten M, Rensen PC, Van Amersfoort ES, Van Eck M, Van Dam AM, Breve JJ, et al. Apolipoprotein E protects against bacterial lipopolysaccharide-induced lethality. A new therapeutic approach to treat gram-negative sepsis. J Biol Chem 2001;276(12):8820–4.

[40] Read TE, Harris HW, Grunfeld C, Feingold KR, Calhoun MC, Kane JP, et al. Chylomicrons enhance endotoxin excretion in bile. Infect Immun 1993;61(8):3496–502.

[41] Fu P, Wang AM, He LY, Song JM, Xue JC, Wang CQ. Elevated serum ApoE levels are associated with bacterial infections in pediatric patients. J Microbiol Immunol Infect 2014;47(2):122–9.

[42] Wang C, Wang Y, Wang A, Fu P, Yang Y. The diagnostic value of apolipoprotein E in pediatric patients with invasive bacterial infections. Clin Biochem 2012;45(3):215–8.

[43] Ali K, Middleton M, Pure E, Rader DJ. Apolipoprotein E suppresses the type I inflammatory response in vivo. Circ Res 2005;97(9):922–7.

[44] Lynch JR, Tang W, Wang H, Vitek MP, Bennett ER, Sullivan PM, et al. APOE genotype and an ApoE-mimetic peptide modify the systemic and central nervous system inflammatory response. J Biol Chem 2003;278(49):48529–33.

[45] Wang H, Christensen DJ, Vitek MP, Sullivan PM. Laskowitz DT. APOE genotype affects outcome in a murine model of sepsis: implications for a new treatment strategy. Anaesth Intensive Care 2009;37(1):38–45.

[46] de Bont N, Netea MG, Demacker PN, Kullberg BJ, van der Meer JW, Stalenhoef AF. Apolipoprotein E-deficient mice have an impaired immune response to *Klebsiella pneumoniae*. Eur J Clin Invest 2000;30(9):818–22.

[47] de Bont N, Netea MG, Demacker PN, Verschueren I, Kullberg BJ, van Dijk KW, et al. Apolipoprotein E knock-out mice are highly susceptible to endotoxemia and *Klebsiella pneumoniae* infection. J Lipid Res 1999;40(4):680–5.

[48] Roselaar SE, Daugherty A. Apolipoprotein E-deficient mice have impaired innate immune responses to *Listeria monocytogenes* in vivo. J Lipid Res 1998;39(9):1740–3.

[49] Vonk AG, De Bont N, Netea MG, Demacker PN, van der Meer JW, Stalenhoef AF, et al. Apolipoprotein-E-deficient mice exhibit an increased susceptibility to disseminated candidiasis. Med Mycol 2004;42(4):341–8.

[50] Chuang K, Elford EL, Tseng J, Leung B, Harris HW. An expanding role for apolipoprotein E in sepsis and inflammation. Am J Surg 2010;200(3):391–7.

[51] Kattan OM, Kasravi FB, Elford EL, Schell MT, Harris HW. Apolipoprotein E-mediated immune regulation in sepsis. J Immunol 2008;181(2):1399–408.

[52] Gale SC, Gao L, Mikacenic C, Coyle SM, Rafaels N, Murray Dudenkov T, et al. APOepsilon4 is associated with enhanced in vivo innate immune responses in human subjects. J Allergy Clin Immunol 2014;134(1):127–34.

[53] Moretti EW, Morris RW, Podgoreanu M, Schwinn DA, Newman MF, Bennett E, et al. APOE polymorphism is associated with risk of severe sepsis in surgical patients. Crit Care Med 2005;33(11):2521–6.

[54] Gerard HC, Fomicheva E, Whittum-Hudson JA, Hudson AP. Apolipoprotein E4 enhances attachment of *Chlamydophila* (*Chlamydia*) *pneumoniae* elementary bodies to host cells. Microb Pathog 2008;44(4):279–85.

[55] Gerard HC, Wang GF, Balin BJ, Schumacher HR, Hudson AP. Frequency of apolipoprotein E (APOE) allele types in patients with *Chlamydia*-associated arthritis and other arthritides. Microb Pathog 1999;26(1):35–43.

[56] Campbell LA, Yaraei K, Van Lenten B, Chait A, Blessing E, Kuo CC, et al. The acute phase reactant response to respiratory infection with *Chlamydia pneumoniae*: implications for the pathogenesis of atherosclerosis. Microbes Infect 2010;12(8–9):598–606.

[57] Nazzal D, Therville N, Yacoub-Youssef H, Garcia V, Thomsen M, Levade T, et al. Apolipoprotein E-deficient mice develop an anti-*Chlamydophila pneumoniae* T helper 2 response and resist vascular infection. J Infect Dis 2010;202(5):782–90.

[58] Burt TD, Agan BK, Marconi VC, He W, Kulkarni H, Mold JE, et al. Apolipoprotein (apo) E4 enhances HIV-1 cell entry in vitro, and the APOE epsilon4/epsilon4 genotype accelerates HIV disease progression. Proc Natl Acad Sci U S A 2008;105(25):8718–23.

[59] Kuhlmann I, Minihane AM, Huebbe P, Nebel A, Rimbach G. Apolipoprotein E genotype and hepatitis C, HIV and herpes simplex disease risk: a literature review. Lipids Health Dis 2010;9:8.

[60] Zhang L, Yesupriya A, Chang MH, Teshale E, Teo CG. Apolipoprotein E and protection against hepatitis E viral infection in American non-Hispanic blacks. Hepatology 2015;62(5):1346–52.

[61] Madenspacher JH, Azzam KM, Gong W, Gowdy KM, Vitek MP, Laskowitz DT, et al. Apolipoproteins and apolipoprotein mimetic peptides modulate phagocyte trafficking through chemotactic activity. J Biol Chem 2012;287(52):43730–40.

[62] Yamashita CM, Fessler MB, Vasanthamohan L, Lac J, Madenspacher J, McCaig L, et al. Apolipoprotein E-deficient mice are susceptible to the development of acute lung injury. Respiration 2014;87(5):416–27.

[63] Jacobsen NR, Moller P, Jensen KA, Vogel U, Ladefoged O, Loft S, et al. Lung inflammation and genotoxicity following pulmonary exposure to nanoparticles in ApoE$^{-/-}$ mice. Part Fibre Toxicol 2009;6:2.

[64] Arunachalam G, Sundar IK, Hwang JW, Yao H, Rahman I. Emphysema is associated with increased inflammation in lungs of atherosclerosis-prone mice by cigarette smoke: implications in comorbidities of COPD. J Inflamm 2010;7:34.

[65] Chen B, Moore TV, Li Z, Sperling AI, Zhang C, Andrade J, et al. Gata5 deficiency causes airway constrictor hyperresponsiveness in mice. Am J Respir Cell Mol Biol 2014;50(4):787–95.

[66] Yao X, Dai C, Fredriksson K, Lam J, Gao M, Keeran KJ, et al. Human apolipoprotein E genotypes differentially modify house dust mite-induced airway disease in mice. Am J Physiol Lung Cell Mol Physiol 2012;302(2):L206–15.

[67] Golder-Novoselsky E, Forte TM, Nichols AV, Rubin EM. Apolipoprotein AI expression and high density lipoprotein distribution in transgenic mice during development. J Biol Chem 1992;267(29):20787–90.

[68] Dai C, Yao X, Keeran KJ, Zywicke GJ, Qu X, Yu ZX, et al. Apolipoprotein A-I attenuates ovalbumin-induced neutrophilic airway inflammation via a granulocyte colony-stimulating factor-dependent mechanism. Am J Respir Cell Mol Biol 2012;47(2):186–95.

[69] Kim TH, Lee YH, Kim KH, Lee SH, Cha JY, Shin EK, et al. Role of lung apolipoprotein A-I in idiopathic pulmonary fibrosis: antiinflammatory and antifibrotic effect on experimental lung injury and fibrosis. Am J Respir Crit Care Med 2010;182(5):633–42.

[70] Davidson WS, Thompson TB. The structure of apolipoprotein A-I in high density lipoproteins. J Biol Chem 2007;282(31):22249–53.

[71] Leman LJ, Maryanoff BE, Ghadiri MR. Molecules that mimic apolipoprotein A-I: potential agents for treating atherosclerosis. J Med Chem 2014;57(6):2169–96.

[72] Provost PR, Boucher E, Tremblay Y. Apolipoprotein A-I, A-II, C-II, and H expression in the developing lung and sex difference in surfactant lipids. J Endocrinol 2009;200(3):321–30.

[73] Bloedon LT, Dunbar R, Duffy D, Pinell-Salles P, Norris R, DeGroot BJ, et al. Safety, pharmacokinetics, and pharmacodynamics of oral apoA-I mimetic peptide D-4F in high-risk cardiovascular patients. J Lipid Res 2008;49(6):1344–52.

[74] Dunbar RL, Bloedon LT, Duffy D, Norris RB, Movva R, Navab M, et al. Daily oral administration of the apolipoprotein A-I mimetic peptide D-4F in patients with heart disease or equivalent risk improves high-density lipoprotein anti-inflammatory function (abstract). J Am Coll Cardiol 2007;366A:49.

[75] Getz GS, Reardon CA. ApoA-I mimetics: tomatoes to the rescue. J Lipid Res 2013;54:995–1010.

[76] Sethi AA, Stonik JA, Thomas F, Demosky SJ, Amar M, Neufeld E, et al. Asymmetry in the lipid affinity of bihelical amphipathic peptides. A structural determinant for the specificity of ABCA1-dependent cholesterol efflux by peptides. J Biol Chem 2008;283(47):32273–82.

[77] Yao X, Gordon EM, Barochia AV, Remaley AT, Levine SJ. The A's have it: developing apolipoprotein A-I mimetic peptides into a novel treatment for asthma. Chest 2016;150(2):283–8.

[78] Tall AR, Yvan-Charvet L. Cholesterol, inflammation and innate immunity. Nat Rev Immunol 2015;15(2):104–16.

[79] Catapano AL, Pirillo A, Bonacina F, Norata GD. HDL in innate and adaptive immunity. Cardiovasc Res 2014;103(3):372–83.

[80] Yvan-Charvet L, Pagler T, Gautier EL, Avagyan S, Siry RL, Han S, et al. ATP-binding cassette transporters and HDL suppress hematopoietic stem cell proliferation. Science 2010;328(5986):1689–93.

[81] Westerterp M, Murphy AJ, Wang M, Pagler TA, Vengrenyuk Y, Kappus MS, et al. Deficiency of ATP-binding cassette transporters A1 and G1 in macrophages increases inflammation and accelerates atherosclerosis in mice. Circ Res 2013;112(11):1456–65.

[82] Wang SH, Yuan SG, Peng DQ, Zhao SP. HDL and ApoA-I inhibit antigen presentation-mediated T cell activation by disrupting lipid rafts in antigen presenting cells. Atherosclerosis 2012;225(1):105–14.

[83] Kim KD, Lim HY, Lee HG, Yoon DY, Choe YK, Choi I, et al. Apolipoprotein A-I induces IL-10 and PGE2 production in human monocytes and inhibits dendritic cell differentiation and maturation. Biochem Biophys Res Commun 2005;338(2):1126–36.

[84] Murphy AJ, Woollard KJ, Suhartoyo A, Stirzaker RA, Shaw J, Sviridov D, et al. Neutrophil activation is attenuated by high-density lipoprotein and apolipoprotein A-I in in vitro and in vivo models of inflammation. Arterioscler Thromb Vasc Biol 2011;31(6):1333–41.

[85] Peshavariya H, Dusting GJ, Di Bartolo B, Rye KA, Barter PJ, Jiang F. Reconstituted high-density lipoprotein suppresses leukocyte NADPH oxidase activation by disrupting lipid rafts. Free Radic Res 2009;43(8):772–82.

[86] Ito A, Hong C, Oka K, Salazar JV, Diehl C, Witztum JL, et al. Cholesterol accumulation in CD11c+ immune cells is a causal and targetable factor in autoimmune disease. Immunity 2016;45(6):1311–26.

[87] Wilhelm AJ, Zabalawi M, Owen JS, Shah D, Grayson JM, Major AS, et al. Apolipoprotein A-I modulates regulatory T cells in autoimmune LDLr$^{-/-}$, ApoA-I$^{-/-}$ mice. J Biol Chem 2010;285(46):36158–69.

[88] Tiniakou I, Drakos E, Sinatkas V, Van Eck M, Zannis VI, Boumpas D, et al. High-density lipoprotein attenuates Th1 and th17 autoimmune responses by modulating dendritic cell maturation and function. J Immunol 2015;194(10):4676–87.

[89] Black LL, Srivastava R, Schoeb TR, Moore RD, Barnes S, Kabarowski JH. Cholesterol-independent suppression of lymphocyte activation, autoimmunity, and glomerulonephritis by apolipoprotein A-I in normocholesterolemic lupus-prone mice. J Immunol 2015;195(10):4685–98.

[90] Yin K, Chen WJ, Zhou ZG, Zhao GJ, Lv YC, Ouyang XP, et al. Apolipoprotein A-I inhibits CD40 proinflammatory signaling via ATP-binding cassette transporter A1-mediated modulation of lipid raft in macrophages. J Atheroscler Thromb 2012;19(9):823–36.

[91] Yin K, Deng X, Mo ZC, Zhao GJ, Jiang J, Cui LB, et al. Tristetraprolin-dependent post-transcriptional regulation of inflammatory cytokine mRNA expression by apolipoprotein A-I: role of ATP-binding membrane cassette transporter A1 and signal transducer and activator of transcription 3. J Biol Chem 2011;286(16):13834–45.

[92] Iqbal AJ, Barrett TJ, Taylor L, McNeill E, Manmadhan A, Recio C, et al. Acute exposure to apolipoprotein A1 inhibits macrophage chemotaxis in vitro and monocyte recruitment in vivo. Elife 2016;5.

[93] van der Vorst EP, Theodorou K, Wu Y, Hoeksema MA, Goossens P, Bursill CA, et al. High-density lipoproteins exert pro-inflammatory effects on macrophages via passive cholesterol depletion and PKC-NF-kappaB/STAT1-IRF1 signaling. Cell Metab 2017;25(1):197–207.

[94] Nguyen SD, Maaninka K, Lappalainen J, Nurmi K, Metso J, Oorni K, et al. Carboxyl-terminal cleavage of apolipoprotein A-I by human mast cell Chymase impairs its anti-inflammatory properties. Arterioscler Thromb Vasc Biol 2016;36(2):274–84.

[95] Dinnes DL, White MY, Kockx M, Traini M, Hsieh V, Kim MJ, et al. Human macrophage cathepsin B-mediated C-terminal cleavage of apolipoprotein A-I at Ser228 severely impairs antiatherogenic capacity. FASEB J 2016;30(12):4239–55.

[96] Wang W, Xu H, Shi Y, Nandedkar S, Zhang H, Gao H, et al. Genetic deletion of apolipoprotein A-I increases airway hyperresponsiveness, inflammation, and collagen deposition in the lung. J Lipid Res 2010;51(9):2560–70.

[97] Emancipator K, Csako G, Elin RJ. In vitro inactivation of bacterial endotoxin by human lipoproteins and apolipoproteins. Infect Immun 1992;60(2):596–601.

[98] Henning MF, Herlax V, Bakas L. Contribution of the C-terminal end of apolipoprotein AI to neutralization of lipopolysaccharide endotoxic effect. Innate Immun 2011;17(3):327–37.

[99] Ma J, Liao XL, Lou B, Wu MP. Role of apolipoprotein A-I in protecting against endotoxin toxicity. Acta Biochim Biophys Sin (Shanghai) 2004;36(6):419–24.

[100] Van Linthout S, Spillmann F, Graiani G, Miteva K, Peng J, Van Craeyveld E, et al. Down-regulation of endothelial TLR4 signalling after apo A-I gene transfer contributes to improved survival in an experimental model of lipopolysaccharide-induced inflammation. J Mol Med (Berl) 2011;89(2):151–60.

[101] Wurfel MM, Kunitake ST, Lichenstein H, Kane JP, Wright SD. Lipopolysaccharide (LPS)-binding protein is carried on lipoproteins and acts as a cofactor in the neutralization of LPS. J Exp Med 1994;180(3):1025–35.

[102] Jiao YL, Wu MP. Apolipoprotein A-I diminishes acute lung injury and sepsis in mice induced by lipoteichoic acid. Cytokine 2008;43(1):83–7.

[103] Wang Y, Zhu X, Wu G, Shen L, Chen B. Effect of lipid-bound apoA-I cysteine mutants on lipopolysaccharide-induced endotoxemia in mice. J Lipid Res 2008;49(8):1640–5.

[104] Yan YJ, Li Y, Lou B, Wu MP. Beneficial effects of ApoA-I on LPS-induced acute lung injury and endotoxemia in mice. Life Sci 2006;79(2):210–5.

[105] Li Y, Dong JB, Wu MP. Human ApoA-I overexpression diminishes LPS-induced systemic inflammation and multiple organ damage in mice. Eur J Pharmacol 2008;590(1–3):417–22.

[106] Sharifov OF, Xu X, Gaggar A, Grizzle WE, Mishra VK, Honavar J, et al. Anti-inflammatory mechanisms of apolipoprotein A-I mimetic peptide in acute respiratory distress syndrome secondary to sepsis. PLoS One 2013;8(5):e64486.

[107] Kwon WY, Suh GJ, Kim KS, Kwak YH, Kim K. 4F, apolipoprotein AI mimetic peptide, attenuates acute lung injury and improves survival in endotoxemic rats. J Trauma Acute Care Surg 2012;72(6):1576–83.

[108] Tu J, Zhang B, Chen Y, Liang B, Liang D, Liu G, et al. Association of apolipoprotein A1 −75 G/A polymorphism with susceptibility to the development of acute lung injury after cardiopulmonary bypass surgery. Lipids Health Dis 2013;12:172.

[109] Hao J, He XD. Haplotype analysis of ApoAI gene and sepsis-associated acute lung injury. Lipids Health Dis 2014;13:79.

[110] Van Lenten BJ, Wagner AC, Navab M, Anantharamaiah GM, Hui EK, Nayak DP, et al. D-4F, an apolipoprotein A-I mimetic peptide, inhibits the inflammatory response induced by influenza A infection of human type II pneumocytes. Circulation 2004;110(20):3252–8.

[111] Van Lenten BJ, Wagner AC, Nayak DP, Hama S, Navab M, Fogelman AM. High-density lipoprotein loses its anti-inflammatory properties during acute influenza a infection. Circulation 2001;103(18):2283–8.

[112] Van Lenten BJ, Wagner AC, Anantharamaiah GM, Garber DW, Fishbein MC, Adhikary L, et al. Influenza infection promotes macrophage traffic into arteries of mice that is prevented by D-4F, an apolipoprotein A-I mimetic peptide. Circulation 2002;106(9):1127–32.

[113] Yao X, Dai C, Fredriksson K, Dagur PK, McCoy JP, Qu X, et al. 5A, an apolipoprotein A-I mimetic peptide, attenuates the induction of house dust mite-induced asthma. J Immunol 2011;186(1):576–83.

[114] Nandedkar SD, Weihrauch D, Xu H, Shi Y, Feroah T, Hutchins W, et al. D-4F, an apoA-1 mimetic, decreases airway hyperresponsiveness, inflammation, and oxidative stress in a murine model of asthma. J Lipid Res 2011;52(3):499–508.

[115] Park SW, Lee EH, Lee EJ, Kim HJ, Bae DJ, Han S, et al. Apolipoprotein A1 potentiates lipoxin A4 synthesis and recovery of allergen-induced disrupted tight junctions in the airway epithelium. Clin Exp Allergy 2013;43(8):914–27.

[116] Barochia AV, Kaler M, Cuento RA, Gordon EM, Weir NA, Sampson M, et al. Serum apolipoprotein A-I and large high-density lipoprotein particles are positively correlated with FEV1 in atopic asthma. Am J Respir Crit Care Med 2015;191(9):990–1000.

[117] Barochia AV, Gordon EM, Kaler M, Cuento RA, Theard P, Figueroa DM, et al. High density lipoproteins and type 2 inflammatory biomarkers are negatively correlated in atopic asthmatics. J Lipid Res 2017;58:1713–21.

[118] Kim C, Lee JM, Park SW, Kim KS, Lee MW, Paik S, et al. Attenuation of cigarette smoke-induced emphysema in mice by apolipoprotein A-1 overexpression. Am J Respir Cell Mol Biol 2016;54(1):91–102.

[119] Nicholas BL, Skipp P, Barton S, Singh D, Bagmane D, Mould R, et al. Identification of lipocalin and apolipoprotein A1 as biomarkers of chronic obstructive pulmonary disease. Am J Respir Crit Care Med 2010;181(10):1049–60.

[120] Lee E, Lee EJ, Kim H, Jang A, Koh E, Uh ST, et al. Overexpression of apolipoprotein A1 in the lung abrogates fibrosis in experimental silicosis. PLoS One 2013;8(2):e55827.

[121] Sharma S, Umar S, Potus F, Iorga A, Wong G, Meriwether D, et al. Apolipoprotein A-I mimetic peptide 4F rescues pulmonary hypertension by inducing microRNA-193-3p. Circulation 2014;130(9):776–85.

[122] de Silva HV, Harmony JA, Stuart WD, Gil CM, Robbins J. Apolipoprotein J: structure and tissue distribution. Biochemistry 1990;29(22):5380–9.

[123] Carnevali S, Luppi F, D'Arca D, Caporali A, Ruggieri MP, Vettori MV, et al. Clusterin decreases oxidative stress in lung fibroblasts exposed to cigarette smoke. Am J Respir Crit Care Med 2006;174(4):393–9.

[124] Garcia-Aranda M, Tellez T, Munoz M, Redondo M. Clusterin inhibition mediates sensitivity to chemotherapy and radiotherapy in human cancer. Anticancer Drugs 2017;28(7):702–16.

[125] de Silva HV, Stuart WD, Duvic CR, Wetterau JR, Ray MJ, Ferguson DG, et al. A 70-kDa apolipoprotein designated ApoJ is a marker for subclasses of human plasma high density lipoproteins. J Biol Chem 1990;265(22):13240–7.

[126] Hong GH, Kwon HS, Moon KA, Park SY, Park S, Lee KY, et al. Clusterin modulates allergic airway inflammation by attenuating CCL20-mediated dendritic cell recruitment. J Immunol 2016;196(5):2021–30.

[127] Sol IS, Kim YH, Park YA, Lee KE, Hong JY, Kim MN, et al. Relationship between sputum clusterin levels and childhood asthma. Clin Exp Allergy 2016;46(5):688–95.

[128] Sol IS, Kim YH, Lee KE, Hong JY, Kim MN, Kim YS, et al. Serum clusterin level in children with atopic dermatitis. Allergy Asthma Proc 2016;37(4):335–9.

[129] Heller AR, Fiedler F, Braun P, Stehr SN, Bodeker H, Koch T. Clusterin protects the lung from leukocyte-induced injury. Shock 2003;20(2):166–70.

[130] Liu X, Meng L, Li J, Meng J, Teng X, Gu H, et al. Secretory clusterin is upregulated in rats with pulmonary arterial hypertension induced by systemic-to-pulmonary shunts and exerts important roles in pulmonary artery smooth muscle cells. Acta Physiol (Oxf) 2015;213(2):505–18.

Index

Note: Page numbers followed by *f* indicate figures and *t* indicate tables.